Concepts and Problems in Physics

Dr. Sanjay Kumar

LAWS OF MOTION
&
FRICTION

FOR JEE (MAIN & ADVANCED) & NEET

Dr. Sanjay Kumar

M.Tech, PhD CMJU Meghalaya
Managing Director- Quanta Classes Lucknow, Mo. 9453763058
Ex. Sr. Faculty of Physics: Imagine Point Kanpur, Jain Classes Jhansi, Bansal Classes MP, TATA Aarambh Engineering and Medical Simplified Lucknow.

Copyright © 2020 Sanjay Kumar
All rights reserved.
No part of this book may be reproduced or distributed in any form or by any means, electronic, mechanical, photocopying, recording, or otherwise or stored in a database or retrieval system without the prior written permission of the author.

To JEE (MAIN & ADVANCED) & NEET aspirants
With the hope that this work will stimulate
an interest in Physics
and provide an acceptable guide to its understanding.

PREFACE

This text book is the product of more than twenty years of teaching and innovation experience in physics for JEE and NEET aspirants. It is primarily intended for students who are preparing for the entrance tests of IIT-JEE/NEET/AIIMS and other esteemed colleges in same fields. This text is equally useful to the students preparing for their school exams.

Our main goals in writing this text book are

- to present the basic concepts and principles of physics that students need to know for their competitive exams.
- to provide a balance of quantitative reasoning and conceptual understanding, with special attention to concepts that have been causing difficulties to student in understanding the concepts.
- to develop students' problem-solving skills and confidence in a systematic manner.
- to motivate students by integrating real-world examples that build upon their everyday experiences.

Main Features of the Book-

1. Every concept is up to the mark and given in student friendly language with various solved problems. The solution is provided with problem solving approach and discussion.
2. Checkpoint questions have been added to applicable sections of the text to allow students to pause and test their understanding of the concept explored within the current section. The answers and solutions to the Checkpoints are given in answer keys, at the end of the chapter, so that students can confirm their knowledge without jumping too quickly to the provided answer.
3. Special attention is given to all tricky topics (like- constrained relations, fictitious (or pseudo) force, condition of contact of two blocks on inclined plane, block over block friction problems, force of friction in case of a block fixed with a string and a lot more) so that student can easily solve them with fun.
4. To test the understanding level of students, multiple choice questions, conceptual questions, practice problems with previous years JEE Main and Advanced problems are provided at the end of the whole discussion. Number of dots indicates level of problem difficulty. Straightforward problems (basic level) are indicated by single dot (•), intermediate problems (JEE mains and NEET level) are indicated by double dots (••), whereas challenging problems (advanced level) are indicated by thee dots (•••). Answer keys with hints and solutions are provided at the end of the chapter.

We have kept these goals in mind while developing the main themes of our physics book.

Dr. Sanjay Kumar

Online/Offline

Physics Classes

Dr. Sanjay Kumar

JEE(Main+Advanced)/NEET/Foundation (IX–XII)

Quanta Classes: K 423A Sector K Ashiyana Colony Lucknow, Mo. +919453763058

Email: spphysicsworld@gmail.com

CONTENTS

1. **FORCE** 1
   1.1. COMBINATION OF FORCES 1
2. **CONTACT AND FIELD FORCES** 1
   2.1. CONTACT FORCES 1
      2.1.1. A SHORT CATALOG OF CONTACT FORCES 1
   2.2. FIELD FORCES 2
      2.2.1. GRAVITATIONAL FORCES 2
      2.2.2. ELECTROMAGNETIC (EM) FORCES 3
      2.2.3. STRONG NUCLEAR FORCES 4
      2.2.4. WEAK NUCLEAR FORCES 4
      2.2.5. HOW FIELD FORCE ACTS THROUGH EMPTY SPACE? 4
3. **FUNDAMENTAL AND NON-FUNDAMENTAL FORCES** 4
   3.1. FUNDAMENTAL FORCES 4
   3.2. NON-FUNDAMENTAL FORCES 4
4. **THE EXISTENCE OF A FIFTH FUNDAMENTAL FORCE** 4
5. **CHECKPOINT 1** 5
6. **INTERNAL VERSUS EXTERNAL FORCES** 5
7. **REFERENCE FRAMES** 5
8. **NEWTON'S LAWS OF MOTION AND FRAME OF REFERENCES** 5
   8.1. NEWTON'S FIRST LAW 5
9. **CHECKPOINT 2** 7
10. **INERTIAL REFERENCE FRAMES** 7
11. **NEWTON'S SECOND LAW OF MOTION ($F = ma$)** 8
    11.1. WORKING WITH NEWTON'S FIRST AND SECOND LAW 9
12. **TRANSLATIONAL EQUILIBRIUM** 10
13. **CHECKPOINT 3** 16
14. **CONCEPT OF INERTIAL MASS AND GRAVITATIONAL MASS** 17
15. **SOME SPECIFIC TERMS** 18
    15.1. FREE FALL 18
    15.2. AT REST 20
    15.3. WEIGHT 20
    15.4. WEIGHING 20
16. **CONTACT FORCES (A MICROSCOPIC EXPLANATION)** 21
    16.1. SPRING FORCE 21
       16.1.1. Equivalent spring constant 22
       16.1.2. Dividing a Spring in to two Parts 23
       16.1.3. How to Solve Massive Spring Problems 24

	16.1.4. Effect on Tension in a spring on sudden Removal of Force	24
17.	CHECKPOINT 4	24
18.	NORMAL FORCE	25
18.1.	DIRECTIONS OF NORMAL REACTIONS IN DIFFERENT SITUATIONS	26
19.	FRICTION (IN BRIEF)	27
20.	TENSION	27
20.1.	SOME IMPORTANT POINTS	28
20.2.	PULLEYS	30
20.3.	TENSIONS ON BOTH SIDES OF A CORD PASSING OVER A PULLEY	30
21.	NEWTON'S THIRD LAW	32
22.	CHECKPOINT 5	36
23.	APPARENT WEIGHT	37
24.	SOLVED EXAMPLES ON NEWTON'S FIRST AND SECOND LAWS	43
24.1.	SYMBOLIC SOLVED EXAMPLES	44
25.	CHECKPOINT 6	47
26.	CONSTRAINED MOTION	49
27.	MATHEMATICAL ANALYSIS OF CONSTRAINED MOTION	50
27.1.	One Degree of Freedom	50
27.2.	Two Degrees of Freedom	51
27.3.	CONSTRAINT WHERE THE DIRECTION OF THE CONNECTING MEMBER CHANGES WITH THE MOTION	52
27.4.	IMPORTANT RESULTS FOR PULLEY CONSTRAINTS	53
27.5.	WEDGE CONSTRAINTS	56
28.	CHECKPOINT 7	58
29.	FICTITIOUS (OR PSEUDO) FORCE	64
29.1.	MOTION IN ACCELERATED FRAMES:	64
30.	CHECKPOINT 8	66
31.	ANALYSIS OF FRICTION	67
31.1.	TYPES OF FRICTION	67
31.2.	TWO TYPES OF DRY FRICTION	67
31.2.1.	KINETIC (SLIDING) DRY FRICTION	68
31.2.2.	STATIC DRY FRICTION	68
31.3.	MICROSCOPIC ORIGIN OF FRICTION	69
32.	LAWS OF STATIC FRICTION	71
33.	HOLDING A BOX AGAINST A ROUGH WALL	73
34.	PUSHING VERSUS PULLING	73
35.	CONDITION OF CONTACT OF TWO BLOCKS ON INCLINED PLANE	76
36.	ANGLE OF FRICTION	77

37.	ANGLE OF REPOSE	77
38.	MINIMUM AND MAXIMUM VALUES OF CONTACT FORCE	77
39.	CHARACTERISTICS OF DRY FRICTION	78
40.	ROLLING FRICTION	78
41.	BLOCK OVER BLOCK FRICTION PROBLEMS	78
41.1.	TWO BLOCK PROBLEM	78
41.2.	MULTI BLOCKS PROBLEM	80
42.	FORCE OF FRICTION IN CASE OF A BLOCK CONNECTED WITH A STRING	84
43.	CHECKPOINT 9	87
44.	DRAG FORCES	89
44.1.	SMALL OBJECTS	89
44.1.1.	Projectile Motion with Air Resistance	91
44.2.	LARGE OBJECTS	92
45.	CHECKPOINT 10	92
46.	CENTRIPETAL & CENTRIFUGAL FORCES	93
47.	FICTITIOUS FORCE IN A ROTATING SYSTEM	93
48.	LIMITATIONS OF NEWTON'S LAWS	94
49.	EXERCISES AND QUESTIONS	94
49.1.	CONCEPTUAL QUESTIONS	94
49.2.	PROBLEMS	97
49.3.	MULTIPLE CHOICE ASSIGNMENTS	112
49.3.1.	NEWTON'S LAWS	112
49.3.1.1.	LEVEL 1	112
49.3.1.2.	LEVEL-2	117
49.3.1.3.	LEVEL-3	120
49.3.1.4.	LEVEL – 4 (Previous Years Questions)	123
49.3.2.	FRICTION	126
49.3.2.1.	LEVEL-1	126
49.3.2.2.	LEVEL-2	128
49.3.2.3.	LEVEL-3	131
49.3.2.4.	LEVEL-4 (Previous Years Questions)	133
50.	ANSWER KEYS AND SOLUTIONS	135
50.1.	CHECKPOINT 1	135
50.2.	CHECKPOINT 2	135
50.3.	CHECKPOINT 3	135
50.4.	CHECKPOINT 4	136
50.5.	CHECKPOINT 5	136
50.6.	CHECKPOINT 6	137
50.7.	CHECKPOINT 7	138
50.8.	CHECK POINT 8	138
50.9.	CHECKPOINT 9	139
50.10.	CHECKPOINT 10	139

- 50.11. PROBLEMS 139
- 50.12. MULTIPLE CHOICE PROBLEMS 147
  - *50.12.1. LAWS OF MOTION* *147*
    - 50.12.1.1. LEVEL 1 147
    - LEVEL 2 147
    - LEVEL 3 148
    - LEVEL 4 148
  - *50.12.2. FRICTION* *148*
    - 50.12.2.1. LEVEL 1 148
    - 50.12.2.2. LEVEL 2 148
    - 50.12.2.3. LEVEL 3 148
    - 50.12.2.4. LEVEL 4 148

5. LAWS OF MOTION AND FRICTION

1. FORCE

In everyday language, a **force** is a push or a pull. A better definition is that a force is an *interaction* between two bodies or between a body and its environment. That's why we always refer to the force that one body *exerts* on other body.

Force is a *vector* quantity; you can push or pull a body in different directions [FIGURE 1].

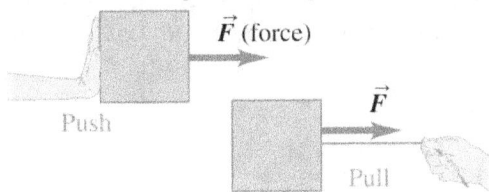

FIGURE 1

For a force, following two points are very important-
1. **A force always acts on an object.** Pushes and pulls are applied *to* something—an object. From the object's perspective, it has a force *exerted* on it. Forces do not exist in isolation from the object that experiences them.
2. **A force requires an external agent.** Every force has an external **agent,** something that acts or exerts force. That is, a force has a specific, identifiable *cause*. As you push a block, the external agent is your hand, which exerts a force on the block. So, block is the object When the block is pulled by a string, the string is the external agent which exerts a force on the block (object) [FIGURE 1].

DYNAMICS AND FORCES

The study of *how* objects move is called **kinematics**. But physicists want to be able to explain *why* the motion occurs. **Dynamics** is the study of why objects move; it is concerned with forces, the central idea of this chapter.

1.1. COMBINATION OF FORCES

when several forces $\vec{F}_1, \vec{F}_2, \vec{F}_3, \ldots \vec{F}_n$ are exerted on an object, they combine to form a **net force** $\vec{F}_{net}$ given by the *vector* sum of *all* the forces:

$$\vec{F}_{net} \equiv \sum_{i=1}^{n} \vec{F}_i = \vec{F}_1 + \vec{F}_2 + \vec{F}_3 + \cdots + \vec{F}_n$$

Note that ≡ is the symbol meaning "is defined as." Mathematically, this summation is called a **superposition of forces.**

2. CONTACT AND FIELD FORCES

In classical mechanics, there are two basic classes of forces, depending on whether the external agent touches the object or not.

2.1. CONTACT FORCES

Contact forces are the forces that act on an object by touching it at a point of contact. The bat must touch the ball to hit it. A string must be tied to an object to pull it.

2.1.1. A SHORT CATALOG OF CONTACT FORCES

Some important contact forces, are given below. A microscopic discussion of these forces will be given later under the article "CONTACT FORCES (A MICROSCOPIC EXPLANATION)".

1. **SPRING FORCE** A spring is a metal coil that can be stretched or compressed.

A spring can either push (when compressed) or pull (when stretched). FIGURE 1 shows the spring force, for which we use the symbol $\vec{F}_{sp}$. In both cases, pushing and pulling, the tail of the force vector is placed on the particle in the force diagram.

FIGURE 1

2. **TENSION FORCE:** When a string or rope or wire pulls on an object, it exerts a contact force that we

call the **tension force,** represented by a capital $\vec{T}$. The direction of the tension force is always along the direction of the string or rope, as you can see in FIGURE 2

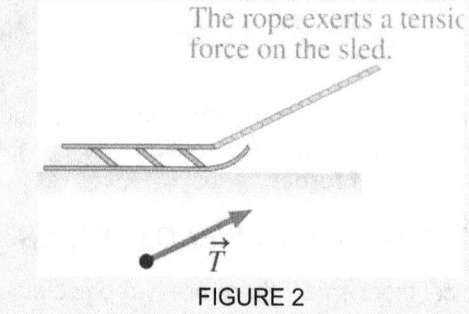

FIGURE 2

3. **NORMAL REACTION** If the two surfaces in contact are perfectly smooth (i.e., frictionless), then the contact force acts only perpendicular (normal) to their surface of contact and is known as 'normal reaction ($\vec{N}$)'

Normal reactions on a block placed on a horizontal and inclined surface, are shown in FIGURE 3. A detailed discussion of normal reaction will be given later, under the article " CONTACT FORCES".

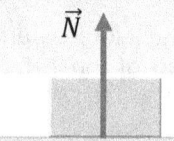

Direction of normal reaction on the block

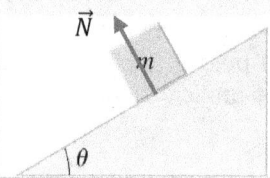

Direction of normal reaction on the block

FIGURE 3

3. **FRICTION: Friction,** like the normal force, is exerted by a surface. But whereas the normal force is perpendicular to the surface, the friction force is always *parallel* to the surface. It is useful to distinguish between two kinds of friction:

(i) *Kinetic friction,* denoted by $\vec{f}_k$, appears when an object slides across a surface. This is a force that "opposes the relative motion of the object," meaning that the friction force vector $\vec{f}_k$ on the object points in a direction opposite the relative velocity of the object.

(ii) *Static friction,* denoted by $\vec{f}_s$, is the force that keeps an object "stuck" on a surface and prevents its motion. Finding the direction of $\vec{f}_s$ is a little trickier than finding it for $\vec{f}_k$. Static friction points opposite the direction in which the object *would* move if there were no friction. That is, it points in the direction necessary to *prevent* motion.

☞ We shall discuss more about it under the heading of friction

4. **DRAG:** Friction at a surface is one example of a *resistive force,* a force that opposes or resists motion. Resistive forces are also experienced by objects moving through fluids—gases and liquids. The resistive force of a fluid is called **drag,** with symbol $\vec{F}_{drag}$. Drag, like kinetic friction, points opposite the direction of relative motion.

Drag can be a significant force for objects moving at high speeds or in dense fluids.

For objects that are heavy and compact, that move in air, and whose speed is not too great, the drag force of air resistance is fairly small. To keep things as simple as possible, **you can neglect air resistance in all problems unless a problem explicitly asks you to include it.**

5. **THRUST** Thrust, occurs when a jet or rocket engine expels gas molecules at high speed. Thrust is a contact force, with the exhaust gas being the agent that pushes on the engine.

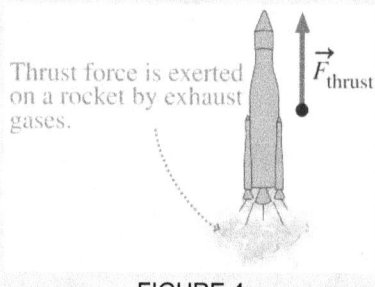

FIGURE 4

The direction of thrust is always opposite to the direction in which the exhaust gas is expelled. There's no special symbol for thrust, we denote it by $\vec{F}_{thrust}$.

2.2. FIELD FORCES

Field forces are the forces that act on an object without physical contact. These forces act through empty space. All known field forces are given below-

2.2.1. GRAVITATIONAL FORCES

The force, between any two objects, by virtue of their masses, is called *gravitational force*. It is *always attractive in nature* i.e., each object is pulled towards the other's center.

According to **Newton's law of universal gravitation**, the gravitational forces between two point masses m_1 and m_2 separated by distance r, is given by

$$F = G \frac{m_1 m_2}{r^2} \quad \ldots (1)$$

where the constant $G = 6.67 \times 10^{-11} \, N.m^2/kg^2$. It is called universal gravitational constant.

The forces on the two objects are equal in magnitude and the directions are opposite, as they must be since they form an interaction pair

Gravitational forces exerted *by* ordinary objects on each other are so small as to be negligible in most

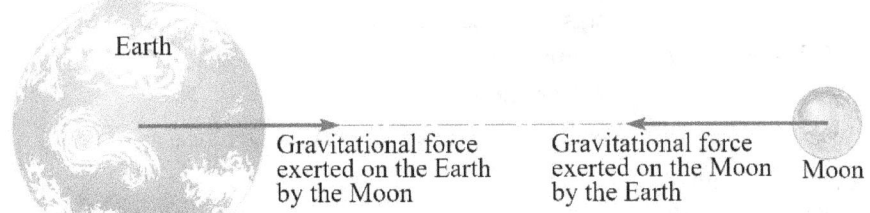

FIGURE 1

cases. For example, considering two persons as point masses of $70\ kg$ each, separated by $1\ m$ apart. The gravitational force between them is given by-

$$F = G\frac{m_1 m_2}{r^2}$$
$$= 6.67 \times 10^{-11} \times \frac{70 \times 70}{1^2} = 32683 \times 10^{-11} N$$
$$\approx 3.3 \times 10^{-7} N$$

It is so small that we can neglect it as compared to other forces experienced in daily life.

Now, if we consider Earth as spherically symmetric body of mass M and radius R, and a particle of mass m is placed near its surface, then we can write
$$m_1 = M, m_2 = m, r \approx R$$
Using these values in (1), we get
$$F = G\frac{Mm}{R^2} \qquad \ldots (2)$$
The mass of earth is $M \approx 6 \times 10^{24} kg$, radius of Earth $R \approx 6.4 \times 10^6 m$

The quantity, $\frac{GM}{R^2}$ has dimensions of acceleration. It is called acceleration due to gravity and is denoted by g. i.e., $g = \frac{GM}{R^2}$

Now, substituting the values of G, M and R in above expression, we get
$$g = \frac{GM}{R^2} = \frac{6.67 \times 10^{-11} \times 6 \times 10^{24}}{(6.4 \times 10^6)^2}$$
$$= 9.7705\ m/s^2 \approx 9.8\ m/s^2$$

For simplicity of calculations, we often use $g \approx 10\ m/s^2$. In terms of g, the equation (2) can be written as
$$F = mg \qquad \ldots (3)$$
Now if we take $m = 70\ kg$, then $F = mg = (70 kg)(9.8\ m/s^2) = 686$ newton, which is large enough and cannot be neglected.

So, it is clear that if one or both objects are planet-sized or larger, then gravitational force becomes an important force. The gravitational force keeps objects bound to the Earth and the planets in orbit around the Sun.

➤ *Gravitational forces always act between the massive objects and affect everything that has mass.*
➤ *If one or both particles are mass less (as like photons at rest), then, the gravitational force between them becomes zero.*

Gravity has an unlimited range. The force gets weaker as the distance between two objects increases, but it never drops exactly to zero, no matter how far apart the objects get.

☞ In this chapter, we will focus on the gravitational force exerted by the Earth on objects located near its surface. Thus, when we speak of *the* gravitational force $\vec{F}_g$ on a body, we usually mean a force that pulls on it directly toward the center of Earth—that is, directly down towards the ground.

2.2.2. ELECTROMAGNETIC (EM) FORCES

Electric force is a force that one electric charge exerts on another.

When we place a bar magnet near a piece of iron, it exerts a magnetic force on it.

In the nineteenth century, Maxwell developed a theory that *unified* the electric and magnetic forces into a single *electromagnetic force*. According to Maxwell a single force under appropriate conditions, exhibits "electric behavior" or "magnetic behavior."

The electromagnetic force is unlimited in range, like gravity. It acts on particles with electric charge. For example, electromagnetic forces can act between two electrons, between two protons, between electron and proton *but cannot act between two neutrons because neutrons are neutral particles*. However, gravitational forces can act between all these particles due to their masses.

Electromagnetism is the fundamental interaction that binds electrons to nuclei to form atoms and binds atoms together in molecules and solids. It is responsible for the properties of solids, liquids, and gases and forms the basis of the sciences of chemistry and biology.

It is the fundamental interaction behind all macroscopic contact forces such as the frictional and normal forces between surfaces, tension produced in strings, and forces exerted by springs, muscles, and the wind.

☞ At the atomic level, there is no fundamental difference between contact forces and other electromagnetic forces.

The electromagnetic force is *much* stronger than gravitational. For example, the electrical repulsion of two electrons at rest is about 10^{43} times as strong as the gravitational attraction between them.

Why the gravitational interactions are dominant for the bodies of astronomical size?

Macroscopic objects have a nearly perfect balance of positive and negative electric charge, resulting in a nearly perfect balance of attractive and repulsive electromagnetic forces between the objects. Therefore, despite the fundamental strength of the electromagnetic forces, the net electromagnetic force between two macroscopic objects is often negligibly small except when atoms on the two surfaces come very close to each other—what we think of as *in contact*. This is the reason why the gravitational interactions are dominant for the bodies of astronomical size.

➤ *Electromagnetic (EM) forces* affects electrically charged particles only.

2.2.3. STRONG NUCLEAR FORCES

The strong force holds protons and neutrons together in the atomic nucleus. The same force binds quarks (a family of elementary particles) in combinations so they can form protons and neutrons and many more exotic subatomic particles. The strong force is the strongest of the four fundamental forces—hence its name—but its range is short: its effect is negligible at distances much larger than the size of an atomic nucleus (about 10^{-15} m).

Radioactivity, nuclear energy (fission, fusion) etc. results from nuclear forces.

If separation between particles $> 10^{-14} m$, then Nuclear forces $\ll$ Coulomb forces (EM force). But at small separation $\approx 10^{-15} m$, the nuclear force will be very strong as compared to Coulomb force.

2.2.4. WEAK NUCLEAR FORCES

Weak nuclear forces arise in certain radioactive decay processes. These forces are encountered when reactions involving protons, electrons and neutrons take place. In classical physics, we are concerned only with gravitational and electromagnetic forces.

The range of the weak force is even shorter than that of the strong force (about 10^{-17} m)

The electromagnetic force and the weak force are sometimes listed under one force, the electroweak force.

2.2.5. HOW FIELD FORCE ACTS THROUGH EMPTY SPACE?

In quantum mechanics bosons are force carrier particles and function as the "glue" that holds matter together and controls the interactions. The gravitational force is carried by bosons named as gravitons(mass = 0), the strong nuclear force is carried by gluons (mass = 0), the electromagnetic force is carried by particles of light, or photons (mass = 0), and the weak force is carried by W and Z bosons (mass of W boson $\approx$ 80.34 GeV/c^2 and mass of Z boson $\approx$ 91.19 GeV/c^2).

3. FUNDAMENTAL AND NON-FUNDAMENTAL FORCES

One of the main goals of physics has been to understand the immense variety of forces of nature in terms of the fewest number of fundamental laws. To reduce the number of fundamental laws we have to define all the forces of nature in terms of further irreducible forces of nature. On this basis, we can classify forces on two categories-
1. Fundamental forces
2. Non-fundamental forces.

3.1. FUNDAMENTAL FORCES

All irreducible forces in terms of which all the other forces of nature can be reduced, are called **fundamental forces**.

In our present understanding of physics, the only known four fundamental forces of nature are all field forces. These are listed below (from the weakest to the strongest)
1. *Gravitational forces*
2. *Weak Nuclear forces*
3. *Electromagnetic (EM) forces*
4. *Strong Nuclear forces*

The electromagnetic force and the weak force are sometimes listed under one force, the electroweak force.

3.2. NON-FUNDAMENTAL FORCES

The **non-fundamental forces** depend on one or more of the fundamental forces.

All contact forces are non-fundamental forces, because they can be reduced to electromagnetic forces between atoms. For example, the normal force exerted by a surface is a result of electromagnetic repulsive forces at the atomic level. Friction, tension, spring force, drag, thrust etc., can also be reduced to electromagnetic forces between atoms. So, we can say that *all contact forces are electromagnetic in nature*.

In this chapter, we will mostly focus on contact forces (non-fundamental forces), with the exception of the force of gravity, which is a fundamental force. Ordinary forces, other than gravity, such as pushes, pulls (for example tension in a string), and other contact forces like the normal force, friction, spring forces, surface tension and viscous forces, are today considered to be due to the electromagnetic force acting at the atomic level. For example, the force your fingers exert on a pencil is the result of electrical repulsion between the outer electrons of the atoms of your finger and those of the pencil.

☞ Out of so many natural forces, for distance $\approx 10^{-15} m$, nuclear force is strongest while gravitational force is weakest, i.e., $F_{nuclear} > F_{electromagnetic} > F_{gravitational}$. The nuclear force will not act for distance greater than $10^{-15} m$.

4. THE EXISTENCE OF A FIFTH FUNDAMENTAL FORCE

We have already discussed the four fundamental forces of nature. Recently, scientists in Hungary's Atomki Nuclear Research Institute, have found a solid evidence of a previously unknown fifth fundamental force of nature, but they are expecting more, independent experimental results to come for it in the coming years.

5. CHECKPOINT 1

1. • The fundamental forces
   (A) can be referred to as field forces
   (B) result from the fundamental interactions
   (C) are the basis of all the non-fundamental forces
   (D) all of the above
2. • A bar magnet is used to lift keys that have fallen into a sewer. The keys fly up a few millimeters to meet the bar magnet. Is the magnetic force a contact force or a field force? Explain
3. • When Neils Bohr shook hand with Werner Heisenberg, what kind of force they exerted?
   (A) gravitational (B) electromagnetic
   (C) nuclear (D) weak
4. • A fish swims in the ocean. Does the fish exert a force on the water? If so, is it a field force or a contact force? Explain.
5. • You blow a small piece of paper through the air. Is the force on the paper a contact force or a field force? Explain.
6. • Let E, G and N represent the magnitudes of electromagnetic, gravitational and nuclear forces between two electrons at a given separation. Then
   (A) $N > E > G$ (B) $E > N > G$
   (C) $G > N > E$ (D) $E > G > N$
7. •The sum of all electromagnetic forces between different particles of a system of charged particles is zero
   (A) only if all the particles are positively charged
   (B) only if all the particles are negatively charged
   (C) only if half the particles are positively charged and half are negatively charged
   (D) irrespective of the signs of the charges

6. INTERNAL VERSUS EXTERNAL FORCES

In addition to distinguishing between contact forces and field forces, we must also distinguish between internal and external forces. Any collection of two or more objects is known as a **system** provided that they are moving with same acceleration. An **internal force** is any force that acts *inside a system*, that is, any force exerted by one object in a system on another object in the same system. To determine which forces are internal to the system and which are external, we must first decide what we want to call the "system." For example, is the system in FIGURE 1 the whole car? Only some part of it? One or more occupants? Suppose we decide the whole car and its occupants are the system as indicated by the red loop. Then, any force exerted on the car (the system) by anything *outside* the system—such as the road, the Earth, or another car— is an **external force**. Any forces that are inside the system, such as the force exerted by the driver on the steering wheel or the force exerted by one child on the other child—are internal forces. In this chapter, we continue to use the particle model for objects and systems, so we represent the whole car and its occupants as a particle.

SYSTEM

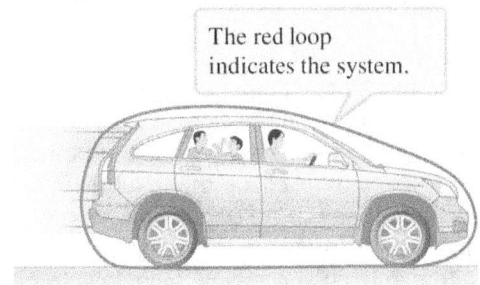

FIGURE 1. The people and the car have been chosen as the system (encircled in red). External forces due to the road or the Earth can accelerate the system. Internal forces such as the two children pushing on each other cannot.

7. REFERENCE FRAMES

A **frame of reference**, or a **reference frame**, consists of a coordinate system that includes the position of the origin; the direction of the axes; and the units of measurement, such as metres, seconds, etc. Notice that a frame of reference must include both the spatial coordinates, such as x, y, and z, and a temporal coordinate, such as time, t. Therefore, in the case of one-dimensional motion, to describe the location of an object at a certain moment of time, we need two coordinates: its position and time (x, t). In general, the number of coordinates needed to define the reference frame precisely for the n-dimensional case is $n + 1$. This explains the four-dimensional space-time (three spatial dimensions and one temporal dimension) described when we discuss Einstein's theory of relativity. We use this theory, when we talk about objects moving with speeds comparable to the speed of light. For now, we will discuss the principles of relativity as applied to slow-moving (as compared to the speed of light) objects, such as many of the objects we encounter in everyday life.

8. NEWTON'S LAWS OF MOTION AND FRAME OF REFERENCES

8.1. NEWTON'S FIRST LAW

What is the relationship between force and motion? Aristotle (384–322 B.C.) believed that a force was required to keep an object moving along a horizontal plane. To Aristotle, the natural state of an object was at rest, and a force was believed necessary to keep an object in motion. Furthermore, Aristotle argued, the greater the force on the object, the greater its speed.

Some 2000 years later, Galileo disagreed: he maintained that it is just as natural for an object to be in motion with a constant velocity as it is for it to be at rest.

Explanation of Galileo's idea: To understand Galileo's idea, consider the motion of an object on a horizontal plane. To push an object, with a rough surface, along a tabletop at constant speed a certain amount of force is required. To push an equally heavy object with a very smooth surface across the table at the same speed, comparatively, a less force is required. Now, if a layer of oil or other lubricant is placed between the surface of the object and the table, then almost no force is required to keep the object moving. Notice that in each successive step, less force is required. As the next step, we imagine that the object does not rub against the table at all or there is a perfect lubricant between the object and the table, then once started, the object would move across the table at constant speed with *no* force applied. A steel ball bearing rolling on a hard horizontal surface approaches this situation. So, does a puck on an air table, in which a thin layer of air reduces friction almost to zero.

Thus, we can conclude that that *if no force is applied to a moving object, it will continue to move with constant speed in a straight line. An object slows down only if a force is exerted on it.* Galileo thus interpreted friction as a force similar to ordinary pushes and pulls.

To push an object across a table at constant speed requires a force from your hand that can balance out the force of friction. When the object moves at constant speed, your pushing force is equal in magnitude to the friction force, but these two forces are in opposite directions, so the *net* force on the object (the vector sum of the two forces) is zero. This is consistent with Galileo's viewpoint, for the object moves with constant speed when no net force is exerted on it.

Upon this foundation laid by Galileo, Isaac Newton built his great theory of motion. Newton's analysis of motion is summarized in his famous "three laws of motion." In his great work, the *Principia* (published in 1687), Newton readily acknowledged his debt to Galileo. In fact, **Newton's first law of motion** is close to Galileo's conclusions. It states that-

Every object continues in its state of rest, or of uniform velocity in a straight line, as long as no net force acts on it.

The tendency of an object to maintain its state of rest or of uniform velocity in a straight line is called **inertia**. As a result, Newton's first law is often called the ***law of inertia**.*

*In **physics**, inertia means resistance to changes in velocity. It does not mean resistance to the continuation of motion (or the tendency to come to rest).*

EXAMPLE 1. Because Newton's first law is counterintuitive, it is important to take some time to think about what the law says and about how and why it differs from our intuition.

a. Why did the unavoidable presence of friction make it difficult for earlier scientists to come to the conclusion expressed in Newton's first law?

b. What is the natural state of an object?

c. How much force does it take to keep an object moving at constant velocity?

SOLUTION a. Because friction cannot be completely eliminated, any sliding object has at least one force acting on it. So, friction made it appear that rest was a natural state and that it takes a force to maintain motion.

b. Newton's first law implies rest is **not** a natural state and replaces the idea of a "natural state of rest" with "constant velocity."

c. No force is required to maintain constant velocity.

EXAMPLE 2. A school bus comes to a sudden stop, and all of the backpacks on the floor start to slide forward. What force causes them to do that?

SOLUTION It isn't "force" that does it. By Newton's first law, the backpacks continue their state of motion, maintaining their velocity. The backpacks slow down if a force is applied, such as friction with the floor.

EXAMPLE 3: ZERO NET FORCE MEANS CONSTANT VELOCITY

In the classic 1950 science fiction film *Rocketship X-M*, a spaceship is moving in the vacuum of outer space, far from any planet, when its engine dies. As a result, the spaceship slows down and stops. What does Newton's first law say about this event?

SOLUTION In this situation there are no forces acting on the spaceship, so according to Newton's first law, it will *not* stop. It continues to move in a straight line with constant speed. Some science fiction movies have made use of very accurate science; this was not one of them.

EXAMPLE 4: CONSTANT VELOCITY MEANS ZERO NET FORCE

You are driving a Porsche Carrera GT on a straight testing track at a constant speed of 150 km/h. You pass a 1971 Volkswagen Beetle doing a constant 75 km/h. For which car is the net force greater?

SOLUTION The key word in this question is 'net'. Both cars are in dynamic equilibrium because their velocities are constant; therefore, the *net* force on each car is *zero*.

This conclusion seems to contradict the 'common sense' idea that the faster car must have a greater force pushing it. It's true that the forward force on your Porsche is much greater than that on the Volkswagen. But there is also a *backward* force acting on each car due to road friction and air resistance. The only reason these cars need engines is to counteract this backward force so that the vector sum of the forward and backward forces will be zero and the car will travel with constant velocity. The backward force on your Porsche is greater because of its greater speed, so its engine has to be more powerful than the Volkswagen's.

9. CHECKPOINT 2

1. • Which of the following statements is correct?
   (A) It is possible for an object to have motion in the absence of forces on the object.
   (B) It is possible to have forces on an object in the absence of motion of the object.
   (C) Neither statement (A) nor statement (B) is correct.
   (D) Both statements (A) and (B) are correct.
2. • A rocket is being launched to place a new satellite in orbit. Air resistance is not negligible. What forces are being exerted on the rocket?
3. • A person stands on a spring scale in an elevator car as shown in adjoining figure. Which of these sources—the Earth, spring scale, elevator car, and cable—exert an external force if the system consists of:

   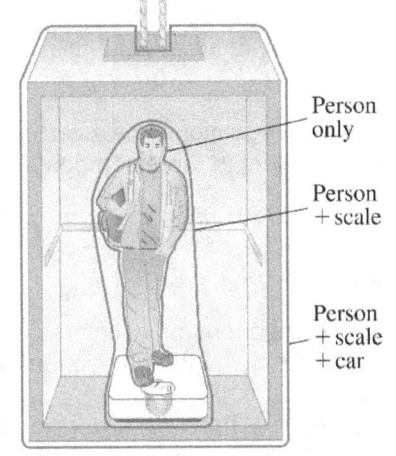

   a. Only the person?
   b. The person and the spring scale?
   c. The person, the spring scale, and the elevator car?
4. • You've just kicked a rock, and it is now sliding across the ground 2 m in front of you. Which of these forces act on the rock? List all that apply.
   (A) Gravity, acting downward.
   (B) The normal force, acting upward.
   (C) The force of the kick, acting in the direction of motion.
   (D) Friction, acting opposite the direction of motion.
5. • Newton's first law of motion describes the following
   (A) Energy   (B) Work
   (C) Inertia   (D) Moment of inertia
6. • An object is subject to a single constant force. Under what circumstances does the object travel in
   (A) a straight line
   (B) along a curved path,
   (C) along a circular path?
7. • Which of the following statements can be explained by Newton's first law?
   (I): When your car suddenly comes to a halt, you lunge forward.
   (II): When your car rapidly accelerates, you are pressed backward against the seat.
   (A) Neither I nor II   (B) Both I and II
   (C) I but not II   (D) II but not I
8. • A car moving at constant speed is suddenly braked. The occupants, all wearing seat belts, are thrown forward. The instant the car stops, however, the occupants are all jerked backward. Why? Is it possible to stop an automobile without this "jerk"?
9. • Three forces act on an object with $\vec{F}_1 = (6.03\hat{\imath} - 10.64\hat{\jmath})N$ and $\vec{F}_2 = (-3.71\hat{\imath} - 12.93\hat{\jmath})N$. If the net force on the object is zero, what is the unknown force, $\vec{F}_3$?
10. • What has more inertia, a bowling ball at rest or the same ball when it is rolling? Explain.
11. • you have more inertia when you are sitting or when you are running? Explain.
12. • Who has more inertia, a child sitting on the sofa or a heavy man jogging?

10. INERTIAL REFERENCE FRAMES

An object can be observed from a number of different perspectives, which are called *reference frames* in physics. The most important difference between reference frames is their relative motion.

Newton's first law does not hold in every reference frame. For example, if your reference frame is fixed in an accelerating car, an object such as a book resting on the dashboard may begin to move toward you (it stayed at rest as long as the car's velocity remained constant). The book accelerated toward you, but neither you nor anything else exerted a force on it in that direction. In accelerating reference frames, Newton's first law does not hold. Reference frames, in which Newton's first law holds, are called **inertial reference frames** (the law of inertia is valid in them). For most purposes, we usually make the approximation that a reference frame fixed on the Earth is an inertial frame. This is not precisely true, due to the Earth's rotation, but usually it is close enough.

Any reference frame that moves with constant velocity (say, a car or an airplane) relative to an inertial frame is also an inertial reference frame. Reference frames where the law of inertia does *not* hold, such as the accelerating reference frame discussed above, are called **noninertial** reference frames. How can we be sure a reference frame is inertial or not? By checking to see if Newton's first law holds. Thus, Newton's first law serves as the definition of inertial reference frames.

☞ **Newton's First Law Defines an Inertial Reference Frame**

☞ **As long as no external force acts on an object, it is always possible to find an inertial reference frame in which the velocity of the object is zero.** In other words, there is nothing special or "natural" about a state of rest.

11. NEWTON'S SECOND LAW OF MOTION ($\sum \vec{F} = m\vec{a}$)

Newton's first law explains what happens to an object when no forces act on it. It either remains at rest or moves in a straight line with constant speed. Newton's second law answers the question of what happens to an object that has a nonzero resultant force acting on it.

Consider the force required to push a block of ice when friction is small enough to ignore. (If there is friction, consider the *net* force, which is the force you exert minus the force of friction.) If you push the block horizontally, it moves with an acceleration of, say $2 \, m/s^2$. If you apply a force twice as large, the acceleration doubles to 4 m/s². If you triple the force, the acceleration is tripled, and so on. From such observations, we conclude that, the acceleration of an object is directly proportional† to the net applied force. But the acceleration depends on the mass of the object as well.

Suppose you stack identical blocks of ice on top of each other while pushing the stack with constant force. If the force applied to one block produces an acceleration of $2 \, m/s^2$, then the acceleration drops to half that value, 1 m/s², when two blocks are pushed, to one-third the initial value when three blocks are pushed, and so on. We conclude that **the acceleration of an object is inversely proportional to its mass.** These observations are summarized in **Newton's second law:**

The acceleration $\vec{a}$ of an object is directly proportional to the net force acting on it and inversely proportional to its mass.

[NEWTON'S SECOND LAW OF MOTION]

The constant of proportionality is equal to one, so in mathematical terms the preceding statement can be written as-

$$\vec{a} = \frac{\sum \vec{F}}{m}$$

[NEWTON'S SECOND LAW OF MOTION]

where $\vec{a}$ is the acceleration of the object, m is its mass, and $\sum \vec{F}$ is the vector sum of all forces acting on it. Above equation can also be written as-

$$\sum \vec{F} = m\vec{a} \qquad \ldots (1)$$

[NEWTON'S SECOND LAW OF MOTION]

Newton's second law relates the description of motion to the cause of motion, force. It is one of the most fundamental relationships in physics. From Newton's second law we can make a more precise definition of **force** as *an action capable of accelerating an object.* Above equation is valid only when the speed of the object is much less than the speed of light. In SI units, we define the unit of force that accelerates a standard $1 \, kg$ by $1 \, m/s^2$ as 1 newton (abbreviated to $1 \, N$). Thus, according to Eq.1, we have:

$$1N = (1kg)(1m/s^2) = 1 kg.m/s^2 \qquad \ldots (2)$$

Although we shall use SI units only from now on, other systems like the CGS (centimetre-gram-second) system and the British system are still in use. Table1 compares lists of all systems currently in use.

Table 1 Units in Newton's second law

System	Force[a]	Mass	Acceleration
SI	newton (N)	kilogram (kg)	m/s^2
CGS	dyne[b]	gram (g)	cm/s^2
British	Pound (lb)[c]	slug	ft/s^2

[a] $1N = 10^5 dyne = 0.255 \, lb.$ [b] $1 \, dyne = 1 \, g.cm/s^2$
[c] $1 \, lb = 1 \, slug.ft/s^2$

Note that equation (1) is a vector expression and hence is equivalent to three component equations.

$$\sum F_x = ma_x, \quad \sum F_y = ma_y, \quad \sum F_z = ma_z \qquad \ldots (3)$$

where, $\vec{F} = F_x \hat{\imath} + F_y \hat{\jmath} + F_z \hat{k}$

The component of acceleration in each direction is affected only by the component of the net force in that direction.

Equations (1) and (3) are valid only when the mass m is *constant*. It's easy to think of systems whose masses change, such as a leaking petrol tanker, a rocket ship or a moving railway wagon being loaded with coal. But such systems are better handled by using the concept of momentum; we'll get to that in next chapter.

☞ In $\sum \vec{F} = m\vec{a}$, $\sum \vec{F}$ is the vector sum of all the forces acting on the body, whereas $m\vec{a}$ is not a force but the product of the object's mass and acceleration. The equal sign says that they have the same value, not that they're the same thing. So, never add an extra force ma, when you're applying Newton's second law.

Acceleration is a *result* of a nonzero net force; it is not a force itself. It's 'common sense' to think that there is a 'force of acceleration' that pushes you back into your seat when your car accelerates forward from rest. But *there is no such force*; instead, your inertia causes you to tend to stay at rest relative to the earth, and the car accelerates around you. The 'common sense' confusion arises from trying to apply Newton's second law where it isn't valid, in the noninertial reference frame of an accelerating car. We will always examine motion relative to *inertial* frames of reference only.

☞ When force is written without direction, then positive force means repulsive while negative force means attractive, e.g., for two similarly charged particles (both positive or negative) electric force is positive meaning repulsive while between dissimilar charged particles is negative meaning attractive.

> ☞ Note that, Newton's first law is not just a special case of the second law ($\sum \vec{F} = m\vec{a}$) with zero acceleration. The first law *defines* what kind of reference frame we can use when applying the second law. For the second law to be valid, we must use an *inertial reference frame* to observe the motion of objects.

11.1. WORKING WITH NEWTON'S FIRST AND SECOND LAW

It is helpful to consider the following steps in solving Newton's second law problems. In this chapter, we limit consideration to situations in which we are only interested in the motion of small objects, and not how an extended object might rotate when forces are applied at different points on it.

The laws as stated above may be used even if the object under consideration is an extended body, provided each part of this body has the same acceleration (in magnitude and direction). A systematic algorithm for writing equations from Newton's laws is as follows:

Step 1. Identify the System *and the external forces on that system.*

(i) Identify the System

The first step is to identify the system on which the laws of motion are to be applied. The system may be a single particle, a block, a combination of two blocks one kept over the other, two blocks connected by a string, a piece of string etc. The only restriction is that all parts of the system should have identical acceleration.

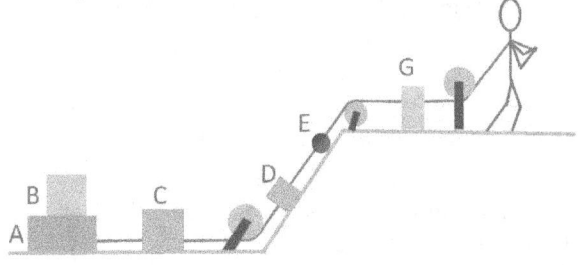

FIGURE 1

Consider the situation shown in FIGURE (1). The block B does not slip over A, the disc E slides over the string and all parts of the string are tight.

A and B move together. C is not in contact with A or B. But as the length of the string between A and C does not change, the distance moved by C in any time interval is same as that by A. The same is true for G. The distance moved by G in any time interval is same as that by A, B or C. The direction of motion is also the same for A, B, C and G. They have identical accelerations. We can take any of these blocks as a system or any combination of the blocks from these as a system. Some of the examples are (A), (B), $(A + B)$, $(B + C)$, $(A + B + C)$, $(C + G)$, $(A + C + G)$, $(A + B + C + G)$ etc. The distance covered by D is also the same as the distance covered by G but their directions are different. D moves along inclined line, whereas G in a horizontal line. Therefore, the accelerations of D and G are not identical and we cannot consider the combination $(D + G)$ as a single system. As the disc E slides over the string, therefore the distance covered by E is not equal to that by D in the same time interval. Therefore, the accelerations of D and E are different and hence, we cannot take $D + E$ as a single system. So, identify the system *carefully*.

(ii) Identify the External Forces that act on the System

Once the system is decided, make a list of the external forces acting *on* the system due to all the objects other than the system. Any force applied *by* the system should not be included in the list of the forces. For example, if you are in a rugby scrum, the external forces on *you* are all the pushes and pulls on *your* body. It does not include any push or pull on another player from you or from anyone else.

By external forces, we mean forces exerted on the body by other bodies in its environment. It's impossible for a body to affect its own motion by exerting a force on itself; if it were possible, you could lift yourself to the ceiling by pulling up on your belt! That's why only external forces are included in the sum $\sum \vec{F}$ in Equations (1) and (3).

Consider the situation shown in adjoining figure. In a circus balancing act, a woman performs a headstand on top of a standing performer's head. The woman weighs W_1, and the standing performer's weigh W_2. The upper woman presses the standing performer, the standing performer pushes the upper woman upward the standing performer presses the floor downward, the floor pushes the

standing performer upward, the earth attracts the upper performer downward, the upper performer attracts the earth upward, the standing performer attracts the earth upward and the earth attracts the standing performer downward. There are many forces operating in this world. List all the forces acting on standing performer and, on the woman, performing the act.

System	Force exerted by	Magnitude of force	Direction of the force	Nature of the force
Standing performer	Earth Floor Woman performing act	W_1 N N_1	Downward Upward Downward	Gravitational EM EM
Woman performing act	Earth Standing performer	W_2 N_1	Downward Upward	Gravitational EM

One may provide as much information as one has about the magnitude and direction of the forces. The contact forces may have directions other than normal to the contact surface if the surfaces are rough. We shall discuss more about it under the heading of friction.

Step 2: Make a Free Body Diagram (FBD)
A powerful tool for analyzing force problems is the **free-body diagram** (**FBD**). The FBD simplifies the problem by isolating the studied object from its environment. On this diagram, we draw *only the forces acting on the object*. Any force generated by the object itself is excluded. The forces on the diagram are always drawn pointing away from the object we are studying. To draw a free-body diagram:
- Draw the object in a simplified way—Almost any object can be represented as a box or a circle, or even a dot.
- Identify all the forces that are exerted on the object. Take care not to omit any forces that are exerted on the object. Consider that everything touching the object may exert one or more contact forces. Then identify long-range forces (for now, just gravity unless electric or magnetic forces are specified in the problem)
- Check your list of forces to make sure that each force is exerted *on* the object of interest *by* some other object. Make sure you have not included any forces that are exerted *on other objects*.
- Draw vector arrows representing all the forces acting on the object. We usually draw the vectors as arrows that start on the object and point away from it. Draw the arrows so they correctly illustrate the directions of the forces. If you have enough information to do so, draw the lengths of the arrows so they are proportional to the magnitudes of the forces.
- While we sometimes put the acceleration direction on the FBD, it should always be made clear that it is not a force. So, its arrow should not be connected with the system under consideration.

The free body diagram for the example discussed above with the standing performer as the system and with the upper acting performer as the system are shown in FIGURE 2.

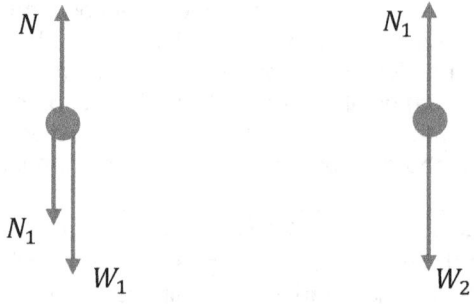

Lower standing performer   Upper acting performer
FIGURE 2

Step 3: Choose Axes and Write Equations
Any three mutually perpendicular directions may be chosen as the X-Y-Z axes. Picking a good coordinate system comes with practice. If the forces are coplanar, only two axes, say X and Y, taken in the plane of forces are needed. One helpful tip is to choose one axis to be parallel to the subject's acceleration.
Here, we use following sub steps-
(i) Establish the directions for the coordinate axes.
(ii) Define the positive direction on these axes.
(iii) Now divide *the forces into components* along the designated axes.
(iv) *Write Newton's second law for each axis direction*. These are called equations of motion.
(v) *Simultaneously solve* these equations. Clearly *state and interpret directions* for all quantities using your sign conventions. As for all problems, ask yourself whether your answer seems reasonable, for example, by considering limiting cases.

12. TRANSLATIONAL EQUILIBRIUM

When the net force acting on an object is zero, the object is said to be in **translational equilibrium**. *Equilibrium* conveys the idea that the forces are in balance; there is as much force upward as there is downward, as much to the right as to the left, and so on. Any object with a constant velocity, whether at rest or moving in a straight line at constant speed, is in translational equilibrium.
Thus, for an object in translational equilibrium,
$$\Sigma \vec{F} = \Sigma (F_x \hat{i} + F_y \hat{j} + F_z \hat{k}) = 0$$
here, F_x, F_y and F_z are, respectively, the x, y and z components of force.
As, a vector can only have zero magnitude if all of its components are zero, so-
For an object in translational equilibrium,
$$\Sigma \vec{F} = 0$$
i.e. $\Sigma F_x = 0, \Sigma F_y = 0,$ and $\Sigma F_z = 0$

If necessary, you can go to step 1, choose another object as the system, repeat steps 2, 3 and 4 to get more equations. This completes the algorithm.

If relevant, draw a vector arrow indicating the direction of the subject's acceleration, but visually different from the arrows representing the force vectors. For example, you might use a dashed line or a different colour. If the subject's acceleration is zero, note that on the diagram.

EXAMPLE 5: DRAWING FBDS

(a) In the adjoining figure, a *duck* sits on the *book* resting on the head of the *woman* who sits on a *chair* resting on the ground. Make FBDs of duck, book, woman and the chair.

(b) In an FBD, show all the forces acting on a crate that is sliding down a frictionless incline.

SOLUTION (a) List all the forces acting on different systems are given in following table

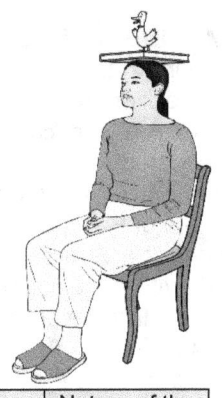

System	Force exerted by	Magnitude of the force	Direction of the force	Nature of the force
Duck	Earth	W_1	Downward	Gravitational
	Book	N	Upward	EM
Book	Earth	W_2	Downward	Gravitational
	Duck	N	Downward	EM
	Woman	N_1	Upward	EM
Woman	Earth	W_3	Downward	Gravitational
	Book	N_1	Downward	EM
	Chair	N_2	Upward	EM
Chair	Earth	W_4	Downward	Gravitational
	Woman	N_2	Downward	EM
	Floor	N_3	Upward	EM

Following figure shows FBDs of different objects given in the problem.

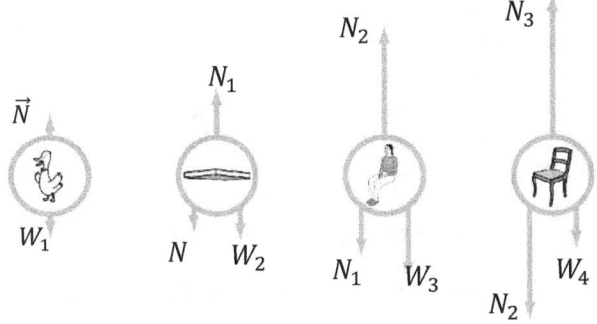

FIGURE 1 FBDs

(b) The forces acting on the crate are the normal force exerted by the inclined plane on the crate, $\vec{N}$, and the force of gravity, $\vec{F}_g = m\vec{g}$. The force of gravity near the surface of Earth acts vertically downward. A normal force must be perpendicular to the surface. Therefore, the FBD for the situation is as shown in FIGURE 1.

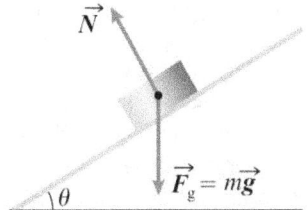

FIGURE 1 The FBD of a crate sliding down a frictionless incline.

MAKING SENSE OF THE RESULT

The acceleration is along the incline in the downward direction, but it is not a part of the FBD. Note that, although, there is no individual force downward along the incline, the net force when the forces $\vec{N}$ and $\vec{F}_g$ are added as vectors, is in downward direction along the incline.

EXAMPLE 6 MOTION ON A FRICTIONLESS INCLINE

A box is free to slide on a frictionless surface. The surface is inclined at an angle θ above the horizontal. Derive an expression for the acceleration of the mass, and calculate the magnitude of the normal force exerted by the inclined surface on the mass.

APPROACH To find acceleration and normal reaction, we will apply the analytical strategy outlined earlier in this section.

SOLUTION **Step 1** *Identify the system and the forces on that system.*

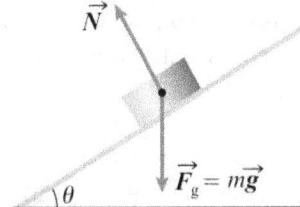

FIGURE 1 The FBD of a crate sliding down a frictionless incline.

The system is the box, and the forces acting on the box are the force of gravity vertically downward, and the normal force perpendicular to the incline (FIGURE 1)

Step 2. *Draw the FBD.*

We did this in the preceding example, with the result shown in FIGURE 1. Notice that we have shown the surface of the incline in the FBD, as well as the box, but the FBD is the box represented by the small circle, and the vectors for the two forces acting on it. Drawing the incline makes it easier to resolve the forces into components and determine the angles and the components.

Step 3. Choose Axes and Write Equations.

(i) Establish the directions for the coordinate axes.

We use the direction of the assumed acceleration to help guide our optimum choice. From common sense we know that the box will accelerate and slide down along the incline. Therefore, one of our axes (we will call it the x-axis) should be along the incline, and the other (the y-axis) will be perpendicular to the incline.

(ii) *Define the positive direction for the coordinate axes.* Which direction is positive is entirely a matter of choice, and as long as you stay consistent with your choice there is no right or wrong answer. Here we will define down along the incline as the positive x-direction, and upward perpendicular to the incline as the positive y-direction. These are demonstrated in FIGURE 2.

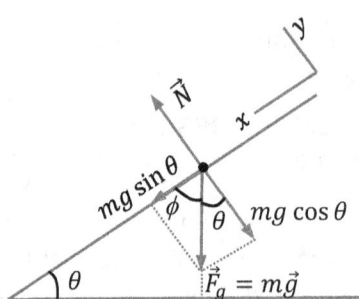

FIGURE 2. This FBD shows the vector components for the gravitational force. For clarity, we have offset the orientation of the axes. The direction of the assumed acceleration is shown, but it is not part of the FBD.

(iii) *Resolve the forces into components along the chosen axes.*

The normal force, $\vec{N}$, from the incline on the block points perpendicular to the plane along the y-axis and does not need to be resolved. To resolve the force of gravity on the block into components, we draw two lines from the tip of the $\vec{F}_g$ vector, one parallel to the incline (along the x-axis) and one perpendicular to the incline (along the y-axis). These lines intersect the axes at the tips of the components of $\vec{F}_g$ (see FIGURE 2).

From the angles of the triangle, we know that $\theta + \phi + 90° = 180°$, which gives $\theta + \phi = 90°$. Therefore, we can see that the angle between the $\vec{F}_g$ vector and the perpendicular to the incline must be θ, as shown, since this angle plus ϕ equals $90°$.

The magnitude of the force of gravity on the box of mass m is mg, so the component of $\vec{F}_g$ perpendicular to the incline must be $mg \cos \theta$, as shown in FIGURE 2, while the component along the incline is $mg \sin \theta$.

(iv) *Write the equations from Newton's second law for each axis.*

Now we will use Newton's second law to write equations for the x-axis and the y-axis directions, keeping in mind that we have chosen down along the incline as the positive x-direction, and upward perpendicular to the incline as the positive y-direction:

X-axis: $\quad mg \sin \theta = ma \quad$ ... (1)
Y-axis: $\quad N - mg \cos \theta = 0 \quad$ ... (2)

Here N without the vector sign means the magnitude of the normal force.

(v) *Simultaneously solve these equations of motion.* We can obtain the magnitude of the normal force directly from (2):

$$N = mg \cos \theta \quad ... (3)$$

From equation (1):

acceleration, $a = mg \sin \theta / m = g \sin \theta \quad$ ... (4)

Since we obtained a positive value for the acceleration, and we defined the positive direction of the x-axis as downward along the incline, as expected the direction of the acceleration is downward along the incline.

MAKING SENSE OF THE RESULT We can use limiting cases to check our result. When the mass is placed on a horizontal surface, we have $\theta = 0$, so $\sin \theta = 0$ and $\cos \theta = 1$. In this case, we obtain the expected results that the acceleration is zero and the magnitude of the normal force is simply mg. An incline angle of $90°$ means that the mass is effectively placed against a vertical wall, in which case it will simply free-fall with an acceleration equal to g. This is what we obtain from relationships (3) and (4), since now $\sin \theta = 1$ and $\cos \theta = 0$. Note that the value of the mass does not enter into the results obtained.

EXAMPLE 7 HORIZONTAL PUSH ON BOX ON AN INCLINED PLANE

FIGURE 1 shows a worker, wearing special boots for added traction, pushing a crate of mass m up an incline with a slope of angle θ using a horizontal force $\vec{F}$. Find the normal force from the incline on the crate, and determine the acceleration of the crate. Assume that the friction between the crate and the incline is negligible. Express your answers in terms of the variables m, g, θ, and $\vec{F}$.

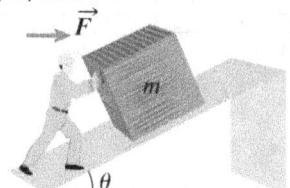

FIGURE 1. The person pushes the box with an entirely horizontal force $\vec{F}$

APPROACH To find the normal force from the incline on the crate, and the acceleration of the crate, we use Newton's second law of motion.

SOLUTION. Here again, we invoke our strategy for solving Newton's second law problems. It is important to keep in mind that the applied force is entirely horizontal (rather than directed along the incline).

Step 1. *Identify the system and the forces on that system.*

We are concerned with the motion of the crate, so that is the system. There are three forces acting on the crate: the force of gravity, the normal force exerted by the incline on the crate, and the horizontal force applied by the worker.

Step 2. *Draw the FBD for crate.*

The FBD is drawn in FIGURE 2, with the three forces acting on the crate being the normal force, $\vec{N}$; the force of gravity, $\vec{F}_g$; and the force applied by the person, $\vec{F}$.

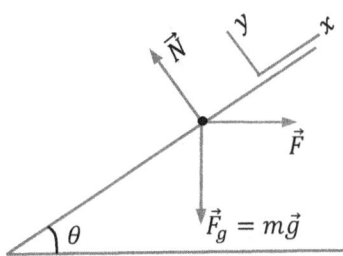

FIGURE 2 This FBD shows the three forces (normal force, force of gravity, and horizontal force applied by the person) acting on the crate.

Step 3. Choose Axes and Write Equations.

(i) *Establish the directions for the coordinate axes.*

We use the direction of the assumed acceleration to help guide our choice of axes. Depending on the strength of the applied force, the crate can either accelerate up along the incline, or if the applied force is too weak, it might slide back down the incline. Therefore, one of our axes (we will call it the x-axis) should be along the incline, and the other (the y-axis) is perpendicular to the incline.

(ii) *Define the positive direction for the coordinate axes.*
Which direction is positive is a matter of choice. We will define up along the incline as the positive x direction, and upward perpendicular to the incline surface as the positive y-direction. These are demonstrated in FIGURE 2 through the offset pair of axes.

(iii) *Resolve the forces into components along the chosen axes.*

The normal force is already in the positive y-axis direction. The gravitational force and the applied force need to be divided into components along the two axes. The technique is similar to that used in the previous example.

The applied force, $\vec{F}$, is divided into components $F\cos\theta$ along the incline and $F\sin\theta$ perpendicular to the incline. The gravitational force has a component $mg\sin\theta$ down along the incline, and a component $mg\cos\theta$ perpendicular to the incline. We show the components in FIGURE 3, along with the angles used to obtain them.

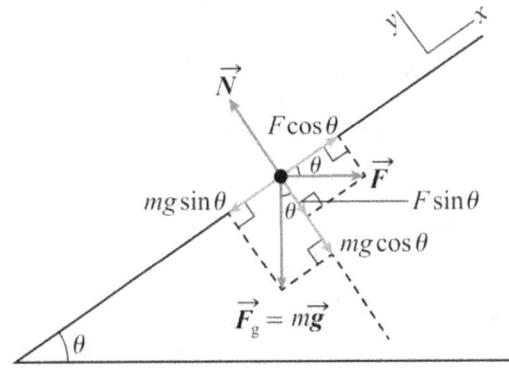

FIGURE 3 Force components, using one axis along the inclined plane and one perpendicular to it.

(iv) *Write the equations from Newton's second law for each axis.*

Keeping in mind that the positive x-direction is up along the incline, and the positive y-direction is up perpendicular to the incline, we obtain the following as we apply Newton's second law for each of the two axis directions.

For the x-axis motion, we have
$$F\cos\theta - mg\sin\theta = ma \qquad \ldots (1)$$

For the y-axis, we have
$$-F\sin\theta - mg\cos\theta + N = 0 \qquad \ldots (2)$$

(v) *Simultaneously solve these equations of motion.*
We can solve (2) for the normal force:
$$N = F\sin\theta + mg\cos\theta \qquad \ldots (3)$$

Note here that the normal force is *not equal* to $mg\cos\theta$. We can find the magnitude of the acceleration of the crate from (1):
$$a = \frac{F\cos\theta - mg\sin\theta}{m} \qquad \ldots (4)$$

Provided that the force was sufficient to make expression (4) positive, the direction of the acceleration is up along the incline, as per our sign convention.

MAKING SENSE OF THE RESULT

Since the worker is pushing the crate partially against the face of the incline, the y-component of the applied force contributes to the normal force. Consequently, the magnitude of the normal force is greater than $mg\cos\theta$. If we consider the limiting cases applied to the acceleration result (4), we see that if there were no incline (i.e., $\theta = 0$), then we would simply have $a = F/m$, as expected.

EXAMPLE 8 HAWK IN EQUILIBRIUM A red-tailed hawk that weighs 8 N is gliding due north at constant speed. What is the force acting on the hawk due to the air? Draw a free-body diagram for the hawk.

APPROACH Since the hawk is gliding at constant velocity (constant speed *and* direction), it is in equilibrium—the net force acting on it is zero. We identify the forces acting on the hawk and determine what the force due to the air must be for the net force to be zero.

SOLUTION Step 1. *Identify the system and the forces on that system.*

The system is the hawk, only two forces are acting on the hawk. One of them is long-range: the downward gravitational pull of the Earth ($\vec{F}_{grav}$) whose magnitude is the bird's weight. The other is an upward air thrust (contact force due to the air) ($\vec{F}_{air}$). Nothing else is in contact with the bird, so there are no other contact forces (FIGURE 1).

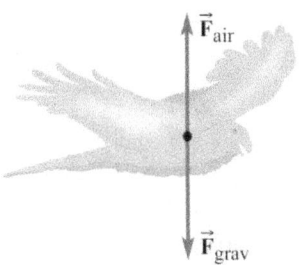

FIGURE 1. FBD for the hawk gliding with constant velocity

Step 2. Draw the FBD.
The FBD is shown by dot in FIGURE 1. Notice that we have shown the hawk in the FBD, although it is not required. The forces on the system are shown by vectors.

Step 3. Choose Axes and Write Equations
(i) Establish the directions for the coordinate axes.
As the problem is one dimensional, so only one axis is sufficient in this case. Now it is also given that the hawk is gliding due north at constant speed. So, the system is in equilibrium along vertical direction, i.e., the magnitude of acceleration $a = 0$.
So, let assume that vertical direction is the direction of Y-axis.

(ii) Define the positive direction for the coordinate axes.
It is a very simple one-dimensional case. Let the upward direction is the positive direction of Y-axis. No need other axes.

(iii) Resolve the forces into components along the chosen axes.
F_{grav} is acting along $-ve$ direction of chosen Y-axis whereas F_{air} is acting along $+ve$ direction of Y-axis.

(iv) Write the equations from Newton's second law for each axis.
Now we will use Newton's second law to write equations for the Y-axis, keeping in mind that we have chosen upward direction as the positive Y-direction:
Y-axis: $F_{air} - F_{grav} = 0$ ... (1)
($\because$ the system is in equilibrium along vertical direction)

(v) Simultaneously solve these equations of motion.
$$F_{air} = F_{grav} = 8\ N$$
From the above equation, it is clear that $\vec{F}_{air}$ and $\vec{F}_{grav}$ both have the same magnitude and opposite directions.
i.e., $\vec{F}_{air} = 8N$ (upward).

DISCUSSION The interaction between the hawk and the air is extremely complex at the microscopic level, but the net effect of all the interactions between air molecules and the hawk produces an upward force of $8\ N$.

EXAMPLE 9 FORCE APPLIED BY THE STRING AND THE TABLE
A toy chest of mass M (with its contents) is pulled on a smooth horizontal table by a string making an angle θ with the horizontal as shown in FIGURE (1a). If the acceleration of the toy chest is a, find the force applied by the string and by the table on the chest.

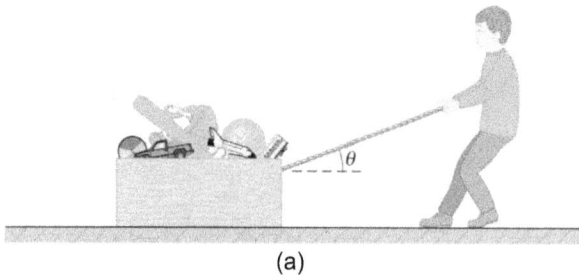
(a)

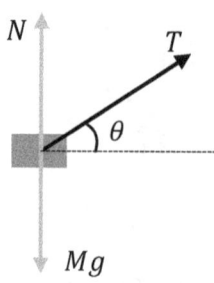
(b) FBD
FIGURE 1

APPROACH To find the force applied by the string and by the table on the system, we apply Newtons second law along horizontal and vertical direction.

SOLUTION. Step 1. Identify the system and the forces on that system.
System is the chest with its contents. The forces on the system are-
(i) pull of the earth, Mg, vertically downward,
(ii) contact force by the table, N, vertically upward,
(iii) pull of the string, T, along the string.

Step 2. Draw the FBD.
The free body diagram for the block is shown in

Step 3. Choose Axes and Write Equations
(i) Establish the directions for the coordinate axes.
The acceleration of the block is horizontal and towards the right. Take this direction as the X-axis and vertically upward direction as the Y-axis.

(ii) Define the positive direction for the coordinate axes.
Since the chest is moving with constant acceleration a horizontally, suppose it is +ve direction of X axis. Also, chest is in vertical equilibrium, therefore
$$\sum F_x = Ma \quad \text{and} \quad \sum F_y = 0$$

(iii) Resolve the forces into components along the chosen axes.

Along X-axis
component of Mg along the X-axis $= 0$
component of N along the X-axis $= 0$
component of T along the X-axis $= T\cos\theta$
$\therefore \quad \sum F_x = 0 + 0 + T\cos\theta = T\cos\theta$
or $\quad \sum F_x = T\cos\theta$

Along Y-axis
component of Mg along the Y-axis $= -Mg$
component of N along the Y-axis $= N$
component of T along the Y-axis $= T\sin\theta$
$\therefore \quad \sum F_y = -Mg + N + T\sin\theta$

(iv) Write the equations from Newton's second law for each axis.
From Newton's second law, along X-axis, we have
$$\sum F_x = Ma$$
$\Rightarrow \quad T\cos\theta = Ma$ ... (1)
From Newton's second law, along Y-axis, we have
$$\sum F_y = 0$$
$\Rightarrow \quad -Mg + N + T\sin\theta = 0$ ... (2)

(v) Simultaneously solve these equations of motion.

From equation (1),
$$T = \frac{Ma}{\cos\theta}$$

From equation (2),
$$N = Mg - T\sin\theta$$
or $\quad N = Mg - \left(\dfrac{Ma}{\cos\theta}\right)\sin\theta \quad \left[\because T = \dfrac{Ma}{\cos\theta}\right]$

or $\quad N = M(g - a\tan\theta)$

EXAMPLE 10 THREE BALANCED FORCES

Two ropes are attached to a ring and exert forces as shown (**FIGURE** 1). The magnitude of these forces is given by $F_1 = 2F_2$ and $f \equiv F_2 = 22.0\ N$. A third force is applied by a rope so that the ring's acceleration is zero. What is the magnitude and direction of the force applied by the third rope?

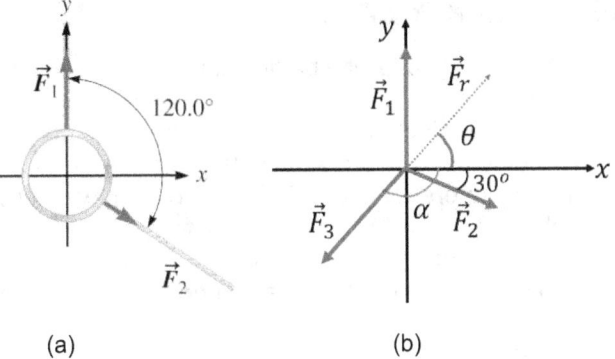

(a) (b)
FIGURE 1

APPROACH In order for the acceleration to be zero the sum of the three forces must be zero. To anticipate the result, we geometrically add $\vec{F}_1$ and $\vec{F}_2$. In order for the net force to be zero, $\vec{F}_3$ must point opposite to the resultant ($\vec{F}_r$) of $\vec{F}_1$ and $\vec{F}_2$. Our goal is to find the magnitude of $\vec{F}_3$ and the angle α from our sketch (FIGURE b) we expect the magnitude of $\vec{F}_3$ is similar to that of $\vec{F}_1$ and absolute value of α is greater than 90°.

SOLUTION. Step 1. *Identify the system and the forces on that system.*
The system is the ring. Three forces are $\vec{F}_1, \vec{F}_2$ and $\vec{F}_3$.
Step 2. *Draw the FBD.*
The free body diagram for the ring is shown in FIGURE (1b).
Step 3. Choose Axes and Write Equations
(i) Establish the directions for the coordinate axes.
   The acceleration of the ring is zero. So, you can consider any two perpendicular directions for coordinate axes. We are considering, X axis in horizontal direction and Y- axis along vertical direction.
(ii) *Define the positive direction for the coordinate axes.*
   +ve directions of axes are shown by arrow tip in FIGURE 1b.
(iii) *Resolve the forces into components along the chosen axes.*
   Along X-axis
   component of F_1 along the X-axis = 0
   component of F_2 along the X-axis = $F_2 \cos 30^\circ$
   component of F_3 along the X-axis = $-F_3 \cos\theta$
   $\therefore \quad \Sigma F_x = 0 + F_2 \cos 30^\circ - F_3 \cos\theta$

or $\quad \Sigma F_x = F_2 \cos 30^\circ - F_3 \cos\theta$

Along Y-axis
component of F_1 along the Y-axis = F_1
component of F_2 along the Y-axis = $-F_2 \sin 30^\circ$
component of F_3 along the Y-axis = $-F_3 \sin\theta$
$\therefore \quad \Sigma F_y = F_1 - F_2 \sin 30^\circ - F_3 \sin\theta$

(iv) *Write the equations from Newton's second law for each axis.*
The ring is not accelerating, so the net force in both the x and y direction is zero.

$\therefore$ From Newton's second law, along X-axis, we have
$$\Sigma F_x = 0$$
$$\Rightarrow F_2 \cos 30^\circ - F_3 \cos\theta = 0 \quad \ldots (1)$$
and from Newton's second law, along Y-axis, we have
$$\Sigma F_y = 0$$
$$\Rightarrow F_1 - F_2 \sin 30^\circ - F_3 \sin\theta = 0 \quad \ldots (2)$$

(v) *Simultaneously solve these equations of motion.*

We have two equations and two unknowns (F_3 and θ).

From, equation (1), we have
$$F_2 \cos 30^\circ = F_3 \cos\theta \quad \ldots (3)$$

and from equation (2), we have
$$F_1 - F_2 \sin 30^\circ = F_3 \sin\theta \quad \ldots (4)$$

Eliminate F_3 by dividing Equation (4) by Equation (3). Use $F_1 = 2F_2$ and $f \equiv F_2$ to simplify. Solve for θ by using $\tan\theta = \dfrac{\sin\theta}{\cos\theta}$.

$$\dfrac{\sin\theta}{\cos\theta} = \dfrac{F_1 - F_2 \sin 30^\circ}{F_2 \cos 30^\circ}$$

or $\quad \tan\theta = \dfrac{F_1 - F_2 \sin 30^\circ}{F_2 \cos 30^\circ}$

or $\quad \tan\theta = \dfrac{2f - f\sin 30^\circ}{f \cos 30^\circ} = \dfrac{2 - \sin 30^\circ}{\cos 30^\circ}$

or $\quad \tan\theta = \dfrac{2 - (1/2)}{\sqrt{3}/2} = \dfrac{3}{\sqrt{3}} = \sqrt{3}$

or $\quad \theta = \tan^{-1} \sqrt{3} = 60^\circ$

Solve either Equation (3) or (4) for F_3 and substitute values. (We choose Equation 3.)

$$F_3 = \dfrac{F_2 \cos 30^\circ}{\cos\theta} = \dfrac{(22\ N)\cos 30^\circ}{\cos 60^\circ}$$

or $\quad F_3 = 38.1\ N$

To find the direction α, we use figure b. We see $\alpha + \theta = 180^\circ$, and because the direction is measured clockwise from the x axis, it must be negative.

$$\alpha = -(180^\circ - 60^\circ) = -120^\circ$$

MAKING SENSE OF THE RESULT

The magnitude $F_1 = 2f = 44.0$ N. So, as expected the magnitude of F_3 is similar to that of F_1. Also, as expected the absolute value of α is greater than 90°.

EXAMPLE 11 FORCE TO ACCELERATE A FAST CAR Estimate the net force needed to accelerate (a) a 1000-kg car at (b) a 200-g apple at the same rate.

APPROACH We use Newton's second law to find the net force needed for each object. This is an estimate (the $\frac{1}{2}$ is not said to be precise) so we round off to one significant figure.

SOLUTION (a) The car's acceleration is $a = \frac{1}{2}g = \frac{1}{2}(9.8 \, m/s^2) \approx 5 \, m/s^2$. We use Newton's second law to get the net force needed to achieve this acceleration:
$$\Sigma F = ma \approx (1000 \, kg)(5 \, m/s^2) = 5000 \, N.$$
(b) For the apple, $m = 200 \, g = 0.2 \, kg$, so
$$\Sigma F = ma \approx (0.2 \, kg)(5 \, m/s^2) = 1 \, N.$$

EXAMPLE 12 FORCE TO STOP A CAR What average net force is required to bring a 1500-kg car to rest from a speed of 100 km/h within a distance of 55 m?

APPROACH We use Newton's second law, to determine the force, but first we need to calculate the acceleration a. We assume the acceleration is constant, so we can use the kinematic equation, to calculate it.

SOLUTION We assume the motion is along the axis (see adjoining **FIGURE**). We are given the initial velocity, the final velocity and the distance travelled We have
$$v^2 = v_0^2 + 2a(x - x_0),$$
so $a = \frac{v^2 - v_0^2}{2(x - x_0)} = \frac{0 - (27.8 \, m/s)^2}{2(55 \, m)} = -7.0 \, m/s^2.$

The net force required is then
$$\Sigma F = ma = (1500 \, kg)(-7 \, m/s^2) = -1.1 \times 10^4 \, N.$$
The force must be exerted in the direction *opposite* to the initial velocity, which is what the negative sign means.

NOTE If the acceleration is not precisely constant, then we are determining an "average" acceleration and we obtain an "average" net force.

☞ **Another statement of Newton's first law (based on inertial frames):** In the absence of external forces, when viewed from an inertial reference frame, an object at rest remains at rest and an object in motion continues in motion with a constant velocity (that is, with a constant speed in a straight line).

If nothing acts to change the object's motion, then its velocity does not change. From the first law, we conclude that any isolated object (one that does not interact with its environment) is either at rest or moving with constant velocity when viewed from an inertial frame. The tendency of an object to resist any attempt to change its velocity is called **inertia**.

$$\Sigma \vec{F} = 0, \quad \text{then} \quad \begin{cases} \vec{v} = 0 \\ \text{or} \\ \vec{v} = \text{constant} \end{cases} \quad \ldots (1)$$
(Newton's first law)

☞ Newton's second law, like the first law, is valid only in inertial reference frames. In the noninertial reference frame of an accelerating car, for example, a cup on the dashboard starts sliding—it accelerates—even though the net force on it is zero; thus $\Sigma \vec{F} = m\vec{a}$ doesn't work in such an accelerating reference frame ($\Sigma \vec{F} = 0$ but $\vec{a} \neq 0$ in this noninertial frame).

13. CHECKPOINT 3

1. • How can you tell if a particular reference frame is an inertial reference frame?
2. • If only a single nonzero force acts on an object, must the object have an acceleration relative to an inertial reference frame? Can it ever have zero velocity?
3. • A box of mass m is placed on an inclined frictionless plane. The magnitude of the normal force
   (A) equals mg
   (B) is greater than mg
   (C) is less than mg
   (D) may be less than, equal to, or greater than mg, depending on the value of θ
4. • Which of the following reference frames are inertial frames?
   (A). An airplane cruising in a straight path at constant speed
   (B) An airplane taking off
   (C) A car taking a sharp turn
   (D) None of above are correct
5. • Two rubber bands stretched to the standard length cause an object to accelerate at 2 m/s². Suppose another object with twice the mass is pulled by four rubber bands stretched to the standard length. The acceleration of this second object is
   (A) 1 m/s²   (B) 2 m/s²
   (C) 4 m/s²   (D) 8 m/s²
6. • (a) Take a moment to be sure that you understand the distinction between Newton's first two laws. How are they different from each other?
   (b) According to Newton's second law, what is the acceleration of an object if there are no forces acting on it? Is your answer consistent with Newton's first law?
7. • A small body of super dense material, whose mass is half the mass of the earth (but whose size is very small compared to the size of the earth), starts from rest at a height H above the earth's surface, and reaches the earth's surface in time t. Calculate time t assuming that H is very small

8. •• A block slides down a frictionless plane inclined at an angle θ. For what value of angle θ the horizontal component of acceleration of the block is maximum? Find this maximum horizontal acceleration.

9. • A particle's acceleration is $\vec{a} = (3.45\hat{i} - 1.84\hat{j})$ m/s², and the total force exerted on the particle is $\vec{F}_{tot} = (10.8\hat{i} - 5.76\hat{j})N$. What is the particle's mass?

10. •• A helicopter of mass $M = 15000$ kg is lifting a cubical box of mass $m = 2000$ kg. The helicopter is going up with an acceleration of $a = 1.2$ m/s². The four strings are tied at mid points of the sides of the square face PQRS of the box. The strings are identical and form a knot at K. Another string KH connects the knot to the helicopter. Neglect mass of all strings and take $g = 10$ m/s². Length of each string AK, BK, CK

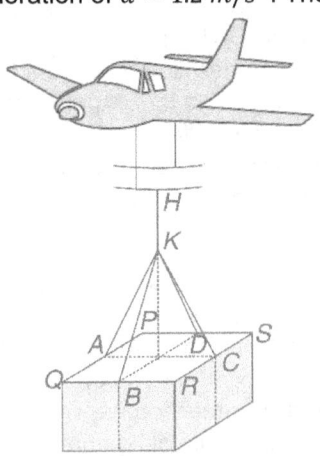

and DK is equal to side length of the cube. (a) Find tension T in string AK. (b) Find tension T_0 in string KH. (c) Find the force (F) applied by the atmosphere on the helicopter. Assume that the atmosphere exerts a negligible force on the box. (d) If the four strings are tied at P, Q, R and S instead of A, B, C & D, how will the quantities T, T_0 and F change? Will they increase or decrease? Assume that length of the four identical strings remains same.

14. CONCEPT OF INERTIAL MASS AND GRAVITATIONAL MASS

It may seem obvious that the gravitational force between any two objects depends on the objects' masses, but there is no theoretical basis for this. That is, no one knows why the gravitational force depends on inertia (mass) and not some other property of the objects, such as volume, density, or composition.

The mass in the law of universal gravity $\left(F_G = G\frac{M_e m}{R_e^2}\right)$ is referred to as the object's **gravitational mass**. It is the property of the particles that creates a gravitational force between them. As we know, the mass in Newton's second law $\left(\sum \vec{F} = m\vec{a}\right)$ is the *inertial mass* of an object. Experimental evidence supports the idea that the gravitational mass of any object equals its inertial mass.

Let us see how this reasoning works-

Case 1. We *assume* the gravitational mass m_g is identical to the inertial mass m_i and then see if this assumption matches experimental evidence.

The magnitude of the gravitational force on an object of gravitational mass m_g near the Earth's surface is

$$F_G = G\frac{M_e m_g}{R_e^2}$$

Suppose our object of inertial mass m_i is in free fall near the Earth's surface. If it is, the only force acting on it is gravity, and we use Newton's second law to find the object's free-fall acceleration g:

$$F_G = m_i g \quad \text{or} \quad G\frac{M_e m_g}{R_e^2} = m_i g$$

Using our assumption that inertial mass and the gravitational mass are identical, i.e., $m_i = m_g \equiv m$ (say), then from above equation, we have

$$G\frac{M_e m}{R_e^2} = mg$$

or or $g = \frac{GM_e}{R_e^2}$ ... (1)

Equation 1 predicts that g is a constant because $\frac{GM_e}{R_e^2}$ is a constant. We have found a testable prediction for the assumption that $m_i = m_g \equiv m$: If the assumption is correct, the gravitational acceleration near the surface of the Earth g has the same value for all objects and does not depend on any property such as size or shape of the object in free fall.

We know from countless experiments carried out over centuries that in the absence of air resistance, all objects near the surface of the Earth fall at the same acceleration g no matter what physical properties they have. Therefore, we infer that the gravitational mass of any object is equal to its inertial mass from the observation that the acceleration due to gravity is constant near the surface of the Earth.

Case 2. In an unaccelerated elevator in the gravitational field of earth (FIGURE 1a), the normal reaction of elevator on the foot of observer

$$N = m_g g$$

Now, if the elevator is accelerated in upward direction with an acceleration g in the absence of any gravitational field (FIGURE 1b), then normal reaction

$$N' = m_i g$$

According to the equivalence principle, these two forces are precisely the same, i.e.,

$$N' = N \quad \text{or} \quad m_i g = m_g g$$

or $m_i = m_g$

so, *the inertial mass must equal the gravitational mass.*

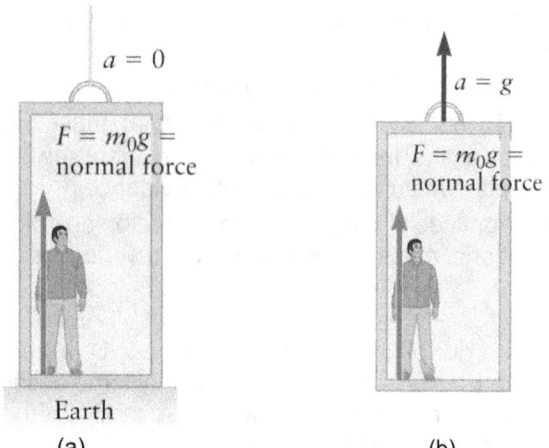

FIGURE 1 A person in an elevator feels a normal force $m_0 g$ on the bottoms of his feet. He cannot tell if he is **(a)** at rest in a gravitational field so that the normal force is the result of a gravitational force or if **(b)** there is no gravitational field and the elevator has an upward acceleration $a = g$. According to the equivalence principle, there is no way to tell the difference between these two situations.

Case 3. As a last thought on this matter, imagine that the gravitational mass did *not* equal the inertial mass. If that were the case, you might weigh two objects on a spring scale and find that they have identical weights, but when you pushed each object horizontally along a frictionless surface, the same force would *not* produce the same acceleration (FIGURE 2).

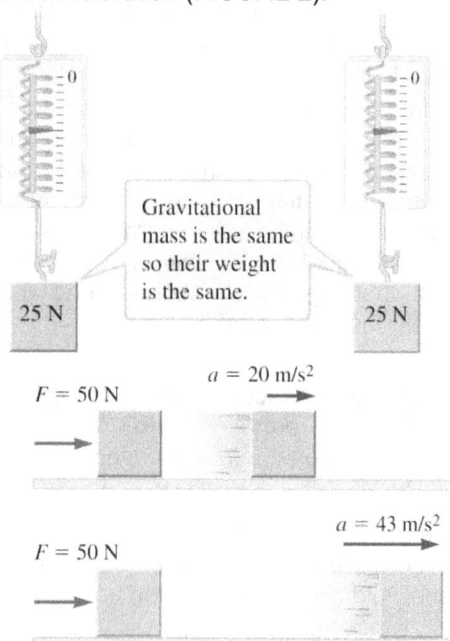

FIGURE 2. If gravitational mass and inertial mass were not the same, two objects that have the same weight may not acquire the same acceleration when the same net force is exerted on each.

☞ The idea that gravitational mass and inertial mass are identical may seem trivial, but this idea is actually very important and led Einstein to discover his theory of *general relativity*.

☞ **Point Mass:** An object with zero dimensions, is considered as a point mass. Practically, if during the motion in a given time, the object covers distance much greater than its own size, the it can be considered as a point mass. Actually, point mass is only a mathematical concept to simplify the problems.

15. SOME SPECIFIC TERMS

15.1. FREE FALL

If you drop a hammer and a feather, you know what will happen. The hammer quickly strikes the ground, and the feather flits and floats and lands some time later. But if you do this experiment on the moon, the result is strikingly different: Both the hammer and the feather experience the exact same acceleration, undergo the exact same motion, and strike the ground at the same time.

The moon lacks an atmosphere, and so objects in motion above its surface experience no air resistance. There is one and only one force that matters—gravity. If an object moves under the influence of gravity only, and no other forces, we call the resulting motion **free fall**. Early investigators concluded, correctly, that any two objects in free fall, regardless of their mass, have the same acceleration. Thus, if you drop two objects and they are both in free fall, they hit the ground at the same time.

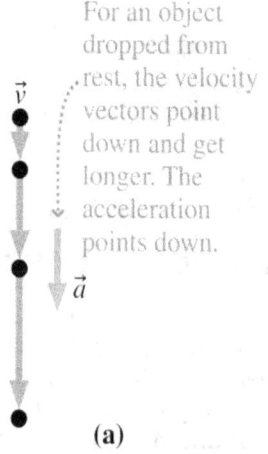

For an object dropped from rest, the velocity vectors point down and get longer. The acceleration points down.

(a)

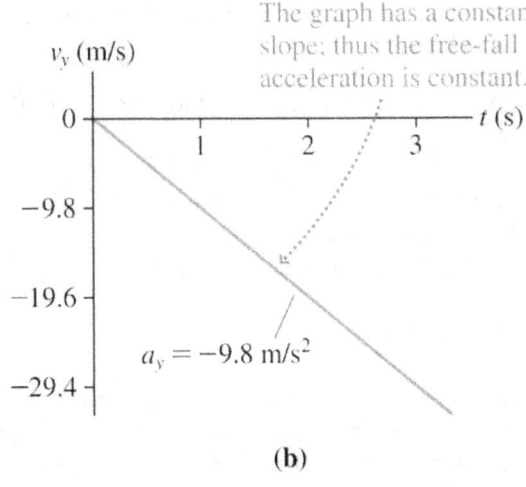

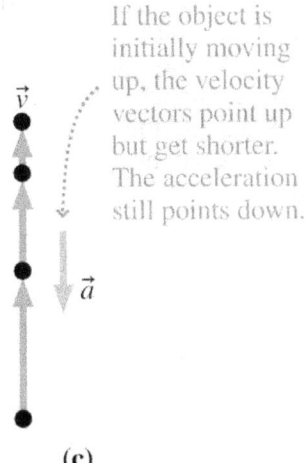

FIGURE 1 Motion of an object in free fall.

FIGURE 1a shows the motion diagram for an object that was released from rest and falls freely. Since the acceleration is the same (in absence of air resistance) for all objects, the diagram and graph would be the same for a falling baseball or a falling boulder! FIGURE 1b shows the object's velocity graph. The velocity changes at a steady rate. The slope of the velocity-versus-time graph is the free-fall acceleration $a_{free\ fall}$. Instead of dropping the object, suppose we throw it upward. What happens then? You know that the object will move up and that its speed will decrease as it rises. This is illustrated in the motion diagram of FIGURE 1c, which shows a surprising result: Even though the object is moving up, its acceleration still points down. In fact, the freefall acceleration always points down, no matter what direction an object is moving.

☞ Despite the name, free fall is not restricted to objects that are literally falling. Any object moving under the influence of gravity only, and no other forces, is in free fall. This includes objects falling straight down, objects that have been tossed or shot straight up, objects in projectile motion (such as a passed football), and, as we will see, satellites in orbit.

Thus, free fall is a situation in which the only force acting on an object is the force due to gravity.

The value of the free-fall acceleration varies slightly at different places on the earth, but for the calculations in this book we will use the following average value:

$$a_{free\ fall} = 9.80\ m/s^2,\ (\text{vertically downward})$$

Standard value for the acceleration of an object in free fall.

The magnitude of the **free-fall acceleration** has the special symbol g: $g = 9.80\ m/s^2$

We will generally work with two significant figures and so will use $g = 9.8\ m/s^2$.

Several points about free fall are worthy of note:

- g, by definition, is *always* positive. There will never be a problem that uses a negative value for g.
- The velocity graph in Figure 1b has a negative slope. Even though a falling object speeds up, it has *negative* acceleration. Alternatively, notice that the acceleration vector $\vec{a}_{\text{free fall}}$ points down. Thus g is *not* the object's acceleration, simply the magnitude of the acceleration. The one-dimensional acceleration is

$$a_y = a_{free\ fall} = -g$$

- Because free fall is motion with constant acceleration, we can use the kinematic equations for constant acceleration with $a_y = -g$.
- g is not called "gravity." Gravity is a force, not an acceleration. g is the *free-fall acceleration*.
- $g = 9.80\ m/s^2$ only on earth. Other planets have different values of g.
- We will sometimes compute acceleration in units of g. An acceleration of $9.8\ m/s^2$ is an acceleration of $1g$; an acceleration of $19.6\ m/s^2$ is $2g$. Generally, we can compute

 acceleration (in units of g)
 $$= \frac{\text{acceleration in units of } m/s^2}{9.8\ m/s^2}$$

 This allows us to express accelerations in units that have a definite physical reference.

Suppose a body of mass m is in free fall with the free-fall acceleration of magnitude g. Then, if we neglect the effects of the air, the only force acting on the body is the gravitational force $\vec{F}_g$. We can relate this downward force and downward acceleration with Newton's second law ($\vec{F} = m\vec{a}$). We place a vertical y axis along the body's path, with the positive direction upward. For this axis, Newton's second law can be written in the form

$$\sum F_y = ma_y$$

which, in our situation, becomes

$$-F_g = m(-g)$$

or $\qquad F_g = mg \qquad \qquad \qquad \ldots (1)$

In words, the magnitude of the gravitational force is equal to the product mg.

Because of its role in Eqn.1, g is also known as the **local acceleration due to gravity**.

15.2. AT REST

This same gravitational force, with the same magnitude, still acts on the body even when the body is not in free fall but is, say, at rest on a pool table or moving across the table. (For the gravitational force to disappear, Earth would have to disappear.)

We can write Newton's second law for the gravitational force in these vector forms:

$$\vec{F}_g = -F_g \hat{\jmath} = -mg\hat{\jmath} = m\vec{g} \qquad \ldots (2)$$

where $\hat{\jmath}$ is the unit vector that points upward along the y axis, directly away from the ground, and $\vec{g}$ is the free-fall acceleration (written as a vector), directed downward.

15.3. WEIGHT

The **weight** W of a body is the magnitude of the net force required to prevent the body from falling freely, as measured by someone on the ground. For example, to keep a ball at rest in your hand while you stand on the ground, you must provide an upward force to balance the gravitational force on the ball from Earth.

Suppose the magnitude of the gravitational force is 3.0 N. Then the magnitude of your upward force must be 3.0 N, and thus the weight W of the ball is 3.0 N. We also say that the ball *weighs* 3.0 N and speak about the ball *weighing* 3.0 N.

A ball with a weight of 4.0 N would require a greater force from you— namely, a 4.0 N force—to keep it at rest. The reason is that the gravitational force you must balance has a greater magnitude—namely, 4.0 N. We say that this second ball is *heavier* than the first ball.

Now let us generalize the situation. Consider a body placed on a table which is unaccelerated with respect to the ground. Two forces act on the body: a downward gravitational force $\vec{F}_g$ and a balancing upward force of magnitude W. This balancing force is applied by table on the body under consideration. We can write Newton's second law for a vertical y axis, with the positive direction upward, as

$$\sum F_y = ma_y$$

In our situation, this becomes

$$W - F_g = m(0) \qquad \ldots (3)$$

or

$$W = F_g \qquad \ldots (4)$$

(weight, with ground as inertial frame)

This equation tells us (assuming the ground is as an inertial frame) that-

The weight W of a body is equal to the magnitude $\vec{F}_g$ of the gravitational force on the body.

Substituting mg for F_g from Eq. 1, we find

$$W = mg \text{ (weight)}, \qquad \ldots (5)$$

which relates a body's weight to its mass.

15.4. WEIGHING

To *weigh* a body means to measure its weight. One way to do this is to place the body on one of the pans of an equal-arm balance (**FIGURE 1**) and then place reference bodies (whose masses are known) on the other pan until we strike a balance (so that the gravitational forces on the two sides match). The masses on the pans then match, and we know the mass of the body. If we know the value of g for the location of the balance, we can also find the weight of the body with Eq. (5).

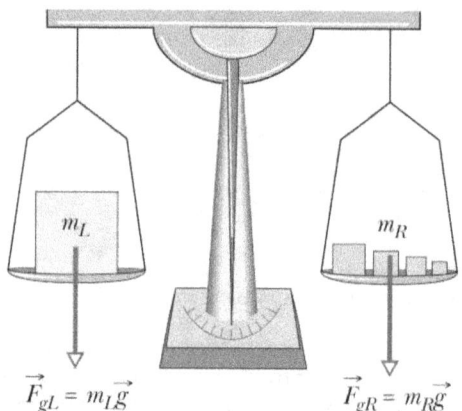

FIGURE 1. An equal-arm balance. When the device is in balance, the gravitational force $\vec{F}_{gL}$ on the body being weighed (on the left pan) and the total gravitational force $\vec{F}_{gR}$ on the reference bodies (on the right pan) are equal. Thus, the mass m_L of the body being weighed is equal to the total mass m_R of the reference bodies.

We can also weigh a body with a spring scale (FIGURE 2). The body stretches a spring, moving a pointer along a scale that has been calibrated and marked in either mass or weight units. (Most bathroom scales in the United States work this way and are marked in the force unit pounds.) If the scale is marked in mass units, it is accurate only where the value of g is the same as where the scale was calibrated.

The weight of a body must be measured when the body is not accelerating vertically relative to the ground. For example, you can measure your weight on a scale in your bathroom or on a fast train. However, if you repeat the measurement with the scale in an accelerating elevator, the reading differs from your weight because of the acceleration. Such a measurement is called an *apparent weight*.

Caution: A body's weight is not its mass. Weight is the magnitude of a force and is related to mass by Eq. 5. If you move a body to a point where the value of g is different, the body's mass (an intrinsic property) is not different but the weight is. For example, the weight of a bowling ball having a mass of 7.2 kg is 71 N on Earth but only 12 N on the Moon. The mass is the same on Earth and Moon, but the free-fall acceleration on the Moon is only $1.6\ m/s^2$.

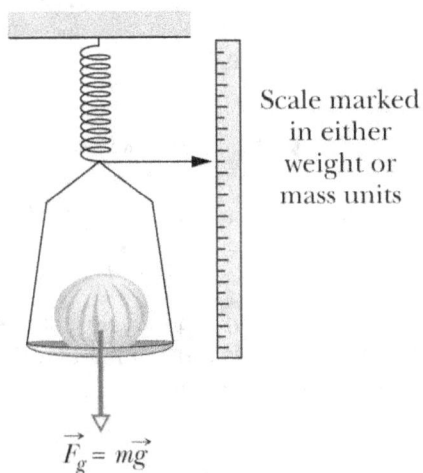

FIGURE 2. A spring scale. The reading is proportional to the *weight* of the object on the pan, and the scale gives that weight if marked in weight units. If, instead, it is marked in mass units, the reading is the object's weight only if the value of g at the location where the scale is being used is the same as the value of g at the location where the scale was calibrated.

16. CONTACT FORCES
(A MICROSCOPIC EXPLANATION)

When we put two bodies in contact with each other, the atoms at the two surfaces come close to each other. These atoms exert great **electromagnetic forces** on each other. The net effect of enormous numbers of **electromagnetic forces** between atoms on the surfaces of the two objects in contact, is called the *Contact force* between the surfaces.

There is no sharp distinction between contact forces and field forces. When examined at the atomic level, all the forces we classify as contact forces turn out to be caused by electric (field) forces.

16.1. SPRING FORCE

A coiled metallic wire is called the spring. The distance between two successive turns in a spring remains the

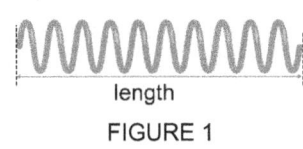

FIGURE 1

same. The straight-lined end to end distance of a spring is called its length (FIGURE 1). If a *spring* is neither compressed nor stretched, then it is said to be the *relaxed spring*. The length of a relaxed spring is called its natural length. Every spring has its own natural length.

FIGURE 2 shows a spring fixed on one end and attached to a block on the other end, which is free to move. A *relaxed spring* does not exert any force on the attached block or on the fixed support. When either compressed or stretched from the relaxed state, however, the spring exerts a contact force on the block and fixed support. Depending on whether the spring is compressed or stretched, the contact force can be either a push or a pull respectively.

The force due to the spring is a **restoring force**; that is, it is directed so as to return ("restore") the spring to its relaxed state. The farther the spring is from being in its relaxed state, the stronger the restoring force. If the extension or the compression is not too large, the restoring force exerted by the spring is proportional to the change in its length. This law is called Hooke's law. Let us chose the x axis to lie along the direction the spring extends or compresses. If x is the position of the block relative to its equilibrium ($x = 0$) position, then by Hooke's law -

$$F_s = -kx \qquad \ldots (1a)$$

In vector form,

$$\vec{F_s} = F_s \hat{\imath} = -kx\hat{\imath} \qquad \ldots (1b)$$

The negative sign in equation (1) signifies that the force exerted by the spring is always directed *opposite* the displacement from equilibrium. The constant k is known as the **spring constant**, a scalar property that differs from one spring to another; it measures the stiffness of the spring. Stiff springs have large k values, and soft springs have small k values. The SI unit of k is newtons per meter (N/m).

In other words, the force required to stretch or compress a spring is proportional to the amount of stretch or compression x. This force law for springs is known as **Hooke's law.**

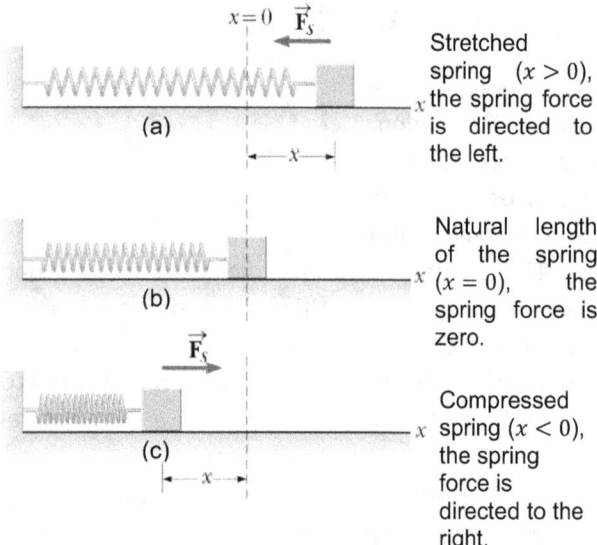

FIGURE 2 The force exerted by a spring on a block varies with the block's position x relative to the equilibrium position $x = 0$. (a) x is positive. (b) x is zero. (c) x is negative.

When $x > 0$ as in FIGURE 2a so that the block is to the right of the equilibrium position and the spring is stretched, the spring force is directed to the left, in the negative x direction. When $x < 0$ as in FIGURE 2c, the block is to the left of equilibrium, the spring is compressed, and the spring force is directed to the right, in the positive x direction. When $x = 0$ as in FIGURE 2b, the spring is unstretched and $F_s = 0$. This force comes into picture due to the electromagnetic forces between the atoms of the material. When spring is compressed or stretched, the subatomic particles (electrons and protons) exert unbalanced electric forces on one another to restore the spring to its equilibrium status.

When we say an ideal spring, we mean a spring that obeys Hooke's law and is also massless.

Since we have assumed spring to be massless, therefore, forces acting on both ends have to be equal and opposite, to have net force on spring to be zero (See FBD shown in FIGURE 3).

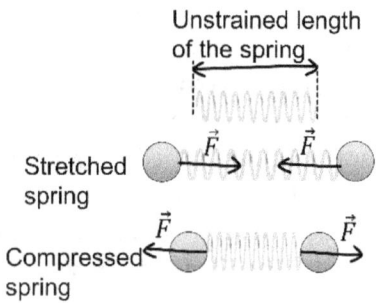

FIGURE 3

Now, suppose, Rahul and Rohit, are pulling a spring from two ends as shown in Fig. 4. Rahul moves x_1 and Rohit moves x_2. In this case, the net extension in spring $x = x_1 + x_2$.

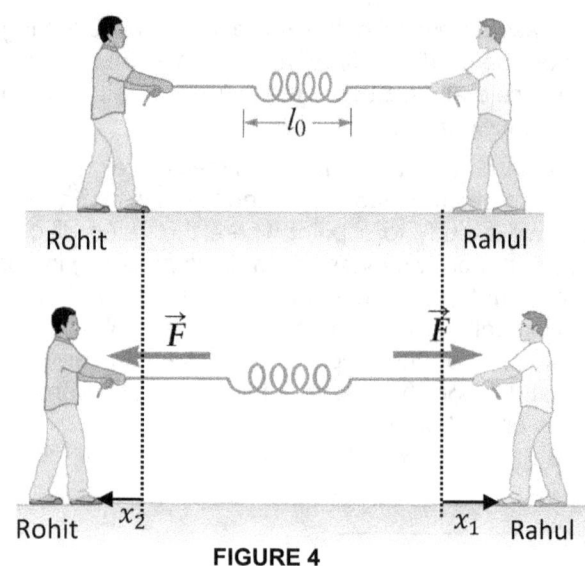

FIGURE 4

Therefore, the force acting on Rahul and Rohit is $k(x_1 + x_2)$, Not kx_1 on Rahul and kx_2 on Rohit. Force due to spring is kx where x is defined as $|l - l_0|$, here, l is present length and l_0 is natural length of spring.

16.1.1. Equivalent spring constant

1. Springs in Parallel: In Fig. 1, two springs having spring constants k_1 and k_2, are connected in parallel with a block in two different ways.

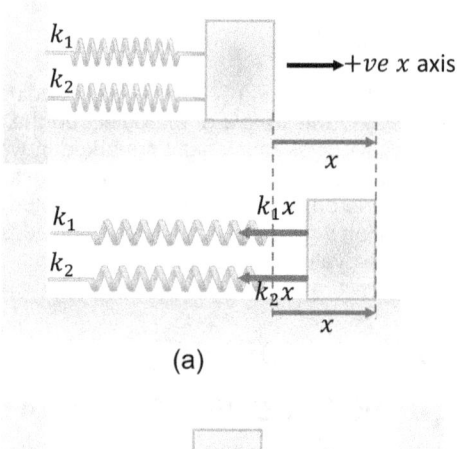

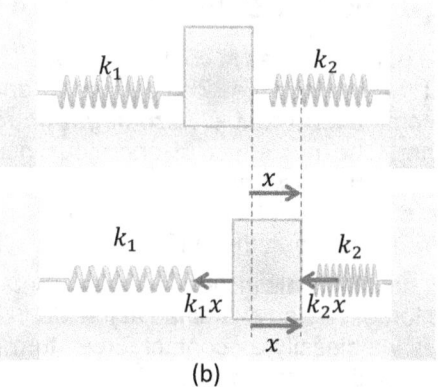

FIGURE 1

We can displace the block either right side (+ve direction of x axis) or left side. If we displace it towards right side by distance x, then in Fig. 1a, both springs get expanded by same amount x and hence apply the restoring forces $k_1 x$ and $k_2 x$ toward left side, whereas in Fig. 1b, left spring k_1 get expanded by length x and right spring k_2 get compressed by length x. As a result, they again apply their restoring forces $k_1 x$ and $k_2 x$ toward left side, i.e., in the same direction.

Similarly, if we displace the block towards left, then in Fig. 1a, both springs get compressed and hence apply the restoring forces toward right side, whereas in Fig. 1b, left spring k_1 get compressed and right spring k_1 get expanded. As a result, they again apply their restoring forces toward right side, i.e., in the same direction.

Now, suppose, the block is shifted by distance x, towards right, then, the net restoring force on the block,

$$\vec{F}_{net} = -(k_1 x + k_2 x)\hat{\imath} \qquad \ldots (1)$$

Here, $\hat{\imath}$ is the unit vector along the positive direction of x axis, i.e., towards right side.

If we replace, the combination of these two springs by an equivalent spring having effective spring constant k, then, corresponding to same expansion x, we can write

$$\vec{F}_{net} = -kx\hat{\imath} \qquad \ldots (2)$$

On comparing equations (1) and (2), we get

$$-kx\hat{\imath} = -(k_1 x\hat{\imath} + k_2 x)\hat{\imath}$$

or $\qquad k = k_1 + k_2 \qquad \ldots (3)$

Similarly, in general, for n springs, having spring constant $k_1, k_2, k_3, \ldots, k_n$ respectively, we have

$$k = k_1 + k_2 + k_3 + \cdots k_n \qquad \ldots (4)$$

2. Springs in Series: In Fig.2, two springs two ideal springs of lengths l_1 and l_2, having spring constants k_1 and k_2 respectively, are connected in series with a block. If we displace the block towards right, then both springs get expanded and if we displace the block towards left, then both springs get compressed simultaneously.

As for a massless spring or string, the spring force or tension at every point on it always remains same (it will be shown in next article. 'TENSION'), therefore if we displace the block by a distance x towards right, then same restoring force will be produced in both the springs.

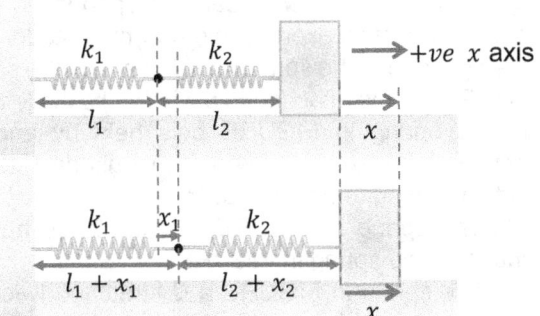

FIGURE 2

Suppose, we displaced the block by distance x along the $+ve$ direction of x axis and corresponding to displacement x of the block towards right, the expansions in springs k_1 and k_2 are x_1 and x_2 respectively, then the restoring force produced in both the springs is given by

$$\vec{F} = -k_1 x_1 \hat{\imath} = -k_2 x_2 \hat{\imath} \qquad \ldots (5)$$

or $\qquad k_1 x_1 = k_2 x_2 \qquad \ldots (6)$

From above it is clear that, in series-
if, $k_1 \neq k_2$, then $x_1 \neq x_2$,

Thus, for two springs in series having different spring constants, the expansions x_1 and x_2 in springs will always be different.

The net expansion, $x = x_1 + x_2$

In vector for, $x\hat{\imath} = x_1\hat{\imath} + x_2\hat{\imath} \qquad \ldots (7)$

Substituting the values of $x_1\hat{\imath}$ and $x_2\hat{\imath}$ from (5) in (7), we get

$$x\hat{\imath} = -\frac{\vec{F}}{k_1} - \frac{\vec{F}}{k_2} \qquad \ldots (8)$$

Now, if we replace the series combination by an equivalent spring having effective spring constant k, then, corresponding to the expansion x, we have

$$\vec{F} = -kx\hat{\imath}$$

$\therefore \qquad x\hat{\imath} = -\frac{\vec{F}}{k} \qquad \ldots (9)$

Substituting, this value of $x\hat{\imath}$, in (8), we get

$$-\frac{\vec{F}}{k} = -\frac{\vec{F}}{k_1} - \frac{\vec{F}}{k_2}$$

or $\qquad \frac{1}{k} = \frac{1}{k_1} + \frac{1}{k_2} \qquad \ldots (10)$

Similarly, in general, for n springs in series, we can show that

$$\frac{1}{k} = \frac{1}{k_1} + \frac{1}{k_2} + \frac{1}{k_3} + \cdots + \frac{1}{k_n} \qquad \ldots (11)$$

If spring of spring constant k and length l is cut into two pieces of length l_1 and l_2, then

$k \propto \frac{1}{l_1 + l_2}$, $k_1 \propto \frac{1}{l_1}$ and $k_2 \propto \frac{1}{l_2}$

Spring constant k is inversely proportional to length of spring i.e., $k \propto \frac{1}{l}$ (l = natural length of spring)

16.1.2. Dividing a Spring in to two Parts

If the spring of spring constant k is cut into two parts in the ratio of lengths $m:n$, then formed new springs have new spring constants given by

$$k_1 = \frac{k(m+n)}{m}, \qquad k_2 = \frac{k(m+n)}{n}$$

The original spring can be considered as a series combination of these two divided parts, therefore the series combination of these two parts will give the original spring constant as shown below-

$$\frac{1}{k_1}+\frac{1}{k_2}=\frac{m}{k(m+n)}+\frac{n}{k(m+n)}=\frac{m+n}{k(m+n)}=\frac{1}{k}$$

16.1.3. How to Solve Massive Spring Problems

A system containing a block of mass M attached with a massive spring of mass m is equivalent to a spring block system with massless spring attached with a block of mass $\left(M+\frac{m}{3}\right)$. In other words, to replace a massive spring (mass $= m$) with a massless spring, in a spring block system, we have to attach an extra mass $m/3$ to the mass of the block M. Here, $m/3$ is the effective mass of spring on the attached body.

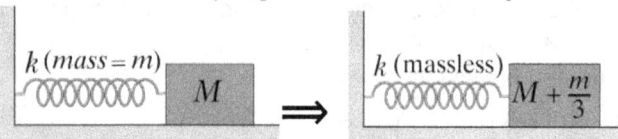

FIGURE 3

16.1.4. Effect on Tension in a spring on sudden Removal of Force

It is important to note that in a string block system, the tension in the string changes immediately on changing the weight of the block attached to it; while in a spring block system, spring need to expand or compress to change tension, i.e., tension does not change immediately on changing the weight attached with spring. The following example is based on this concept.

EXAMPLE 13 THREE SPRING SYSTEM *The system shown in adjoining figure is in equilibrium. Find the initial acceleration of A, B and C just after the spring-2 is cut.*

APPROACH To get elongations and spring forces in equilibrium, first make equilibrium condition equations for all three blocks and spring systems. After that cut spring 2 (in this case spring force of spring 2 will be zero), and form equations of motion for each block by using Newton's second law of motion. To get the acceleration of each block, you have to use the values of spring forces as obtained in equilibrium conditions.

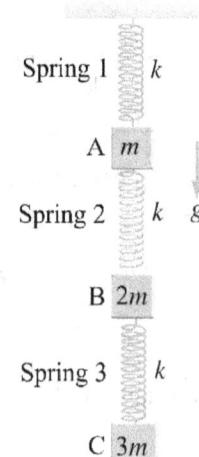

SOLUTION

In equilibrium (**FIGURE** 1)
$3mg = kx_3$ ... (1)
$2mg + kx_3 = kx_2$
$\therefore \quad 2mg + 3mg = kx_2$
$\Rightarrow 5mg = kx_2$ ... (2)
$kx_1 = kx_2 + mg = 5mg + mg$
i.e., $\quad kx_1 = 6mg$ ... (3)

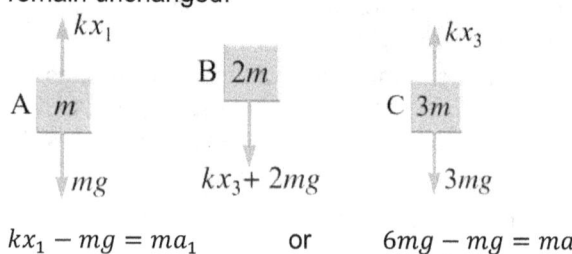

FIGURE 1

When spring 2 is cut spring force in other two strings remain unchanged.

$kx_1 - mg = ma_1$ or $6mg - mg = ma_1$

$\Rightarrow a_1 = 5g \uparrow$ (Since at instant when the spring 2 is cut, the spring force on spring 1 is kx_1.
$kx_3 + 2mg = 2ma_2$
or $\quad 3mg + 2mg = 2ma_2$
$\Rightarrow a_2 = 5g \Rightarrow a_2 = \frac{5g}{2} \downarrow$
$kx_3 - 3mg = 3ma_3$
$\Rightarrow 3mg - 3mg = 3ma_3 \Rightarrow a_3 = 0$
Thus, acceleration of $3m$ will be zero.

17. CHECKPOINT 4

1. • The net force on a moving object is suddenly reduced to zero and remains zero. As a consequence, the object (a) stops abruptly, (b) stops during a short time interval, (c) changes direction, (d) continues at constant velocity, (e) changes velocity in an unknown manner.
2. • FIGURE shows two identical springs. In both cases, a person exerts a force of magnitude F on the right end of the spring. The left end of the spring in FIGURE a is attached to a wall, while the left end of the spring in FIGURE b is held by another person, who exerts a force of magnitude F to the left. Which statement is true?
(A) The spring in FIGURE a is stretched half as much as the spring in FIGURE b.
(B) The spring in FIGURE a is stretched twice as much as the spring in FIGURE b.
(C) The two springs are stretched the same amount.

(D) The spring in FIGURE b is stretched greater amount as compared to the spring in FIGURE a.

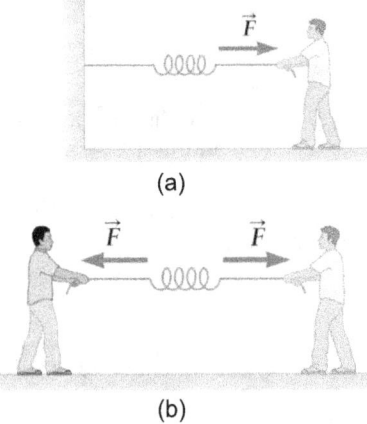

3. •• An ideal spring is in its natural length (L) with two objects A and B connected to its ends. A point P on the unstretched spring is at a distance $2L/3$ from B. Now the objects A and B are moved by 4 cm to the left and 8 cm to the right respectively. Find the displacement of point P.

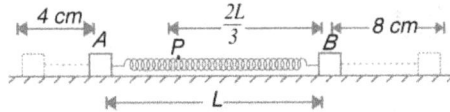

4. •• The figure shows an infinite tower of identical springs each having force constant k. The connecting bars and all springs are massless. All springs are relaxed and the bottom row of springs is fixed to horizontal ground. The free end of the top spring is pulled up with a constant force F. In equilibrium, find (a) The displacement of free end A of the top spring from relaxed position. (b) The displacement of the top bar B1 from the initial relaxed position.

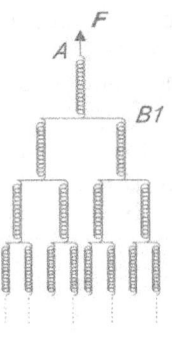

5. •• A uniform light spring has unstretched length of 3.0 m. One of its ends is fixed to a wall. A particle of mass $m = 20\ g$ is glued to the spring at a point 1.0 m away from its fixed end. The free end of the spring is pulled away from the wall at a constant speed of 5 cm/s. Assume that the spring remains horizontal (i.e., neglect gravity). Force constant of spring = 0.6 N /cm. (a) With what speed does the particle of mass m move? (b) Find the force applied by the external agent pulling the spring at time 2.0 s after he started pulling

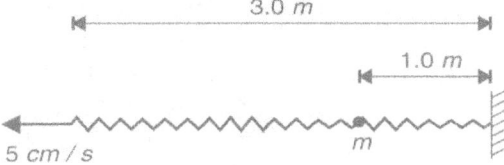

18. NORMAL FORCE

A contact force perpendicular to the contact surface that prevents two objects from passing through one another is called the normal force. (In geometry, the word *normal* means *perpendicular*.) Consider a book resting on a horizontal table surface. The normal force due to the table must have just the right magnitude to keep the book from falling through the table. If no other vertical forces act, the normal force on the book is equal in magnitude to the book's weight because the book is in equilibrium (FIGURE 1a).

According to Newton's third law, a normal force is also exerted on the table by the book; this normal force acts downward and is of equal magnitude. In *everyday* language, we might say that the table "feels the book's weight." That is not an accurate statement in the language of physics. The table cannot "feel" the gravitational force on the book; the table can only feel forces exerted *on the table*. What the table does "feel" is the normal force—a *contact* force—exerted on the table by the book.

If the table's surface is horizontal, the normal force on the book will be vertical and equal in magnitude to the book's weight. If the surface of the table is *not* horizontal, the normal force is not vertical and is not equal in magnitude to the weight of the book. Remember that the normal force is *perpendicular to the contact surface* (FIGURE1b). Even on a horizontal surface, if there are other vertical forces acting on the book, then the normal force is *not* equal in magnitude to the book's weight (FIGURE 1c). Never *assume* anything about the magnitude of the normal force. In general, we can figure out what the magnitude of the normal force must be in various situations if we have enough information about other forces.

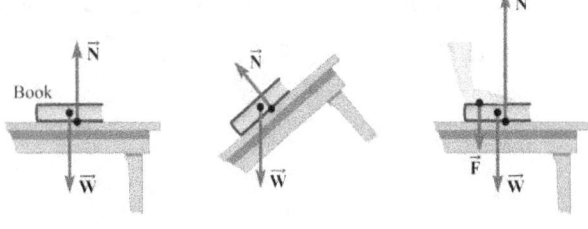

FIGURE 1 (a) The normal force is equal in magnitude to the weight of the book; the two forces sum to zero. (b) On an incline, the normal force is smaller than the weight of the book and is not vertical. (c) If you push down on the book ($\vec{F}$), the normal force on the book due to the table is larger than the book's weight

How does the table "know" how hard to push on the book? First imagine putting the book on a bathroom scale instead of the table. A spring inside the scale provides the upward force. The spring "knows" how hard to push because, as it is compressed, the force it exerts increases. When the book reaches equilibrium, the spring is exerting just the right amount of force, so there is no tendency to compress it further. The spring is compressed until it pushes up with a force equal to the book's weight. If the spring were stiffer, it would exert the same upward force but with less compression.

The forces that bind atoms together in a rigid solid, like the table, act like extremely stiff springs that can provide large forces with little compression—so little that it's usually not noticed. The book makes a tiny indentation in the surface of the table (FIGURE 2); a heavier book would make a slightly larger indentation. If the book were to be placed on a soft foam surface, the indentation would be much more noticeable.

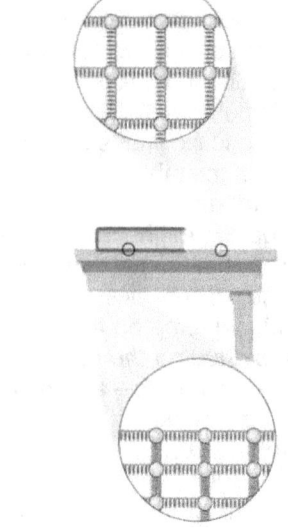

FIGURE 2 The book compresses the "atomic springs" in the table until they push up on the book to hold it up. The slight decrease in the distance between atoms is greatly exaggerated here.

Note: *The direction of normal reaction is always perpendicular to the common normal between the bodies under contact.*

18.1. DIRECTIONS OF NORMAL REACTIONS IN DIFFERENT SITUATIONS

(i)
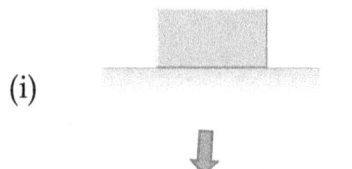

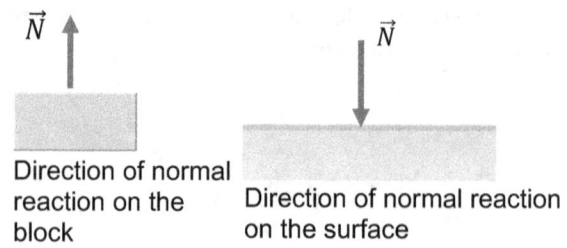

Direction of normal reaction on the block

Direction of normal reaction on the surface

(ii) The number of normal reaction vectors is equal to number of contact surfaces.

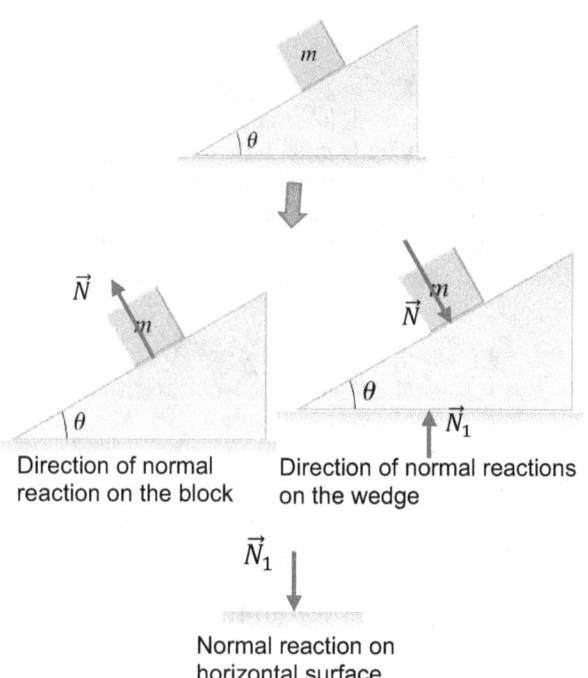

Direction of normal reaction on the block

Direction of normal reactions on the wedge

Normal reaction on horizontal surface

Here, four surfaces (one surface of the block, two surfaces of the wedge, and one surface of plane ground), are in contact, so there are four normal reaction vectors.

(iii) In following figure, the normal reaction of lower block A on upper block B is in upward direction and normal reaction of upper block B on lower block A is in downward direction.

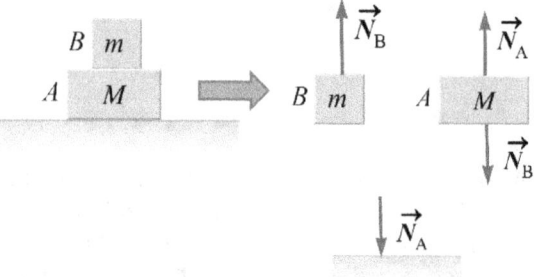

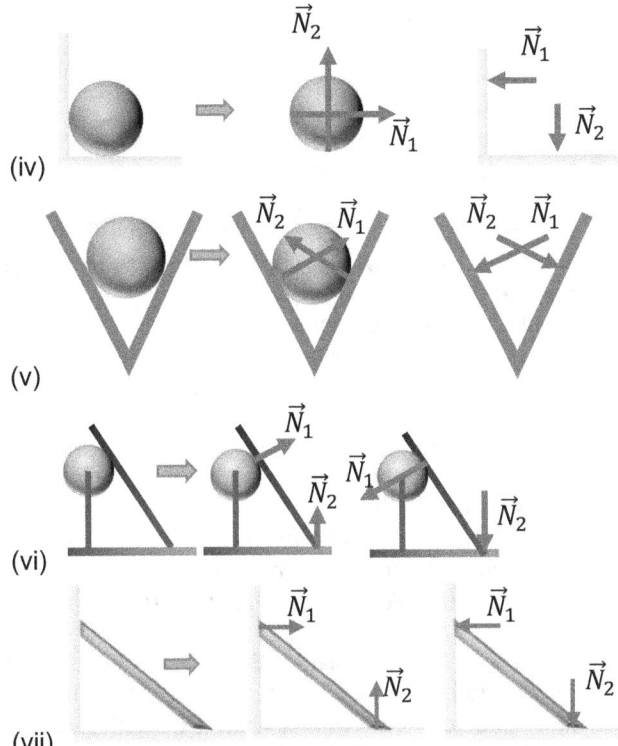

19. FRICTION (in brief)

A contact force *parallel* to the contact surface is called **frictional force** or simply **friction**. We will discuss this *electromagnetic bonding force* a bit later in the chapter. This force is directed along the surface, opposite the direction of the intended motion (FIGURE 1). Sometimes, to simplify a situation, friction is assumed to be negligible (the surface, or even the body, is said to be *frictionless*).

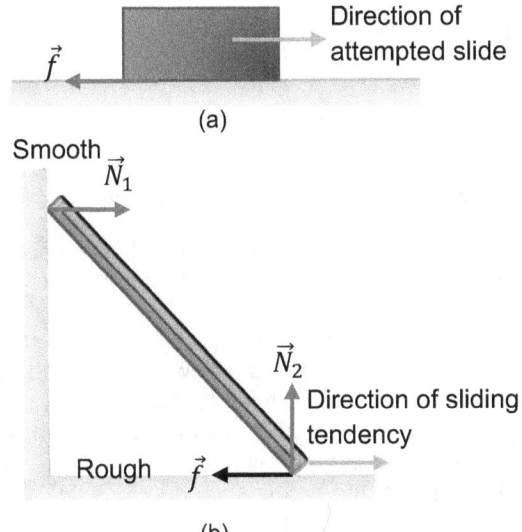

FIGURE 1 (a) A frictional force opposes the attempted slide of a body over a surface. (b) A smooth wall supports the ladder by pushing on the ladder with a normal force $\vec{N}_1$ while floor applies normal force ($\vec{N}_2$) in upward direction and a frictional force $\vec{f}$ opposite to its sliding tendency.

The bodies with smooth surfaces can exert only small amount of forces parallel to the surface and hence are close to frictionless surface. We shall often use the word smooth to mean frictionless.

CONNECTION

Normal force and *frictional force* are just names given to two perpendicular components of a surface contact force.

20. TENSION

Imagine picking up a light string and holding it with one end in each hand. If you pull to the right with your right hand with a force T and to the left with your left hand with a force T, the string becomes taut. In such a case, we say that there is a **tension** T in the string. To be more specific, if your friend were to cut the string at some point, the tension T is the force pulling the ends apart, as illustrated in FIGURE 1—that is, T is the force your friend would have to exert with each hand to hold the cut ends together. At any given point in the string, the tension pulls equally to the right and to the left.

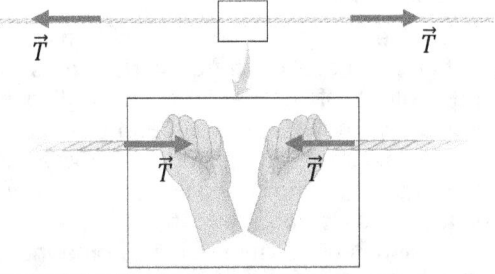

FIGURE 1 Tension in a string A string, pulled from either end, has a tension, T. if the string were to be cut at any point, the force required to hold the ends together is T.

It is important to note that the string must be in contact with the object to exert a tension force. Because the string pulls on the object, the tension force is always directed along the string.

An ideal string is one which is unstretchable and massless, where by "massless" we mean that the mass of any string in our discussion is so small that we can ignore the effect of gravity on the string. For such an ideal string, the tension is the same everywhere in it. Thus, if the spring connects two objects, the magnitude of the tension force on each object is the same.

A tension force T is exerted only a taut string. For a loose string, tension $T = 0$.

Let us consider bananas hanging by a string from the trunk of a tree (FIGURE 2). The bananas are is in equilibrium, so the upward force on it due to the string is equal in magnitude to the bananas weight. Like the normal force, the tension force can be modelled as a spring force (**electromagnetic in nature**) exerted by molecular bonds between molecules (close-up, FIGURE 3). In the case of the tension force, a taut string stretches the molecular bonds. Like very stiff springs, these molecular bonds pull back. If the "springs" are very stiff, the string does not stretch perceptibly, and we say the string is *unstretchable*.

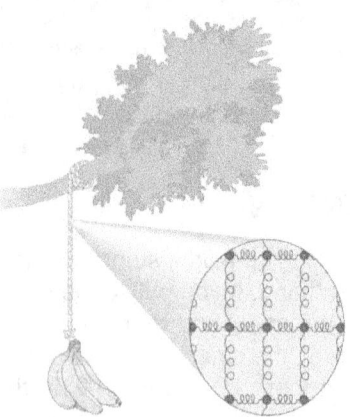

FIGURE 2. Bananas hang from a string. The tension force on the bananas is upward. Molecular bonds in the rope act like very, very stiff springs (close-up). When those bonds are stretched, they pull back.

In simple words we can say that the electrons and protons near the end point of string exerts forces on electrons and protons of the bananas. According to Newton's third law of motion electrons and protons of the bananas exerts same magnitude of force on electrons and protons near point *end point* of the string. These forces are cause of tension in the string. This is why, force of tension is an example of electromagnetic force.

20.1. SOME IMPORTANT POINTS

> Let us consider a segment of string of mass m. suppose T_1 and T_2 are the tension forces acting on this string segment as shown in adjoining figure. If $\vec{a}$ is the acceleration of the string, then, by Newton's second law, we have-

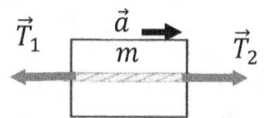

FIGURE 1. FBD of a segment of string

$$\sum F = T_2 - T_1 = ma \qquad \ldots (1)$$

If magnitude of acceleration $a \neq 0$, then from above equation (1), we have

$$T_2 - T_1 \neq 0$$
or $\quad T_2 \neq T_1$

☞ Thus, for an accelerated massive string, tension is always different at different points.

But for a massless string, $m = 0$
∴ From equation (1),

$$T_2 - T_1 = (0)a$$
i.e., $\quad T_2 - T_1 = 0$
or $\quad T_2 = T_1$

☞ Thus, for an accelerated massless string, tension is always same at every point on the string.

> For a vertically suspended massive string, tension varies from top to bottom because of its mass (see example 14).
> Unless stated, always assume the given string massless.
> If a string is slacked, then tension in it will be zero.
> The direction of tension on a body or pulley is always away from the body or pulley.
> If a free pulley is massless then tension remains same even if there is friction between pulley and rope Also it is important to remember that ropes can change tension instantaneously.
> All contact forces (Normal force, friction, spring force, tension force, viscous force and force of buoyancy) are electromagnetic in nature.

EXAMPLE 14 TENSION IN A HEAVY ROPE Consider a rope of length $5\ m$ that is attached to the ceiling at one end, and to a box with a weight of $105\ N$ at the other end, as shown in **FIGURE 1**. In addition, suppose that the rope is uniform and has a total weight of $2.00\ N$. What is the tension in the rope (a) where it attaches to the box, (b) at any general distance x from the lower end of rope, and (c) at its midpoint, and (d) where it attaches to the ceiling?

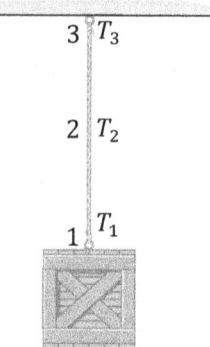

FIGURE 1 Tension in a heavy rope Because of the weight of the rope, the tension is noticeably different at points 1, 2, and 3. In the limit of a rope of zero mass, the tension is the same throughout the rope.

APPROACH The box and rope system is in equilibrium under gravity. So, we use Newtons second law with zero acceleration in vertical direction.

SOLUTION Suppose the upward direction of y-axis is the positive direction of all forces.

(a) The FBD of the box is shown in FIGURE 2.

LAWS OF MOTION AND FRICTION 29

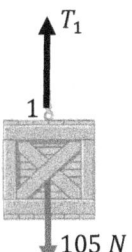

FIGURE 2. FBD of the box

The forces acting on the box are-

(i) The tension T_1 (applied by the rope on the box) in vertically upward direction.

(ii) The gravitational force $105\,N$ (applied by the earth on the box) in vertically downward direction.

∴ Net force on the box in vertical direction

$$\Sigma F_y = T_1 - 105\,N$$

As the system is in equilibrium, therefore

$$\Sigma F_y = 0$$

or $\quad T_1 - 105\,N = 0$

or $\quad T_1 = 105\,N$

(b) If m is the mass of rope and L is its length, then its linear mass density (i.e., mass per unit length)

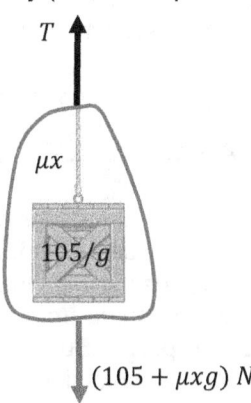

FIGURE 2. FBD of box and rope (up to length x) system.

$$\mu = \frac{m}{L} = \frac{\text{weight of rope}/g}{L},$$

or $\quad \mu = \dfrac{2}{5g} = \dfrac{0.4}{g}\,kg/m$

∴ Mass of length x of rope $= \mu x$

Since the accelerations of rope up to distance x from bottom and the box, are same (both are in equilibrium, therefore zero in this case), therefore we can consider this part of rope and the box in a single system.

Now from FBD (FIGURE 3), the forces acting on the system are-

(i) The tension T (applied by the rope above the distance x) in vertically upward direction.

(ii) The gravitational force = weight of the box + weight of the part of rope up to distance x from bottom end

$$= 105\,N + \mu x g$$

(applied by the earth on the x part of rope plus box system) in vertically downward direction.

∴ Net force on the box in vertical direction

$$\Sigma F_y = T - (105\,N + \mu x g)$$

As the system is in equilibrium, therefore

$$\Sigma F_y = 0$$

or $\quad T - (105\,N + \mu x g) = 0$

or $\quad T = (105\,N + \mu x g),$

Here $\mu = \dfrac{0.4}{g}\,kg/m,$

∴ $\quad T = \left(105 + \dfrac{0.4}{g} x g\right) N$

(c) The FBD of the box and half string system is shown in FIGURE 4.

The forces acting on the system are-

(i) The tension T_2 (applied by the rope above the mid point) in vertically upward direction.

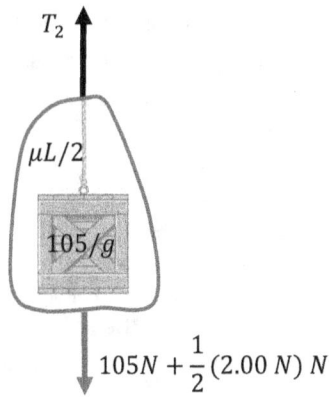

FIGURE 2. FBD of box and half rope system.

(ii) The gravitational force = weight of the box + weight of the half rope $= 105\,N + \dfrac{1}{2}(2.00\,N) = 106\,N$ (applied by the earth on the half rope plus box system) in vertically downward direction.

∴ Net force on the box in vertical direction

$$\Sigma F_y = T_2 - 106\,N$$

As the system is in equilibrium, therefore

$$\Sigma F_y = 0$$

or $\quad T_1 - 106\,N = 0$

or $\quad T_1 = 106\,N$

(d) This part is similar to part c, except the half rope is replaced by the complete rope.

In this case, the gravitational force on the system = weight of the box + weight of the rope = $105\,N +$

$2.00\ N = 107\ N$. If T_3 is the tension in the rope where it attaches to the ceiling, then

The forces acting on the system will be-

(i) The tension T_3 (in rope at top) in vertically upward direction.

(ii) The gravitational force $= 107\ N$ (applied by the earth on the rope plus box system) in vertically downward direction.

∴ Net force on the box in vertical direction
$$\Sigma F_y = T_y - 107\ N$$
As the system is in equilibrium, therefore
$$\Sigma F_y = 0$$
or $\qquad T_3 - 107\ N = 0$
or $\qquad T_3 = 107\ N$

This problem is so simple, that we can also solve it orally as follows-

First, the rope holds the box at rest; thus, the tension where the rope attaches to the box is simply the weight of the box, $T_1 = 105\ N$. At the midpoint of the rope, the tension supports the weight of the box, plus half the weight of the rope. Thus, $T_2 = 105\ N + \frac{1}{2}(2.00\ N) = 106\ N$. Similarly, at the ceiling the tension supports the box plus all of the rope, giving a tension of $T_3 = 107\ N$. Parts (a), (c) and (d) can also be solved by directly putting $x = 0, \frac{5}{2}$ and 5 respectively in the result of part (b).

DISCUSSION
We can see that the tension in the rope changes slightly from top to bottom because of the mass of the rope. If the rope had less mass, the difference in tension between its two ends would also be less. In particular, if the rope's mass were to be vanishingly small, the difference in tension would vanish as well.

20.2. PULLEYS

Strings and ropes often pass over pulleys. The application might be as simple as lifting a heavy weight or as complex as the internal cable-and-pulley arrangement that precisely moves a robot arm. FIGURE 1 shows a simple situation in which a person pulls a box by applying force on the string passing over a pulley. As the string moves, static friction between the string and pulley causes the pulley to turn. If we assume that

- The string *and* the pulley are both massless, and
- There is no friction where the pulley turns on its axle,

then no net force is needed to accelerate the string or turn the pulley. Thus, *the tension in a massless string remains constant as it passes over a massless, frictionless pulley.*

☞ In the ideal case, a pulley has no mass and no friction in its bearings

PULLEYS CHANGE THE DIRECTION OF THE TENSION

Pulleys are often used to redirect a force (without changing its magnitude) exerted by a string, as indicated in FIGURE 1.

If a system contains more than one pulley, however, it's possible to arrange them in such a way as to "magnify a force," even if each pulley itself merely redirects the tension in a string. The traction device considered in the Example 15 shows one way this can be accomplished in a system that uses four ideal pulleys.

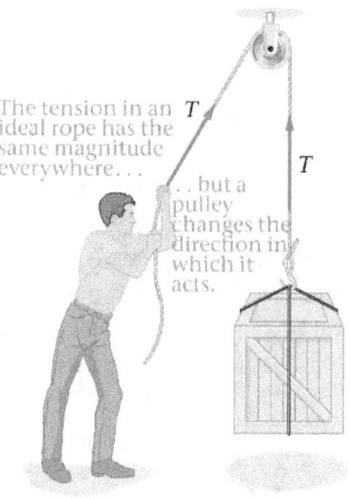

FIGURE 1 A pulley changes the direction of a tension

20.3. TENSIONS ON BOTH SIDES OF A CORD PASSING OVER A PULLEY

We have already shown that the tension at every point of an ideal string is always same weather it is at rest or accelerated. So, the tension for either side of string passing over a pulley is same from lower end to the point of contact of the pulley. But still, we don't know the relation in tensions on both sides of the pulley.

Fig. 1 shows an ideal cord passing around an ideal pulley and the FBD for a short segment of the cord at the top of the pulley. Choosing the x-axis to be horizontal, the normal force ($\vec{N}$) has no x-component.

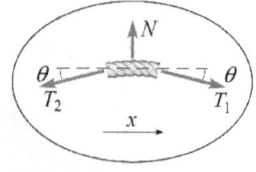

Applying Newton's second law along the x-axis:

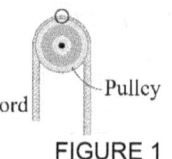

FIGURE 1

$$\Sigma F_x = T_1 \cos\theta - T_2 \cos\theta = ma_x.$$
If, $m = 0$, then $T_1 \cos\theta - T_2 \cos\theta = 0$
i.e., $T_1 = T_2$.

The same reasoning can be applied to any segment of cord in contact with the massless pulley to show that the tensions are the same on either side of the pulley and it is **independent on the friction between ideal pulley and the ideal cord**.

☞ Forces T_1 and T_2 act *as if* they are an action/reaction pair, even though they are not opposite in direction because the tension force gets "turned" by the pulley.

Now, suppose, an ideal pulley is connected with strings as shown in Fig.1 and its acceleration is $\vec{a}$. Let T_1 is the tension in lower string passing over the pulley and T_2 is the tension in upper string, then from Newton's second law, we have-

$$T_2 - 2T_1 = ma$$

As for an ideal pulley, $m = 0$, therefore-

$$T_2 - 2T_1 = (0)a$$

or $\quad T_2 = 2T_1$

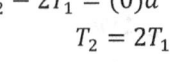

FIGURE 2

Thus, we can say that tension forces on two opposite sides of an ideal pulley are always same.

EXAMPLE 15 SETTING A BROKEN LEG WITH TRACTION

A young boy with a broken leg is undergoing traction. (a) Find the magnitude of the total force of the traction apparatus applied to the leg, assuming the weight of the leg is 22 N and the weight hanging from the traction apparatus is also 22 N. (b) What is the horizontal component of the traction force acting on the leg? (c) What is the magnitude of the force exerted on the femur by the lower leg?

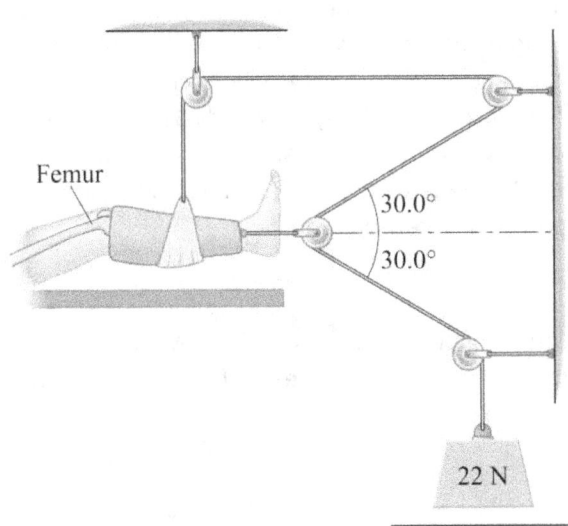

(a) Physical picture

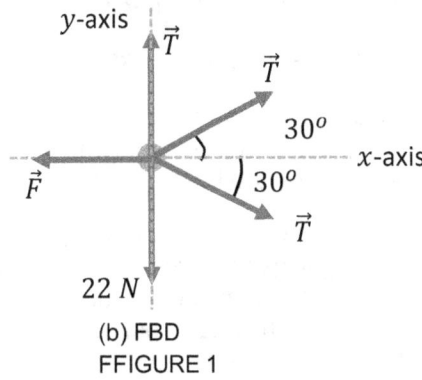

(b) FBD
FFIGURE 1

APPROACH The tension T is the same along the length of the cord. Its magnitude is equal to the hanging weight, $22\ N$. i.e., $T = 22\ N$. Therefore, we don't need the equilibrium condition of leg, and we can directly obtain the components of traction force in various directions. Let the $+y$-direction be up and the $+x$-direction be to the right.

SOLUTION Let us consider the banded part of the leg as the system

The components of traction forces are-

(i) Along X-axis

Component of tension T of string above the x- axis in the right side to banded part $= T \cos 30^o$

Component of tension T of string below the x- axis in the right side to banded part $= T \cos 30^o$

∴ Net force of traction along x-axis $\sum F_x = T \cos 30^o + T \cos 30^o = 2T \cos 30^o = 2T \frac{\sqrt{3}}{2} = T\sqrt{3}$

(ii) Along Y-axis

Component of tension T of string along the y- axis $= T$

Component of tension T of string above the x- axis in the and right side to the banded part $= T \sin 30^o$

Component of tension T of string below the x- axis in the right side to the banded part $= -T \sin 30^o$

∴ Net traction force on the leg along y-axis $\sum F_y = T + T \sin 30 - T \sin 30 = T$

(a) Magnitude of the total force of the traction apparatus applied to the leg

$$\sum F = \sqrt{(\sum F_x)^2 + (\sum F_y)^2}$$

$$= \sqrt{(T\sqrt{3})^2 + (T)^2} = 2T$$

$$= 2 \times 22\ N = 44\ N \qquad [\because T = 22\ N]$$

(b) The horizontal component of traction

$$\sum F_x = T\sqrt{3} = 22\,N \times \sqrt{3} \approx 38\,N$$

(c) The magnitude of the horizontal force acting on the femur is equal to the horizontal component of the traction force acting on the leg ≈ 38 N

DISCUSSION As pointed out earlier, this pulley arrangement "magnifies the force" in the sense that a 22 N weight attached to the rope produces a 44-N force exerted on the foot by the middle pulley. Notice that the tension in the rope always has the same value—$T = 22\,N$—as expected with ideal pulleys, but because of the arrangement of the pulleys the force applied to the foot by the rope is 44 N.

EXAMPLE 16 ACCELEROMETER A small mass m hangs from a thin string and can swing like a pendulum. You attach it above the window of your car as shown in FIGURE 1a. When the car is at rest, the string hangs vertically. What angle does the string make (a) when the car accelerates at a constant $a = 1.20$ m/s², and (b) when the car moves at constant velocity, $v = 90$ km/h? (c) Calculate tension in string in both cases (a) and (b).

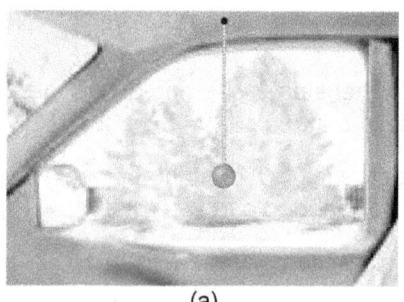

(a)

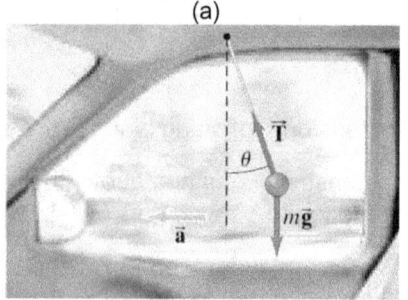

(b)
FIGURE 1.

APPROACH The free-body diagram of FIGURE 1b shows the pendulum at some angle θ and the forces on it: $m\vec{g}$ downward, and the tension $\vec{T}$ in the cord. These forces do not add up to zero if $\theta \neq 0$, and since we have an acceleration a, we therefore expect $\theta \neq 0$. Note that θ is the angle relative to the vertical.

SOLUTION (a) The acceleration is horizontal, so from Newton's second law,

$$ma = T\sin\theta \qquad \ldots (1)$$

for the horizontal component, whereas the vertical component gives

$$0 = T\cos\theta - mg, \qquad \ldots (2)$$

Dividing these two equations, we obtain

$$\tan\theta = \frac{T\sin\theta}{T\cos\theta} = \frac{ma}{mg} = \frac{a}{g}$$

or

$$\tan\theta = \frac{1.20\,m/s^2}{9.80\,m/s^2} = 0.122,$$

so,

$$\theta = 7.0°$$

(b) The velocity is constant, so $a = 0$ and $\tan\theta = 0$. Hence the pendulum hangs vertically ($\theta = 0°$).

(c) In case (a), tension $T = \frac{mg}{\cos\theta}$ and in the case (b), tension $\theta = 0$, therefore $T = mg$

NOTE This simple device is an **accelerometer**—it can be used to measure acceleration.

21. NEWTON'S THIRD LAW

Nature never produces just one force at a time—*forces always come in pairs*. In addition, the forces in a pair always act on *different objects*, are *equal in magnitude*, and point in *opposite directions*. This is Newton's third law of motion.

NEWTON'S THIRD LAW

For every force that acts on an object, there is a reaction force acting on a different object that is equal in magnitude and opposite in direction.

For example, if you lean against a wall with a certain force, the wall reacts and pushes back on you with a force of equal magnitude. Another example of two interacting bodies is shown in **FIGURE 1**, where body 1 exerts an action force $\vec{F}_{21}$ (a pull) on body 2. ($\vec{F}_{21}$ is read: force exerted on body 2 by body 1). Experiments show that body 2 would also exert a reaction force $\vec{F}_{12}$ on body 1. These two forces are equal in magnitude and opposite in direction. That is:

$$\vec{F}_{21} = -\vec{F}_{12} \quad \text{(Newton's third law)} \qquad \ldots (4)$$

Equation.4 implies that $F_{21} = F_{12}$. Moreover, this equation holds true regardless of whether the two bodies move or remain stationary.

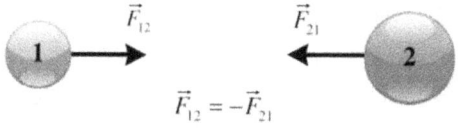

FIGURE. 1 The force exerted by body 1 on 2 is equal in magnitude but opposite to the force exerted by body 2 on 1

This law, more commonly written as, "For every action there is an equal and opposite reaction," completes Newton's laws of motion.

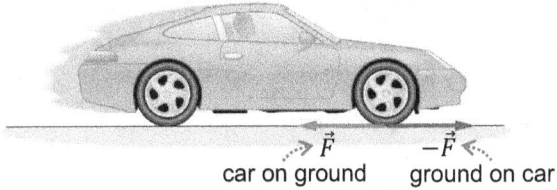

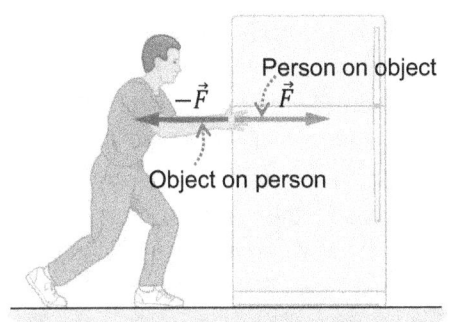

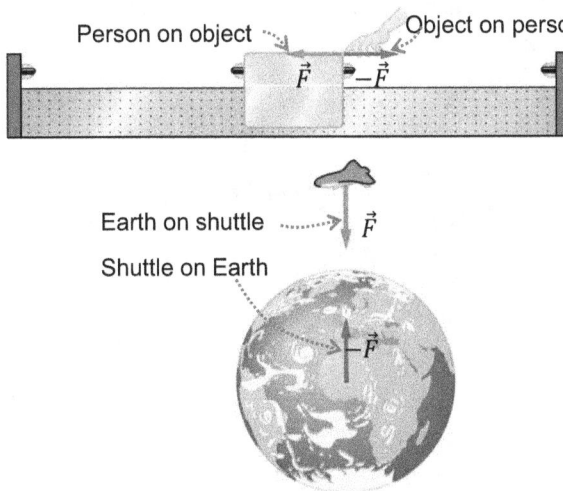

FIGURE 2 Examples of action-reaction force pairs

FIGURE 2 illustrates some action-reaction pairs. Notice that there is always a reaction force, whether the action force pushes on something hard to move, like a refrigerator, or on something that moves with no friction, like an air-track cart. In some cases, the reaction force tends to be overlooked, as when the Earth exerts a *downward* gravitational force on the space shuttle, and the shuttle exerts an equal and opposite *upward* gravitational force on the Earth. Still, the reaction force always exists.

Third-Law Forces Act on Different Objects

Another important aspect of the third law is that the action-reaction forces always act on *different* objects.

This, again, is illustrated in FIGURE 2. Thus, in drawing a free-body diagram, we draw only one of the action-reaction pair of forces for a given object. The other force in the pair would appear in the free-body diagram of a different object. As a result:

The two forces in an action-reaction pair do not cancel in the free-body diagram for an object.

For example, consider a car accelerating from rest, as in FIGURE 2. As the car's engine turns the wheels, the tires exert a force on the road. By the third law, the road exerts an equal and opposite force on the car's tires. It is this second force—which acts on the car through its tires—that propels the car forward. The force exerted by the tires on the road does not accelerate the car.

Because the action-reaction forces act on different objects, they generally produce different accelerations.

EXAMPLE 17 APPLYING NEWTON'S THIRD LAW: OBJECTS AT REST- I An apple sits on a table in equilibrium. What forces act on it? What is the reaction force to each of the forces acting on the apple? What are the action–reaction pairs?

APPROACH Action and reaction forces always acts simultaneously between two interacting bodies. To find action and reaction pair, always consider the given bodies in pair.

SOLUTION FIGURE 1a shows the forces acting on the apple. In the diagram, $\vec{F}_{earth\ on\ apple}$ is the weight of the apple—that is, the downward gravitational force exerted *by* the earth (first subscript) *on* the apple (second subscript). Similarly, $\vec{F}_{table\ on\ apple}$ is the upward force exerted *by* the table (first subscript) *on* the apple (second subscript).

As the earth pulls down on the apple, the apple exerts an equally strong upward pull $\vec{F}_{apple\ on\ earth}$ on the earth, as shown in **FIGURE** 1b. $\vec{F}_{apple\ on\ earth}$ and $\vec{F}_{earth\ on\ apple}$ are an action–reaction pair, representing the mutual interaction of the apple and the earth, so

$$\vec{F}_{apple\ on\ earth} = -\vec{F}_{earth\ on\ apple}$$

The two forces on the apple cannot be an action–reaction pair because they act on the same object.

(a) The forces acting on the apple

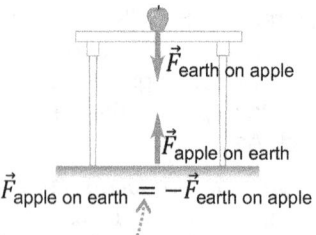

An action–reaction pair is a mutual interaction between two objects. The two forces act on two different objects.

(b) The action–reaction pair for the interaction between the apple and the earth

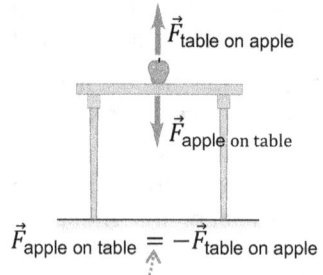

An action–reaction pair is a mutual interaction between two objects. The two forces act on two different objects.

(c) The action–reaction pair for the interaction between the apple and the table

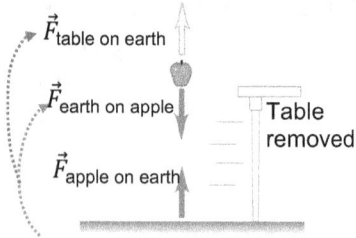

When we remove the table, $\vec{F}_{table\ on\ apple}$ becomes zero but $\vec{F}_{earth\ on\ apple}$ is unchanged. Hence these forces (which act on the same object) cannot be an action–reaction pair.

(d) We eliminate the force of the table on the apple.

FIGURE 1 The two forces in an action–reaction pair always act on different bodies.

Also, as the table pushes up on the apple with force $\vec{F}_{table\ on\ apple}$ the corresponding reaction is the downward force $\vec{F}_{apple\ on\ table}$ exerted by the apple on the table (**FIGURE 1c**). So, we have

$$\vec{F}_{apple\ on\ table} = -\vec{F}_{table\ on\ apple}$$

The two forces acting on the apple are $\vec{F}_{table\ on\ apple}$ and $\vec{F}_{earth\ on\ apple}$. Are they an action–reaction pair? No, they aren't, despite being equal and opposite. They do not represent the mutual interaction of two bodies; they are two different forces acting on the *same* body. *The two forces in an action–reaction pair **never** act on the same body.* Here's another way to look at it. Suppose we suddenly yank the table out from under the apple (FIGURE 1d). The two forces $\vec{F}_{apple\ on\ table}$ and $\vec{F}_{table\ on\ apple}$ then become zero, but $\vec{F}_{apple\ on\ table}$ and $\vec{F}_{earth\ on\ apple}$ are still there (the gravitational interaction is still present). Since $\vec{F}_{table\ on\ apple}$ is now zero, it can't be the negative of $\vec{F}_{earth\ on\ apple}$ and these two forces can't be an action–reaction pair.

EXAMPLE 18 APPLYING NEWTON'S THIRD LAW: OBJECTS AT REST- II A block of weight W sits on a table in equilibrium. Make FBD of the block and find the normal reaction of table on it.

APPROACH Apply the methods discussed in FBD section.

SOLUTION The reaction of a block of weight $\vec{W}$ is the force exerted on the Earth $\vec{W}'$, see **FIGURE** 1a. When this block rests on a table, the table exerts an upward action force, $\vec{N}$, called the normal force, see **FIGURE** 1b. The normal force can have any value up to the point of breaking the table. The reaction to $\vec{N}$ is the force that the block exerts on the table, $\vec{N}'$, see **FIGURE** 1c. Therefore, from Newton's third law

$$\vec{W} = -\vec{W}', \vec{N} = -\vec{N}' \qquad \ldots (1)$$

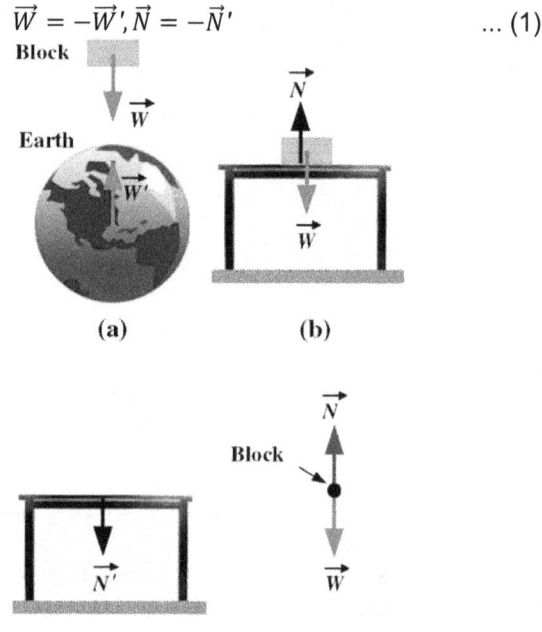

FIGURE 1 (a) The reaction of a block of weight $\vec{W}$ is the force $\vec{W}'$ (b) A block resting on a table experiences a normal force $\vec{N}$ perpendicular to the table. (c) The reaction force $\vec{N}'$ exerted on the table. (d) The free-body diagram used to solve the block problem

Now, from FBD of block placed on table, we have

$$\Sigma \vec{F} = 0 \qquad \Rightarrow \qquad \vec{N} + \vec{W} = 0$$
$$\Rightarrow \qquad N - W = 0$$

Thus, $N = W$ ... (2)

EXAMPLE 19 APPLYING NEWTON'S THIRD LAW: OBJECTS IN MOTION (one block pushes another)

Two blocks of masses m_1 and m_2, with $m_1 > m_2$, are placed in contact with each other on a frictionless, horizontal surface as in FIGURE 1. A constant horizontal force $\vec{F}$ is applied to m_1 as shown.

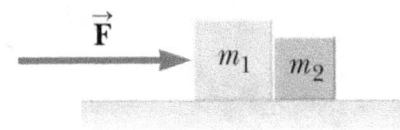

FIGURE 1 A force is applied to a block of mass m_1, which pushes on a second block of mass m_2.

(A) Find the magnitude of the acceleration of the system.

(B) Determine the magnitude of the contact force between the two blocks.

PART (A) APPROACH

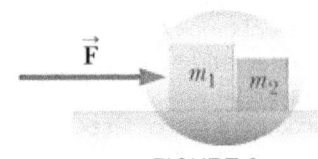

FIGURE 2

Conceptualize the situation by using FIGURE 1 and realize that both blocks must experience the *same* acceleration because they are in contact with each other and remain in contact throughout the motion.

Categorize We categorise this problem as one involving a *particle under a net force* because a force is applied to a system of blocks and we are looking for the acceleration of the system.

SOLUTION First model the combination of two blocks as a single particle under a net force (FIGURE 2). Apply Newton's second law to the combination in the x direction to find the acceleration:

$$\sum F_x = F = (m_1 + m_2)a_x$$

$$a_x = \frac{F}{(m_1 + m_2)} \quad ...(1)$$

Finalize The acceleration given by Equation (1) is the same as that of a single object of mass $(m_1 + m_2)$ and subject to the same force.

PART-B APPROACH The contact force is internal to the system of two blocks. Therefore, we cannot find this force by modelling the whole system (the two blocks) as a single particle.

Categorize Now consider each of the two blocks individually by categorizing each as a *particle under a net force*.

SOLUTION FBD's of blocks m_1 and m_2 are shown in Figures 3a and 3b respectively, where the contact forces are denoted by $\vec{N}_{ij}$. From FIGURE 3b, we see that the only horizontal force acting on m_2 is the contact force $\vec{N}_{12}$. (the force exerted by m_1 on m_2), which is directed to the right.

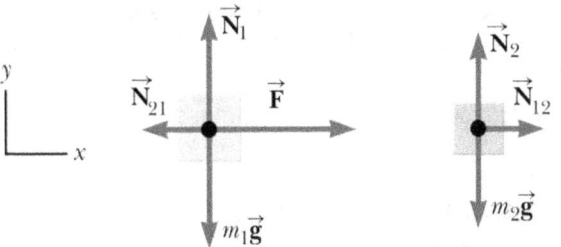

FIGURE 3 (a) The forces acting on m_1. (b) The forces acting on m_2.

Apply Newton's second law to m_2:

$$\sum F_x = N_{12} = m_2 a_x \quad ...(2)$$

Substitute the value of the acceleration a_x given by Equation (1) into Equation (2):

$$N_{12} = \left(\frac{m_2}{m_1 + m_2}\right) F \quad ...(3)$$

Finalize This result shows that the contact force N_{12} is *less* than the applied force F. The force required to accelerate block 2 alone must be less than the force required to produce the same acceleration for the two-block system.

To finalize further, let us check this expression for N_{12} by considering the forces acting on m_1, shown in **FIGURE 3a**. The horizontal forces acting on m_1 are the applied force $\vec{F}$ to the right and the contact force $\vec{N}_{21}$ to the left (the force exerted by m_2 on m_1). From Newton's third law, $\vec{N}_{21}$ is the reaction force to $\vec{N}_{12}$, so $N_{21} = N_{12}$.

Apply Newton's second law to m_1:

$$\sum F_x = F - N_{21} = F - N_{12} = m_1 a_x \quad ...(4)$$

Solve for N_{12} and substitute the value of a_x from Equation (1):

$$N_{12} = F - m_1 a_x = F - m_1\left(\frac{F}{m_1 + m_2}\right) = \left(\frac{m_2}{m_1 + m_2}\right)F$$

This result agrees with Equation (3), as it must.

WHAT IF?

Imagine that the force $\vec{F}$ in **FIGURE 1** is applied toward the left on the right-hand block of mass m_2. Is the magnitude of the force $\vec{N}_{12}$ the same as it was when the force was applied toward the right on m_1?

ANSWER When the force is applied toward the left on m_2, the contact force must accelerate m_1. In the original situation, the contact force accelerates m_2. Because $m_1 > m_2$, more force is required, so the magnitude of $\vec{N}_{12}$ is greater than in the original situation. To see this mathematically, modify Equation (4) appropriately and solve for $\vec{N}_{12}$.

EXAMPLE 20 A force $\vec{F}$ is applied at the free end of the rod of mass M and length L. Find the tension at distance x from the other end of the rod.

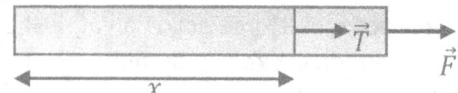

SOLUTION If we pull the rod from one end then at any point on rod there will be tension (internal force) driving the remaining body.

Acceleration of the rod $a = \frac{F}{M}$, mass density of the rod $\lambda = \frac{M}{L}$

∴ Mass of rod of length x, $m = \lambda x = \frac{M}{L}x$

∴ tension at distance x from free end, $T = ma = \left(\frac{M}{L}\right) x \left(\frac{F}{M}\right)$

22. CHECKPOINT 5

1. • True or false. (a) If two external forces that are both equal in magnitude and opposite in direction act on the same object, the two forces can never be an action-reaction force pair. (b) Action equals reaction only if the objects are not accelerating.

2. • In the photo, two children are pulling on a toy. If they are exerting equal and opposite forces on the toy, are these two forces interaction partners (action-reaction pair)?

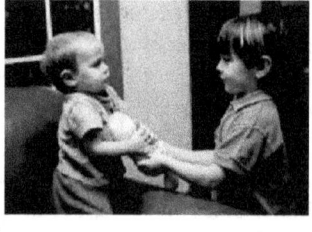

3. •• N identical carts are connected to each other using strings of negligible mass. A pulling force F is applied on the first cart and the system moves without friction along the horizontal ground. The tension in the string connecting 4th and 5th cart is twice the tension in the string connecting 8th and 9th cart. Find the total number of carts (N) and tension in the last string.

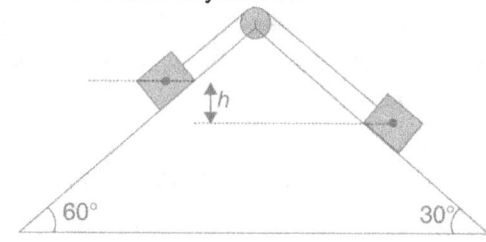

4. •• Two strange particles A and B in space, exert no force on each other when they are at a separation greater than $x_0 = 1.0$ m. When they are at a distance less than x_0, they repel one another along the line joining them. The repulsion force is constant and does not depend on the distance between the particles. This repulsive force produces an acceleration of 6 ms^{-2} in A and 2 ms^{-2} in B when the particles are at separation less than x_0. In one experiment particle B is projected towards A with a velocity of 2 ms^{-1} from a large distance so as to hit A head on. The particle A is originally at rest and the system of two particles do not experience any external force. (a) Find the ratio of mass of A to that of B. (b) Find the minimum distance between the particles during subsequent motion. (c) Find the final velocity of the two particles.

5. • Will hanging a magnet in front of the iron cart in adjoining figure make it go? Explain.

6. • A small car is pushing a large truck. They are speeding up. Is the force of the truck on the car larger than, smaller than, or equal to the force of the car on the truck?

7. • A young boy is at rest in the middle of a pond on perfectly frictionless ice. How can he get himself to the shore of the pond? Explain.

8. •• The hand in adjoining figure is pushing on the back of block A. Blocks A and B, with $m_B > m_A$, are connected by a massless string and slide on a frictionless surface. Is the force of the string on B larger than, smaller than, or equal to the force of the hand on A? Explain.

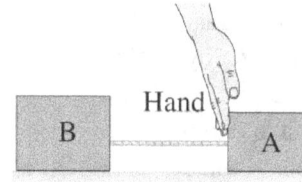

9. •• A heavy block of mass M hangs in equilibrium at the end of a rope of mass m and length l connected to a ceiling. Determine the tension in the rope at a distance x from the ceiling.

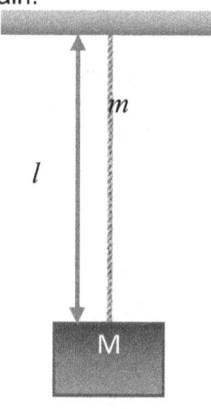

10. •• Two blocks of equal mass have been placed on two faces of a fixed wedge as shown in figure. The blocks are released from position where centre of one block is at a height h above the centre of the other block. Find the time after which the centre of the two blocks will be at same horizontal level. There is no friction anywhere.

11. •• A disc of mass m lies flat on a smooth horizontal table. A light string runs halfway around it as shown in figure. One end of the string is attached to a particle of mass m and the other end is being pulled with a force F. There is no friction between the disc and the string. Find acceleration of the end of the string to which force is being applied.

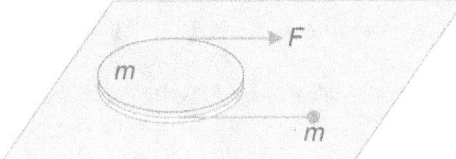

23. APPARENT WEIGHT

Imagine being in an elevator when the cable snaps. Assume that some safety mechanism brings you to rest after you have been in free fall for a while. While you are in free fall, you *seem* to be "weightless," but your weight has not changed; the Earth still pulls downward with the same gravitational force. In free fall, gravity gives the elevator and everything in it a downward acceleration equal to $\vec{g}$. If you jump up from the elevator floor, you seem to "float" up to the ceiling of the elevator. Your *weight* hasn't changed, but your *apparent* weight is zero while you are in free fall.

Similarly, astronauts in a space station in orbit around the Earth are in free fall (their acceleration is equal to the local value of $\vec{g}$). Earth exerts a gravitational force on them so they are not weightless; their *apparent* weight is zero.

Imagine an object that appears to be resting on a bathroom scale. The scale measures the object's *apparent* weight W', which is equal to the true weight only if the object and the scale have zero acceleration. Newton's second law requires that

$$\sum \vec{F} = \vec{N} + m\vec{g} = m\vec{a}$$

where $\vec{N}$ is the normal force of the scale pushing up. The apparent weight W' is the reading of the scale—that is, the magnitude of $\vec{N}$:

$$W' = |\vec{N}| = N$$

In FIGURE 1a, the acceleration of the elevator is a_y and it is in upward direction. The normal force must be larger than the weight for the net force to be upward (FIGURE 1b). Writing the forces in component form where the $+y$-direction is upward

$$\sum F_y = N - mg = ma_y$$

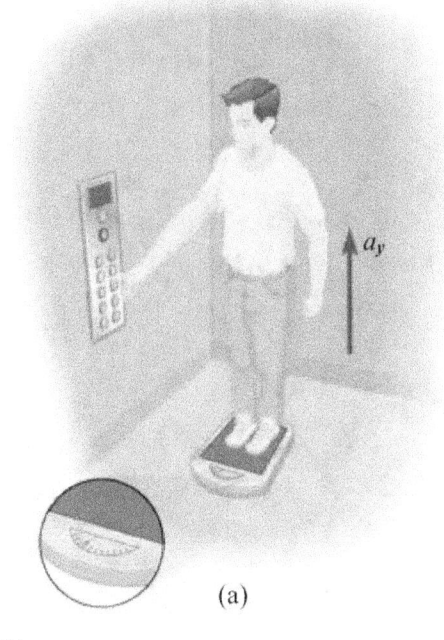

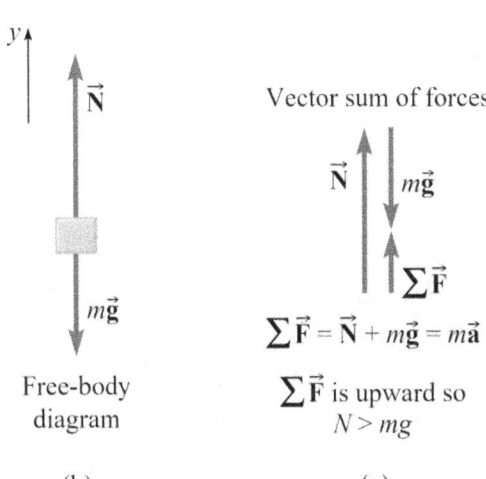

FIGURE 1 (a) Apparent weight in an elevator with acceleration upward. (b) FBD for the passenger. (c) The normal force must be greater than the weight to have an upward net force.

or $\qquad N = mg + ma_y$

Therefore, $\qquad W' = N = m(g + a_y) \qquad \ldots (1)$

If actual weight of the person is W, then $W = mg$, therefore $m = \frac{W}{g}$

Substituting this value in (1), we get

$$W' = \frac{W}{g}(g + a_y) = W\left(1 + \frac{a_y}{g}\right) \qquad \ldots (2)$$

Clearly, in this case, $W' > W$, i.e., the apparent weight is greater than the true weight (FIGURE 1c).

(a)

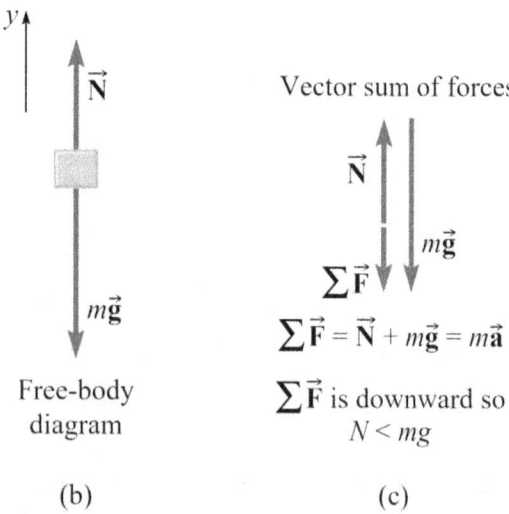

(b) (c)

FIGURE 2 (a) Apparent weight in an elevator with acceleration downward. (b) FBD for the passenger. (c) The normal force must be less than the weight to have a downward net force.

In FIGURE 2a, the acceleration is downward. Therefore, the net force must also point downward. The normal force is still upward, but it must be smaller than the weight in order to produce a downward net force (FIGURE 2b).

In this case, $\sum \vec{F} = \vec{N} + m\vec{g} = m\vec{a}$

$\Rightarrow \quad N - mg = -ma_y$

or $\quad N = mg - ma_y$

∴ apparent weight

$$W' = N = m(g - a_y) \qquad \ldots (3)$$

or $\quad W' = \frac{W}{g}(g - a_y)$

or $\quad W' = W\left(1 - \frac{a_y}{g}\right) \qquad \ldots (4)$

Clearly, in this case, apparent weight is less than the true weight, i.e., $W' < W$ (FIGURE 2c).

In general, we can write $W' = W\left(1 + \frac{a}{g}\right) \qquad \ldots (5)$

For upward acceleration $a > 0$ while for downward acceleration $a < 0$.

If the elevator is in free fall, then in equation (5), $a = -g$ and the apparent weight of the unfortunate passenger is zero.

If the magnitude of downward acceleration of the elevator is greater than the magnitude of $\vec{g}$, i.e., $|\vec{a}| > |\vec{g}|$, then elevator will move faster than you in downward direction and then you seem to "float" up to the ceiling of the elevator.

> The apparent weight does not depend on the direction of velocity, it only depends on the direction of acceleration of elevator.

EXAMPLE 21. APPARENT WEIGHT IN AN ELEVATOR

A passenger weighing 600 N rides in an elevator. What is the apparent weight of the passenger in each of the following situations? In each case, the magnitude of the elevator's acceleration is 0.500 m/s².

(a) The passenger is on the first floor and has pushed the button for the 15th floor; the elevator is beginning to move upward. (b) The elevator is slowing down as it nears the 15th floor. (c) If cable of elevator break what will be the reading shown by scale. Explain your result qualitatively.

APPROACH *Use the formula discussed in section 'apparent weight'. If the cable of elevator breaks, then the elevator will be in the situation of free fall i.e., the acceleration of elevator $a = g$.*

SOLUTION Apparent weight, $W' = W\left(1 + \frac{a}{g}\right)$

here, $W = 600 N$

(a) Since, acceleration is in vertically upward direction, therefore, $a = +0.500 \ m/s^2$

∴ $W' = 600N\left(1 + \frac{0.500 \ m/s^2}{10 \ m/s^2}\right) = 630 \ N$

(b) When the elevator approaches the 15th floor, it is slowing down while still moving upward; its acceleration is downward $a = -0.500 \ m/s^2$

∴ $W' = 600N\left(1 - \frac{0.500 \ m/s^2}{10 \ m/s^2}\right) = 570 \ N$

(c) Put $a = g$ in the above case, you get $N = W\left(1 - \frac{g}{g}\right) = 0$

Thus, the reading of the scale will be zero. This is the condition of weightlessness.

EXAMPLE 22 WEIGHING A FISH IN AN ELEVATOR
A person weighs a fish of mass m on a spring scale attached to the ceiling of an elevator as illustrated in FIGURE 1.

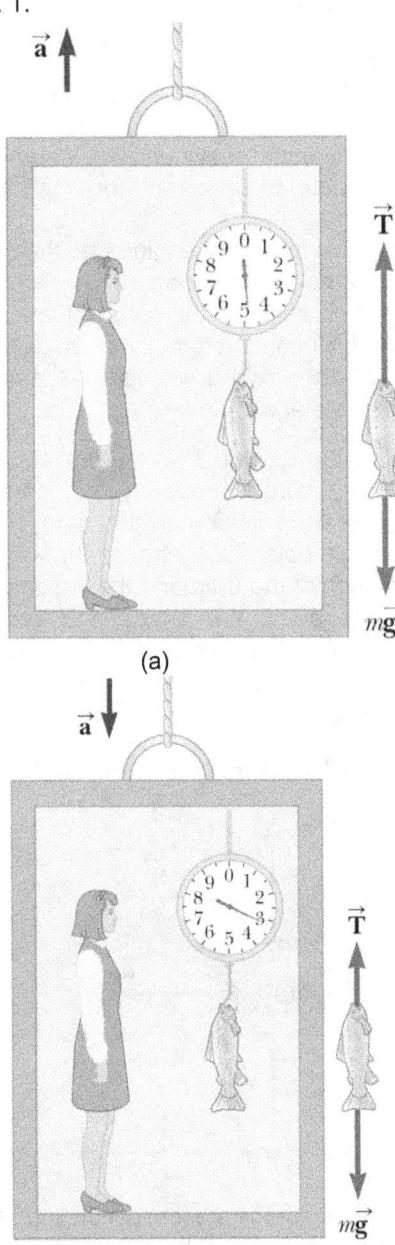

FIGURE. 1. (a) When the elevator accelerates upward, the spring scale reads a value greater than the weight of the fish. (b) When the elevator accelerates downward, the spring scale reads a value less than the weight of the fish.

(A) Show that if the elevator accelerates either upward or downward, the spring scale gives a reading that is different from the weight of the fish.
(B) Evaluate the scale readings for a 40.0-N fish if the elevator moves with an acceleration $a_y = \pm 2.00$ m/s².

PART A APPROACH The reading on the scale is related to the extension of the spring in the scale, which is related to the force on the end of the spring. Imagine that the fish is hanging on a string attached to the end of the spring. In this case, the magnitude of the force exerted on the spring is equal to the tension T in the string. Therefore, we are looking for T. The force $\vec{T}$ pulls down on the spring and pulls up on the fish.

We can categorize this problem by identifying the fish as a *particle in equilibrium* if the elevator is not accelerating or as a *particle under a net force* if the elevator is accelerating.

SOLUTION Inspect the diagrams of the forces acting on the fish in FIGURE 1 and notice that the external forces acting on the fish are the downward gravitational force $\vec{F}_g = m\vec{g}$ and the force $\vec{T}$ exerted by the string. If the elevator is either at rest or moving at constant velocity, the fish is a particle in equilibrium, so
$$\sum F_y = T - F_g = 0$$
or $\quad T = F_g = mg$ (Remember that the scalar mg is the weight of the fish.)

Now suppose the elevator is moving with an acceleration $\vec{a}$ relative to an observer standing outside the elevator in an inertial frame. The fish is now a particle under a net force.

Apply Newton's second law to the fish:
$$\sum F_y = T - mg = ma_y$$
Solve for T: $\quad T = ma_y + mg = mg\left(\frac{a_y}{g} + 1\right)$

or $\quad T = F_g\left(\frac{a_y}{g} + 1\right)$ ... (1)

where we have chosen upward as the positive y direction. We conclude from Equation (1) that the scale reading T is greater than the fish's weight mg if $\vec{a}$ is upward, so a_y is positive (FIGURE 1a), and that the reading is less than mg if $\vec{a}$ is downward, so a_y is negative (FIGURE 1b).

PART B Evaluate the scale reading from Equation (1) if $\vec{a}$ is upward: $T = (40.0\ N)\left(\frac{2.00\ m/s^2}{9.80\ m/s^2} + 1\right) = 48.2\ N$

Evaluate the scale reading from Equation (1) if $\vec{a}$ is downward: $T = (40.0\ N)\left(\frac{-2.00\ m/s^2}{9.80\ m/s^2} + 1\right) = 31.8\ N$

Finalize Take this advice: if you buy a fish by weight in an elevator, make sure the fish is weighed while the elevator is either at rest or accelerating downward! Furthermore, notice that from the information given here, one cannot determine the direction of the velocity of the elevator.

WHAT IF? Suppose the woman in FIGURE 1 tires of watching the scale and exits the elevator. Then the elevator cable breaks and the elevator and its remaining contents are in free fall. What happens to the reading on the scale?

ANSWER If the elevator falls freely, the fish's acceleration is $\vec{a}_y$. We see from Equation (1) that the scale reading T is zero in this case; that is, the fish *appears* to be weightless.

EXAMPLE 23 TIGHTENING A HAMMER

The metal head of a hammer is loose. To tighten it, you drop the hammer down onto a table. Should you (a) drop the hammer with the handle end down, (b) drop the hammer with the head end down, or (c) do you get the same result either way?

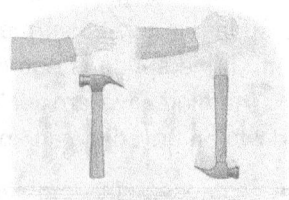

FIGURE 1.

APPROACH It might seem that since the same hammer hits against the same table in either case, there shouldn't be a difference. Actually, there is.
In case (a) the handle of the hammer comes to rest when it hits the table, but the head—with its large inertia—continues downward until a force acts on it to bring it to rest. The force that acts on it is supplied by the handle, which results in the head being wedged more tightly onto the handle. Because the metal head is heavy, the force wedging it onto the handle is great.
In case (b) the head of the hammer comes to rest, but the handle continues to move until a force brings it to rest. The handle is lighter than the head, however; thus, the force acting on it is less, resulting in less tightening.

ANSWER (a) Drop the hammer with the handle end down.

REAL WORLD PHYSICS APPLICATIONS A similar effect occurs when you walk—with each step you compress your head down onto your spine, as when you drop a hammer handle end down. This causes you to grow shorter during the day! Try it. Measure your height first thing in the morning, then again before going to bed. If you're like many people, you'll find that you have shrunk by an inch or so during the day.

EXAMPLE 24 COUPLING FORCE ON FREIGHT CARS

(a) A train engine pulls out of a station along a straight horizontal track with five identical freight cars behind it, each of which weighs $90.0 \, kN$. The train reaches a speed of $15.0 \, m/s$ within $5.00 \, min$ of starting out. Assuming the engine pulls with a constant force during this interval, with what magnitude of force does the coupling between cars pull forward on the first and last of the freight cars? Ignore air resistance and friction on the freight cars.
(b) With what force does the coupling between the first and second cars pull forward on the second car?

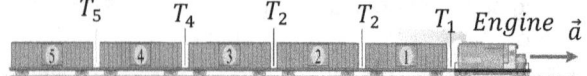

FIGURE 1. An engine pulling five identical freight cars. The entire train has a constant acceleration $\vec{a}$ to the right.

(a) APPROACH A sketch of the situation is shown in FIGURE 1. Here, all freight cars and engine are moving with same acceleration. So, if we want, we can consider all freight cars and engine in a single system. To find the force exerted by the first coupling, we consider all five cars to be one system so we do not have to worry about the force exerted on the first car by the second car. The only *external* forces on the group of five cars are the normal force, gravity, and the pull of the first coupling. To find the force exerted by the fifth coupling, we consider car five by itself to be a system. In each case, once we identify a system, we draw an FBD, choose a coordinate system, and then apply Newton's second law.
As discussed previously, the engine and the cars must all have the same acceleration at any instant. We expect the acceleration to be *constant* because the engine pulls with a constant force. We can calculate the acceleration of the train from the initial and final velocities and the elapsed time.

SOLUTION For the tension T_1 in the first coupling, we consider the five cars as *one system* of mass M. FIGURE 2 shows the FBD in which cars 1 to 5 are treated as a single object. We choose the x-axis in the direction of motion of the train and the y-axis up. Since the train moves along the x-axis, the acceleration vector is along the x-axis. Therefore, $a_y = 0$. Using the y-component of Newton's second law, the vertical forces add to zero:

$$\Sigma F_y = M a_y = N_{1\text{-}5} - W_{1\text{-}5} = 0$$

FIGURE 2 FBD for the system consisting of cars 1 through 5 (but not the engine).

The only external horizontal force is the force $\vec{T}_1$ due to the tension in the first coupling. This force is constant according to the problem statement, so we know that the acceleration a_x is constant:

$$\Sigma F_x = T_1 = M a_x$$

The mass of the system M is five times the mass of one car m. We are given the *weight* of one car ($W = 90.0 \, kN = 9.00 \times 10^4 \, N$). From the relation between mass and weight, $W = mg$, the mass of one car is $m = W/g$ and the mass of five cars is $M = 5W/g$.

The constant acceleration of the train is-

$$a_x = \frac{\Delta v_x}{\Delta t} = \frac{v_{fx} - v_{ix}}{t_f - t_i} = \frac{15.0 \, m/s - 0}{300 \, s - 0} = 0.0500 \, m/s^2$$

Therefore, $T_1 = Ma_x = \frac{5W}{g} \times \frac{\Delta v_x}{\Delta t}$

$$= \frac{5 \times 9.00 \times 10^4 N}{9.80 \, m/s^2} \times \frac{15.0 \, m/s}{300 \, s} = 2.30 \, kN$$

Now consider the last freight car (car 5). If we ignore friction and air resistance, the only external forces acting are the force $\vec{T}_5$ due to the tension in the fifth coupling, the normal force $\vec{N}_5$, and the gravitational force $\vec{W}_5$; the FBD is shown in FIGURE 3. Since $\vec{N}_5 + \vec{W}_5 = 0$, the net force is equal to $\vec{T}_5$. From Newton's second law,

$$\Sigma F_x = T_5 = ma_x = \frac{W}{g} a_x$$

$$T_5 = \frac{W}{g} \times \frac{\Delta v_x}{\Delta t} = \frac{9.00 \times 10^4 N}{9.80 \, m/s^2} \times \frac{15 \, m/s}{300 \, s} = 459 \, N$$

FIGURE 3 FBD for car 5. (Vector lengths are not to the same scale as those in FIGURE 2.)

DISCUSSION We considered two systems (cars 1 to 5 and car 5) that have the same acceleration and different masses. As expected, the net force is proportional to the mass: the net force on five cars is 5 times the net force on one car.

The solution to this problem is much simpler when Newton's second law is applied to a system comprised of all five cars, rather than to each car individually. Although the problem can be solved by looking at individual cars, to find the tension in the first coupling you would have to draw five FBDs (one for each car) and apply Newton's second law five times. That's because each car, except the fifth, is acted on by the unequal tensions in the couplings on either side. You'd have to first find the tension in the fifth coupling, then the fourth, then the third, and so on.

(b) APPROACH Try two methods. One of them is to draw the FBD for the first car and apply Newton's *third* law as well as the second.

ANSWER $1.84 \, kN$.

EXAMPLE 25. FORCES BETWEEN CARS IN A TRAIN
Train cars are connected by *couplers*, which are under tension as the locomotive pulls the train. Imagine you are on a train speeding up with a constant acceleration. As you move through the train from the locomotive to the last car, measuring the tension in each set of couplers, does the tension increase, decrease, or stay the same? When the engineer applies the brakes, the couplers are under compression. How does this compression force vary from the locomotive to the last car? (Assume only the brakes on the wheels of the engine are applied.)

APPROACH While the train is speeding up, tension decreases from the front of the train to the back. The coupler between the locomotive and the first car must apply enough force to accelerate the rest of the cars. As you move back along the train, each coupler is accelerating less mass behind it. The last coupler has to accelerate only the last car, and so it is under the least tension.

When the brakes are applied, the force again decreases from front to back. The coupler connecting the locomotive to the first car must apply a large force to slow down the rest of the cars, but the final coupler must apply a force large enough to slow down only the last car.

ANSWER As you move through the train from the locomotive to the last car, measuring the tension in each set of couplers, the tension decreases.
When the engineer applies the brakes, the force again decreases from front to back.

EXAMPLE 26. THE ATWOOD MACHINE
When two objects of unequal mass are hung vertically over a frictionless pulley of negligible mass as in FIGURE 1a, the arrangement is called an *Atwood machine*. The device is sometimes used in the laboratory to determine the value of g by measuring the acceleration of the objects. Determine the magnitude of the acceleration of the two objects and the tension in the lightweight string.

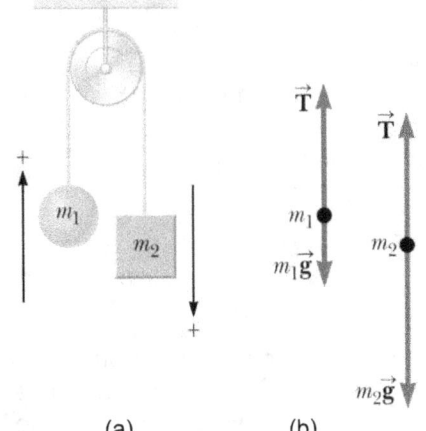

Figure 1 The Atwood machine. (a) Two objects connected by a massless inextensible string over a frictionless pulley. (b) The free-body diagrams for the two objects.

APPROACH Imagine the situation pictured in FIGURE 1a in action: as one object moves upward, the other object moves downward. Because the objects are connected by an inextensible string, the distance one object travels in a given time interval must be the same as the distance the other one travels, and their velocities and accelerations must be of equal magnitude.

The objects in the Atwood machine are subject to the gravitational force as well as to the forces exerted by the strings connected to them. Therefore, we can categorize this problem as one involving two *particles under a net force*.

SOLUTION The free-body diagrams for the two objects are shown in FIGURE 1b. Two forces act on each object: the upward force $\vec{T}$ exerted by the string and the downward gravitational force. In problems such as this one in which the pulley is modelled as massless and frictionless, the tension in the string on both sides of the pulley is the same. If the pulley has mass or is subject to friction, the tensions on either side are not the same and the situation requires techniques we will learn in Chapter "**Rotational Motion**".

We must be very careful with signs in problems such as this one. In FIGURE 1a, notice that if object 1 accelerates upward, object 2 accelerates downward. Therefore, for consistency with signs, if we define the upward direction as positive for object 1, we must define the downward direction as positive for object 2. With this sign convention, both objects accelerate in the same direction as defined by the choice of sign. Furthermore, according to this sign convention, the y component of the net force exerted on object 1 is $T - m_1 g$, and the y component of the net force exerted on object 2 is $m_2 g - T$.

From the particle under a net force model, apply Newton's second law to object 1

$$\sum F_y = T - m_1 g = m_1 a_y \quad \ldots (1)$$

Apply Newton's second law to object 2:

$$\sum F_y = m_2 g - T = m_2 a_y \quad \ldots (2)$$

Add Equation (2) to Equation (1), noticing that T cancels:

$$-m_1 g + m_2 g = m_1 a_y + m_2 a_y$$

Solve for the acceleration:

$$a_y = \left(\frac{m_2 - m_1}{m_1 + m_2}\right) g \quad \ldots (3)$$

Substitute Equation (3) into Equation (1) to find T:

$$T = m_1(g + a_y) = \left(\frac{2 m_1 m_2}{m_1 + m_2}\right) \quad \ldots (4)$$

Finalize The acceleration given by Equation (3) can be interpreted as the ratio of the magnitude of the unbalanced force on the system $(m_2 - m_1)g$ to the total mass of the system $(m_1 + m_2)$, as expected from Newton's second law. Notice that the sign of the acceleration depends on the relative masses of the two objects; if $m_2 > m_1$, the acceleration is positive, corresponding to downward motion for m_2 and upward for m_1. However, if $m_2 < m_1$, Equation (3) gives a negative acceleration, indicating that m_1 moves downward and m_2 moves upward.

WHAT IF? Describe the motion of the system if the objects have equal masses, that is, $m_1 = m_2$.

ANSWER If we have the same mass on both sides, the system is balanced and should not accelerate. Mathematically, we see that if $m_1 = m_2$, Equation (3) gives us $a_y = 0$.

WHAT IF? What if one of the masses is much larger than the other: $m_1 \gg m_2$?

ANSWER In the case in which one mass is infinitely larger than the other, we can ignore the effect of the smaller mass. Therefore, the larger mass should simply fall as if the smaller mass were not there. We see that if $m_1 \gg m_2$, Equation (3) gives us $a_y = g$.

EXAMPLE 27. ACCELERATION OF TWO OBJECTS CONNECTED BY A CORD

A ball of mass m_1 and a block of mass m_2 are attached by a lightweight cord that passes over a frictionless pulley of negligible mass as in FIGURE 1a. The block lies on a frictionless incline of angle θ. Find the magnitude of the acceleration of the two objects and the tension in the cord.

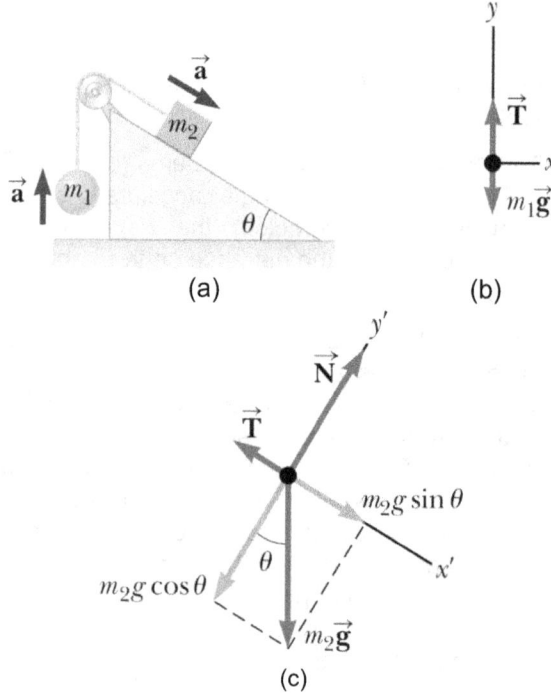

FIGURE 1. (a) Two objects connected by a lightweight cord strung over a frictionless pulley. (b) The free-body diagram for the ball. (c) The free-body diagram for the block. (The incline is frictionless.)

APPROACH Imagine the objects in FIGURE 1 in motion. If m_2 moves down the incline, then m_1 moves upward. Because the objects are connected by a cord (which we assume does not stretch), their accelerations have the same magnitude. Notice the normal coordinate axes in FIGURE 1b for the ball and the "tilted" axes for the block in FIGURE 1c. Just as we chose the positive direction to be different for each of the objects in Example 21, we are free to choose entirely different coordinate axes for the two objects here.

We can identify forces on each of the two objects and we are looking for an acceleration, so we categorize the objects as *particles under a net force*. For the block, this model is only valid for the x' direction. In the y' direction, we apply the *particle in equilibrium* model because the block does not accelerate in that direction.
SOLUTION Consider the free-body diagrams shown in figures 1b and 1c.
Apply Newton's second law in the y direction to the ball, choosing the upward direction as positive:
$$\sum F_y = T - m_1 g = m_1 a_y = m_1 a \qquad \ldots (1)$$
For the ball to accelerate upward, it is necessary that $T > m_1 g$. In Equation (1), we replaced a_y with a because the acceleration has only a y component.

For the block, we have chosen the x' axis along the incline as in FIGURE 1c. For consistency with our choice for the ball, we choose the positive x' direction to be down the incline.

Apply the particle under a net force model to the block in the x' direction and the particle in equilibrium model in

the y' direction:
$$\sum F_{x'} = m_2 g \sin\theta - T = m_2 a_{x'} = m_2 a \qquad \ldots (2)$$
$$\sum F_{y'} = N - m_2 g \cos\theta = 0 \qquad \ldots (3)$$

In Equation (2), we replaced $a_{x'}$ with a because the two objects have accelerations of equal magnitude a.

Solve Equation (1) for T:
$$T = m_1(g + a) \qquad \ldots (4)$$

Substitute this expression for T into Equation (2):
$$m_2 g \sin\theta - m_1(g + a) = m_2 a$$

Solve for a: $\quad a = \left(\dfrac{m_2 \sin\theta - m_1}{m_1 + m_2}\right) g \qquad \ldots (5)$

Substitute this expression for a into Equation (4) to find T:
$$T = \left[\dfrac{m_1 m_2(\sin\theta + 1)}{m_1 + m_2}\right] g$$

FINALIZE The block accelerates down the incline only if $m_2 \sin\theta > m_1$. If $m_1 > m_2 \sin\theta$, the acceleration is up the incline for the block and downward for the ball. Also notice that the result for the acceleration, Equation (5), can be interpreted as the magnitude of the net external force acting on the ball–block system divided by the total mass of the system; this result is consistent with Newton's second law.

(i) What happens in this situation if $\theta = 90°$?

ANSWER If $\theta = 90°$, the inclined plane becomes vertical and there is no interaction between its surface and m_2. Therefore, this problem becomes the Atwood machine of Example 21. Letting $\theta \to 90°$ in Equations (5) and (6) causes them to reduce to Equations (3) and (4) of Example 21!

(ii) What if $m_1 = 0$?

ANSWER If $m_1 = 0$, then m_2 is simply sliding down an inclined plane without interacting with m_1 through the string. Therefore, this problem becomes the sliding block problem in EXAMPLE 6. Letting $m_1 \to 0$ in Equation (5) causes it to reduce to the result of EXAMPLE 6.

24. SOLVED EXAMPLES ON NEWTON'S FIRST AND SECOND LAWS

EXAMPLE 28. A 2.00 kg object is subjected to three forces that give it an acceleration $\vec{a} = -(8.00\ m/s^2)\hat{\imath} + (6.00\ m/s^2)\hat{\jmath}$. If two of the three forces are $\vec{F}_1 = (30.0\ N)\hat{\imath} + (16.0\ N)\hat{\jmath}$ and $\vec{F}_2 = -(12.0\ N)\hat{\imath} + (8.00\ N)\hat{\jmath}$, find the third force.

APPROACH Let third force is $\vec{F}_3$. From Newton's second law, we have
$$\sum \vec{F} = m\vec{a}$$
or $\quad \vec{F}_1 + \vec{F}_2 + \vec{F}_3 = m\vec{a}$
or $\quad \vec{F}_3 = m\vec{a} - \vec{F}_1 - \vec{F}_2 \qquad \ldots (1)$

SOLUTION Here,
$$m\vec{a} = 2[-(8.00\ m/s^2)\hat{\imath} + (6.00\ m/s^2)\hat{\jmath}]$$
$$= -(16.00\ m/s^2)\hat{\imath} + (12.00\ m/s^2)\hat{\jmath},$$
$$\vec{F}_1 = (30.0\ N)\hat{\imath} + (16.0\ N)\hat{\jmath}$$
$$\vec{F}_2 = -(12.0\ N)\hat{\imath} + (8.00\ N)\hat{\jmath}$$

Using these values in (1), we get
$$\vec{F}_3 = -(16.00\ m/s^2)\hat{\imath} + (12.00\ m/s^2)\hat{\jmath} - [(30.0\ N)\hat{\imath} + (16.0\ N)\hat{\jmath}] - [-(12.0\ N)\hat{\imath} + (8.00\ N)\hat{\jmath}]$$
$$\vec{F}_3 = (-34\ N)\hat{\imath} - (12.00\ N)\hat{\jmath}$$

This is the required third force.

EXAMPLE 29. A 0.340 kg particle moves in an xy plane according to $x(t) = -15.00 + 2.00\,t - 4.00 t^3$ and $y(t) = 25.00 + 7.00\,t - 9.00\,t^2$, with x and y in meters and t in seconds. At $t = 0.700\,s$, what are (a) the magnitude of acceleration and force? (b) the angle (relative to the positive direction of the x axis) of the net force on the particle, and (c) what is the angle of the particle's direction of travel?

APPROACH To solve the problem, we note that acceleration is the second time derivative of the position function; it is a vector and can be determined from its components. The net force is related to the acceleration via Newton's second law. Thus, differentiating $x(t) = -15.0 + 2.00\,t + 4.00\,t^3$ twice with respect to t, we get

$$\frac{dx}{dt} = 2.00 - 12.00\,t^2, \quad \frac{d^2x}{dt^2} = -24.0\,t$$

Similarly, differentiating $y(t) = 25.00 + 7.00\,t - 9.00\,t^2$, twice with respect to t yields

$$\frac{dy}{dt} = 7.00 - 18.0\,t, \quad \frac{d^2y}{dt^2} = -18.0$$

The magnitude of the force is given by $F = ma$ and the angle $\vec{F}$ or $\vec{a} = \vec{F}/m$ makes with $+x$ is given by

$$\theta = \tan^{-1}\left(\frac{a_y}{a_x}\right)$$

The direction of travel is the direction of a tangent to the path, which is the direction of the velocity vector.

SOLUTION (a) The acceleration

$$\vec{a} = a_x\hat{\imath} + a_y\hat{\jmath} = \frac{d^2x}{dt^2}\hat{\imath} + \frac{d^2y}{dt^2}\hat{\jmath}$$

$$= (-24.0\,t)\hat{\imath} + (-18.0)\hat{\jmath}$$

At $t = 0.700\,s$, we have $\vec{a} = (-16.8)\hat{\imath} + (-18.0)\hat{\jmath}$, with magnitude

$$a = \sqrt{(-16.8)^2 + (-18.0)^2} = 24.6\,m/s^2$$

Thus, the magnitude of the force is $F = ma = (0.34\,kg)(24.6\,m/s^2) = 8.37\,N$.

(b) The angle $\vec{F}$ or $\vec{a} = \vec{F}/m$ makes with $+x$ is

$$\theta = \tan^{-1}\left(\frac{a_y}{a_x}\right) = \tan^{-1}\left(\frac{-18.0\,m/s^2}{-16.8\,m/s^2}\right)$$

$$= 47.0° \text{ or } -133°$$

We choose the latter ($-133°$) since $\vec{F}$ is in the third quadrant.

(c) The direction of travel is the direction of a tangent to the path, which is the direction of the velocity vector:

$$\vec{v}(t) = v_x\hat{\imath} + v_y\hat{\jmath} = \frac{dx}{dt}\hat{\imath} + \frac{dy}{dy}\hat{\jmath}$$

$$= (2.00 - 12.00\,t^2)\hat{\imath} + (7.00 - 18.0\,t)\hat{\jmath}$$

At $t = 0.700\,s$, we have,
$\vec{v}(t = 0.700\,s) = (-3.88\,m/s)\hat{\imath} + (-5.60\,m/s)\hat{\jmath}$.
Therefore, the angle $\vec{v}$ makes with $+x$ is

$$\theta_v = \tan^{-1}\left(\frac{v_y}{v_x}\right)$$

$$= \tan^{-1}\left(\frac{-5.60\,m/s^2}{-3.88\,m/s^2}\right) = 55.3° \text{ or } -125°.$$

We choose the latter ($-125°$) since $\vec{v}$ is in the third quadrant.

EXAMPLE 30. A $0.150\,kg$ particle moves along an x axis according to $x(t) = -13.00 + 2.00\,t + 4.00\,t^2 - 3.00\,t^3$, with x in meters and t in seconds. In unit-vector notation, what is the net force acting on the particle at $t = 3.40\,s$?

APPROACH To solve the problem, we note that acceleration is the second time derivative of the position function, and the net force is related to the acceleration via Newton's second law. Thus, differentiating

$$x(t) = -13.00 + 2.00\,t + 4.00\,t^2 - 3.00\,t^3$$

twice with respect to t, we get

$$\frac{dx}{dt} = 2.00 + 8.00\,t - 9.00\,t^2,$$

$$\frac{d^2x}{dt^2} = 8.00 - 18.00\,t$$

The net force acting on the particle is given by

$$\vec{F} = m\frac{d^2x}{dt^2}\hat{\imath} = m[8.00 - 18.00\,t]\hat{\imath}$$

SOLUTION At $t = 3.40\,s$ is

$$\vec{F} = m[8.00 - 18.00\,t]\hat{\imath} = (0.150)[8.00 - 18.0)(3.40)]\hat{\imath} = (-0.798\,N)\hat{\imath}$$

EXAMPLE 31. A $2.0\,kg$ particle moves along x axis, being propelled by a variable force directed along that axis. Its position is given by $x = 3.0\,m + (4.0\,m/s)t + ct^2 - (2.0\,m/s^3)t^3$, with x in meters and t in seconds. The factor c is a constant. At $t = 3.0\,s$, the force on the particle has a magnitude of $36\,N$ and is in the negative direction of the axis. What is c?

APPROACH The velocity is the derivative (with respect to time) of given function x, and the acceleration is the derivative of the velocity. Thus, $a = 2c - 3(2.0)(2.0)t$, which we use in Newton's second law: $F = (2.0\,kg)a = 4.0c - 24t$ (with SI units understood).

SOLUTION At $t = 3.0\,s$, we are told that $F = -36\,N$. Thus, $-36 = 4.0c - 24(3.0)$ can be used to solve for c. The result is $c = +9.0\,m/s^2$.

24.1. SYMBOLIC SOLVED EXAMPLES

EXAMPLE 32. (i) Suppose that blocks A and B have masses of 2 and 6kg, respectively, and are in contact on a smooth horizontal surface. If a horizontal force of 6N pushes them, calculate (a) the acceleration of the system and (b) the force that the 2kg block exerts on the other block.

SOLUTION. (a) Considering the blocks to move as unit, $M = m_0 + m_b = 8kg$, $F = Ma = 6N$, $a = 0.75\,m/s^2$.

(b) If we now consider block B to be our system, the only force acting on it is the force due to block A, F_{ab}. Then since the acceleration is the same as in part (a), we have $F_{ab} = M_b a = 4.5$ N

EXAMPLE 33. Three blocks, of masses 2.0, 4.0 and 6.0 kg, arranged in the order lower, middle, and upper, respectively, are connected by strings on a frictionless inclined plane of 30°. A force 120 N is applied upward along the incline to the uppermost block, causing an upward movement of the blocks. The connecting cords are light. What is the acceleration of the blocks?

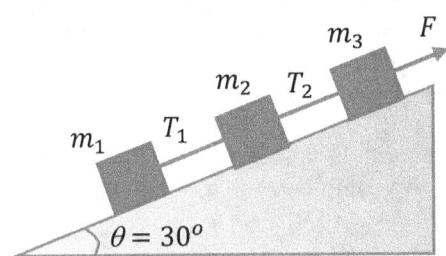

SOLUTION The situation is depicted in adjoining figure with $F = 120$ N

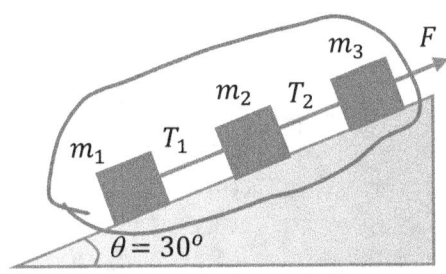

$m_1 = 2.0$ kg, $m_2 = 4.0$ kg and $m_3 = 6.0$ kg

Applying Newton's second law to each block, we have

For block m_3: $F - T_2 - m_3 g \sin 30° = m_3 a$

For block m_2: $T_2 - T_1 - m_2 g \sin 30° = m_2 a$

For block m_1: $T_1 - m_1 g \sin 30° = m_1 a$

Adding these equations,

$F - (m_1 + m_2 + m_3)g \sin 30° = (m_1 + m_2 + m_3)a$

$a = \frac{F-(m_1+m_2+m_3)g \sin 30}{(m_1+m_2+m_3)} = \frac{120-(2+4+6)\times 10 \times \frac{1}{2}}{(2+4+6)} = \frac{60}{12}$

or $a = 5 \, m/s^2$

ALTERNATE METHOD
Since all three blocks are connected with strings, therefore all blocks will move with same acceleration along the incline.

So, we can consider all the blocks in a single system. Now, mass of the system, $m = (m_1 + m_2 + m_3)$
Net force on the system along the incline
$= F - (m_1 + m_2 + m_3)g \sin 30$
$F - (m_1 + m_2 + m_3)g \sin 30 = (m_1 + m_2 + m_3)a$

$a = \frac{F-(m_1+m_2+m_3)g \sin 30}{(m_1+m_2+m_3)} = \frac{120-(2+4+6)\times 10 \times \frac{1}{2}}{(2+4+6)}$
$= \frac{60}{12} = 5 \, m/s^2$

EXAMPLE 34. The device in adjoining FIGURE is called an Atwood's machine. In terms of m_1 and m_2 with $m_2 > m_1$. What is the tension in the light cord that connects the two masses? Assume the pulley to be frictionless and massless.

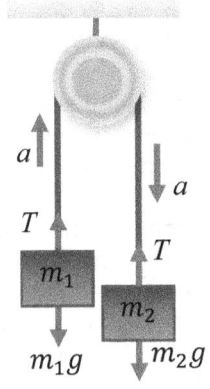

SOLUTION. Isolate the forces on each mass and write Newton's second law

$T - m_1 g = m_1 a$ and $m_2 g - T = m_2 a$.

Eliminating T gives $a = \frac{(m_2-m_1)g}{m_1+m_2}$.

From the above equations,

$T = \frac{2m_1 m_2 g}{m_1+m_2}$

EXAMPLE 35. Two bodies of masses m_1 and m_2 are connected by a light string going over a smooth light pulley at the end of an incline. The mass m_1 lies on the incline and m_2 hangs vertically. (a) find the acceleration of m_1. (b) Find force acting on the pulley due to rod which attaches it to incline.

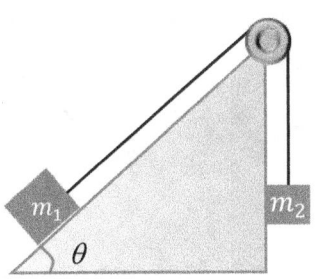

SOLUTION. (a) Adjoining figure shows the situation with the forces on m_1 and m_2. Take the body of mass m_2 as the system. The forces acting on it are:

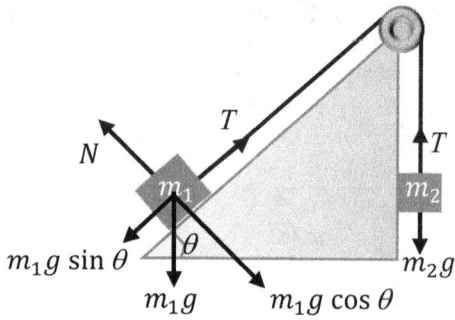

(i) m_2g vertically downward (by the Earth)
(ii) T vertically upward (by the string)
This gives $\quad m_2g - T = m_2a \quad$ ... (i)
Next, consider the body of mass m₁ as the system. The forces acting on this system are
(i) m_1g vertically downward (by the Earth)
(ii) T along the string up the incline (by the string)
(iii) N normal to the incline (by the incline)
As the string and the pulley are all light and smooth, the tension in the string is uniform everywhere.
Taking components parallel to the incline,
$$T - m_1g \sin\theta = m_1 a \quad ... \text{(ii)}$$
Taking components along the normal to the incline,
$$N = m_1 g \cos\theta$$
Eliminating T from (i) and (ii),
$$a = \frac{m_2 g - m_1 g \sin\theta}{(m_1 + m_2)}$$
(b) Suppose, F_x and F_y are the components of force applied by rod on the pulley along X and Y axis respectively. As the pulley is in equilibrium, therefore from FBD of pulley, we have

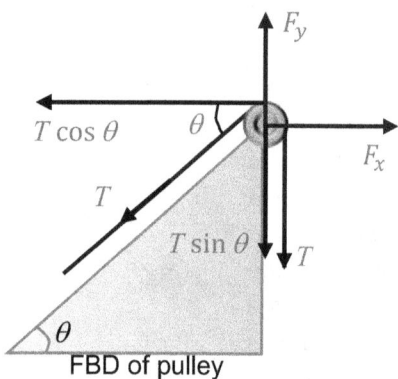

FBD of pulley

Along X-axis-
$F_x - T\cos\theta = 0 \Rightarrow F_x = T\cos\theta$
Along Y-axis-
$F_y - T\sin\theta - T = 0 \Rightarrow F_y = T(1 + \sin\theta)$
$$|\vec{F}| = \sqrt{F_x^2 + F_y^2}$$
or $|\vec{F}| = T\sqrt{2 + 2\sin\theta}$
Direction of force from X axis: If ϕ is the angle of net force from X axis, then $\tan\phi = \frac{F_y}{F_x} = \frac{T(1+\sin\theta)}{T\cos\theta} = \frac{(1+\sin\theta)}{\cos\theta}$
or $\quad \phi = \tan^{-1}\left(\frac{1+\sin\theta}{\cos\theta}\right)$

☞ As the tension in string, along the incline, has both X and Y components of force, therefore, it is always safer to assume that the force applied by the rod on pulley also has components along X and Y axis.

EXAMPLE 36. An inventive child of mass M wants to reach an apple in a tree without climbing the tree. Sitting in a chair, of mass m, connected to a rope that passes over a frictionless pulley (FIGURE 1), the child *raises himself and the chair by pulling the rope downward*.

FIGURE 1

(a) With what minimum force should the child pull the rope so as to prevent himself from falling down.
(b) If the child pulls the rope with a force F greater than the minimum force, then determine the acceleration of the (child + chair) system.
(c) Determine the normal reaction between the child and the chair in part (b).
SOLUTION. Let the whole system moves upward with an acceleration a. Applying Newton's Second law,
$$2F - (M+m)g = (M+m)a$$

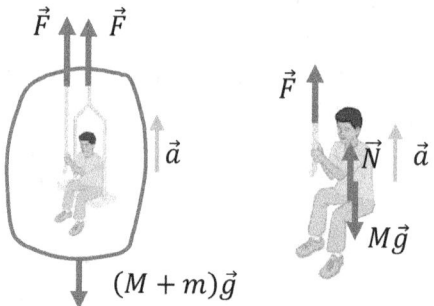

(a) FBD of whole system (b) FBD of child
FIGURE 2

(a) When $F = F_{min}$; $a = 0$, thus $F_{min} = \frac{(M+m)g}{2}$
(b) When $F > F_{min}$, then acceleration of the system is
$$a = \frac{2F}{m+M} - g$$
(c) Considering the free body diagram of the child, we have from Newton's Second Law,
$$F + N - Mg = Ma$$
or $\quad F + N - Mg = m\left[\frac{2F}{M+m} - g\right]$
or $\quad N = (M - m)\left[\frac{F+(m+M)g}{m+M}\right]$

EXAMPLE 37. A block of mass $m = 10$ kg is pulled by a force $F = 100$ N at an angle $\theta = 30°$ with the horizontal along a smooth horizontal surface. What is the acceleration of the block? ($g = 10$ m/s²)
SOLUTION The forces that act on the body can be decomposed along x and y axis.
As there is no acceleration along y-axis, net force acting along vertical or y axis should be zero i.e.
$$\Sigma F_y = N + F\sin\theta - mg = 0 \quad ... (1)$$
The body accelerates along x-axis. Therefore
$$\Sigma F_x = F\cos\theta = ma \quad ... (2)$$
where the acceleration along x-axis is a.
$\Rightarrow \quad F\cos\theta = ma$

$$\Rightarrow \quad a = \frac{F \cos\theta}{m}$$
$$= \frac{100 \cos 30°}{10} = \frac{100\sqrt{3}}{2 \times 10} = 5\sqrt{3} \ m/s^2$$

∴ The acceleration of the block is $5\sqrt{3} \ m/s^2$ Directed towards right. Since $F \sin\theta < mg$ & the surface is rigid, the block remains in equilibrium along y-axis.

EXAMPLE 38. In the adjoining figure, *the tension in the horizontal cord is 30N. Find the weight of the body B.*

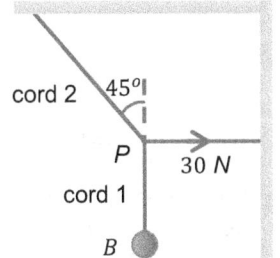

SOLUTION. (i) Isolate the point P
(ii) The forces acting on it are:
Unknown tension T_2 in cord 2
Unknown tension T_1 in cord 1
Known tension of 30N in the horizontal cord.
(iii) T_2 is resolved along x and y axes.
(iv) Condition of equilibrium, at point P, is:
$\sum F_x = 0 \Rightarrow 30 - T_2 \sin 45° = 0$ ... (1)
and $\sum F_y = T_2 \cos 45° - T_1 = 0$ ... (2)

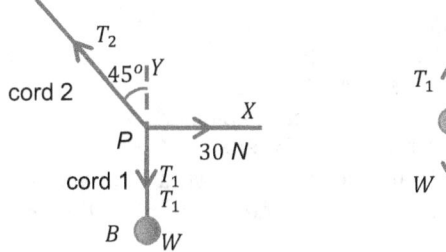

For body B
Since the body B is also in equilibrium,
Hence, $T_1 = W$ ... (3)
(v) After solving these equations, we get $W = 30N$

EXAMPLE 39. *A block of mass m_1 moves on a level frictionless surface. It is connected by a light flexible cord passing over a small frictionless pulley to a second hanging block of mass m_2. What is the acceleration of the system and what is the tension in the string?*

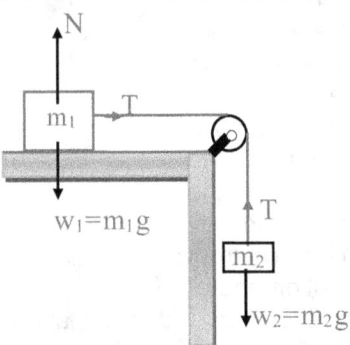

SOLUTION: The diagram shows the forces acting on each block.

For block m_1: $\sum f(x) = T = m_1 a_x$
$\sum f(y) = N - mg = m_1 a_y = 0$
For block m_2: $\sum f(y) = m_2 g - T = m_2 a_y$
$T = m_1 a$ ... (1)
$m_2 g - T = m_2 a$ ... (2)
Adding above two equations, we get
$m_2 g = (m_1 + m_2) a$
$a = \frac{m_2}{m_1 + m_2} g$

Substituting the value of a in equation (1) gives the tension as $T = \frac{m_1 m_2}{m_1 + m_2} g$

EXAMPLE 40. Two blocks of masses $m_1 = 2$ kg and $m_2 = 1$ kg, are in contact on a smooth horizontal surface as shown in the adjoining figure. A horizontal force $F = 3N$ is applied on the block m_1. Find the contact force between the blocks

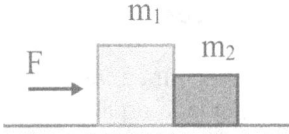

SOLUTION. Let contact force between the blocks be N. Since N is responsible for the acceleration of m_2, therefore, it will be in the direction of acceleration of m_2, and on m_1, it will be opposite to the acceleration
Now, equations of motion
$F - N = m_1 a$ ... (1)
$N = m_2 a$ ... (2)
Where a = acceleration of the blocks

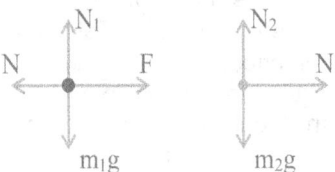

From equations (1) and (2), we get
$a = \frac{F}{m_1 + m_2} \Rightarrow N = \frac{m_2 F}{m_1 + m_2} = 1$ newton

25. CHECKPOINT 6

1. • How would an astronaut in apparent weightlessness be aware of her mass?
2. • Under what circumstances would your apparent weight be greater than your true weight?
3. •• A person of mass 50 kg stands on a weighing scale on a lift. If the lift is descending with a downward acceleration of $9 \ ms^{-2}$, what would be the reading of the weighing scale? ($g = 10 \ ms^{-2}$)

4. •• Two monkeys A and B are holding on the two sides of a light string passing over a smooth pulley. Mass of the two monkeys are $m_A = 8$ kg and $m_B = 10$ kg respectively [$g = 10$ m/s²] (a) Monkey A holds the string tightly and B goes down with an acceleration $a_r = 2$ m/s² relative to the string. Find the weight that A feels of his own body. (b) What is the weight experienced by two monkeys if A holds the string tightly and B goes down with an acceleration $a_r = 4$ m/s² relative to the string

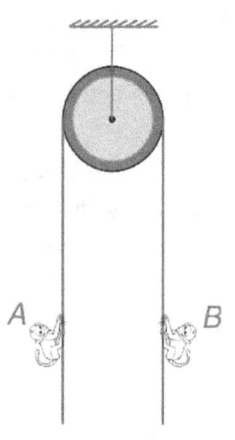

5. ••• Six identical blocks – numbered 1 to 6 – have been glued in two groups of three each and have been suspended over a pulley as shown in fig. The pulley and string are massless and the system is in equilibrium. The block 1, 2, 3, and 4 get detached from the system in sequence starting with block 1. The time gap between separation of two consecutive block (i.e., time gap between separation of 1 and 2 or gap between separation of 2 and 3) is t_0. Finally, blocks 5 and 6 remain connected to the string. (a) Find the final speed of blocks 5 and 6. (b) Plot the graph of variation of speed of block 5 with respect to time. Take $t = 0$ when block 1 gets detached.

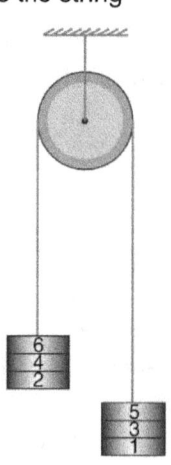

6. ••• A light string passing over a smooth pulley holds two identical buckets at its ends. Mass of each empty bucket is M and each of them holds M mass of sand. The system was in equilibrium when a small leak developed in bucket B (take this time as $t = 0$). The sand leaves the bucket at a constant rate of m kg/s. Assume that the leaving sand particles have no relative speed with respect to the bucket (it means that there is no impulsive force on the bucket like leaving exhaust gases exert on a rocket). Find the speed (v_0) of the two buckets when B is just empty.

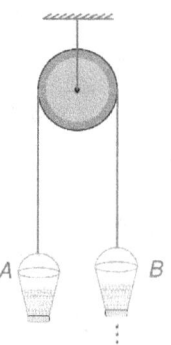

7. ••• A chain is lying on a smooth table with half its length hanging over the edge of the table [fig(a)]. If the chain is released it slips off the table in time t_1. Now, two identical small balls are attached to the two ends of the chain and the system is released [fig(b)]. This time the chain took t_2 time to slip off the table. Which time is larger, t_1 or t_2?

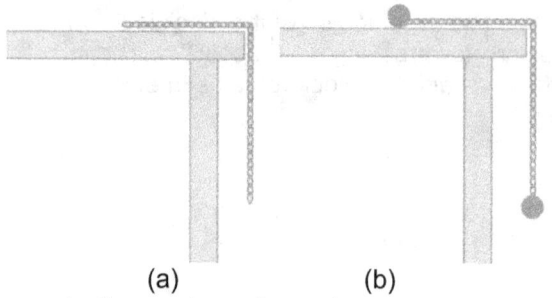

(a)   (b)

8. ••• In the system shown in the figure, AB and CD are identical elastic cords having force constant k. The string connected to the block of mass M is inextensible and massless. The pulley is also massless. Initially, the cords are just taut. The end D of the cord CD is gradually moved up. Find the vertical displacement of the end D by the time the block leaves the ground.

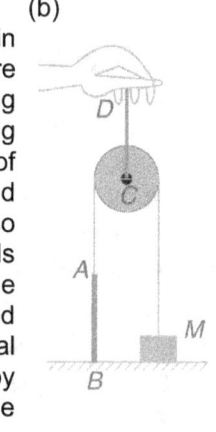

9. •• Blocks A and B have dimensions as shown in the fig. and their masses are 8 kg and 1 kg respectively. A small block C of mass 0.5 kg is placed on the top left corner of block A. All surfaces are smooth. A horizontal force $F = 18$ N is applied to the block B at time $t = 0$. At what time will the block C hit the ground surface? Take $g = 10$ m/s².

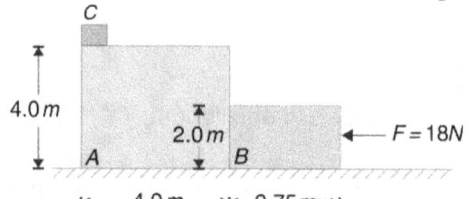

10. ••• In the arrangement shown in the fig. the pulley, the spring and the thread are ideal. The spring is stretched and the two blocks are in contact with a horizontal platform P. When the platform is gradually moved up by 2 cm the tension in the string becomes zero. If the platform is gradually moved down by 2 cm from its original position one of the blocks lose contact with the platform. Given $M = 4$ kg; $m = 2$ kg. (a) Find the force constant (k) of the spring (b) If the platform continues to move

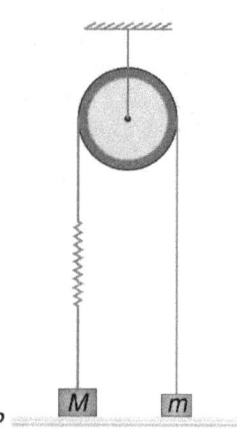

down after one of the blocks loses contact, will the other block also lose contact? Assume that that the platform moves very slowly.

11. ••• In the arrangement shown in the adjoining figure, a monkey of mass M keeps itself as well as block A at rest by firmly holding the rope. Rope is massless and the pulley is ideal. Height of the monkey and block A from the floor is h and 2h respectively [h = 2.5 m]. (a) The monkey loosens its grip on the rope and slides down to the floor. At what height from the ground is block A at the instant the money hits the ground? (b) Another block of mass equal to that of A is stuck to the block A and the system is released. The monkey decides to keep itself at height h above the ground and it allows the rope to slide through its hand. With what speed will the block strike the ground? (c) In the situation described in (b), the monkey decides to prevent the block from striking the floor. The monkey remains at height h till the block crosses it. At the instant the block is crossing the monkey it begins climbing up the rope. Find the minimum acceleration of the monkey relative to the rope, so that the block is not able to hit the floor. Do you think that a monkey can climb with such an acceleration? ($g = 10$ m/s²)

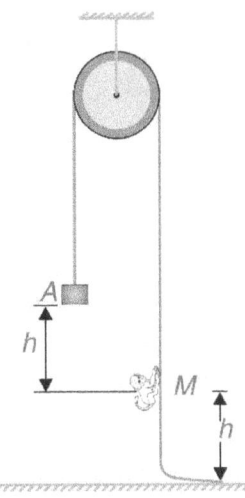

26. CONSTRAINED MOTION

Constrained motion results when an object is forced to move in a restricted way. For example, it may have to move along a curved track, to slide on a table that may accelerate upwards, to stay in contact with an accelerating wedge, etc.

Constraint Forces are the forces that the constraining object exerts on the object to make it follow the motional constraints.

Constraints- These are the relation obtained by using the given restrictions. Some constraints are given below-

1. When two or more bodies are connected & their motion are related to maintain connection, e.g. if we have a block kept on an inclined plane and we want the block to maintain contact with it. The block cannot have velocity and acceleration in direction perpendicular to the incline.

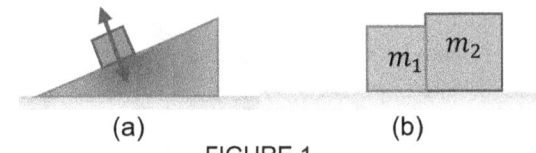

FIGURE 1

If we have two blocks kept touching each other on horizontal surface as shown in FIGURE 1b, then they must have same velocity and acceleration to maintain the contact. i.e.,

$$\vec{v}_1 = \vec{v}_2 \quad \text{and} \quad \vec{a}_1 = \vec{a}_2$$

If we keep a block on wedge which can move over it, then again constraint is defined in reference frame attached to wedge. The block cannot have any acceleration in 'y' direction in reference frame attached to the wedge.

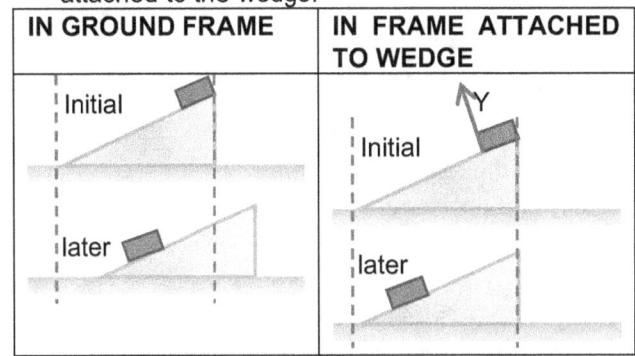

IN GROUND FRAME	IN FRAME ATTACHED TO WEDGE
Initial	Initial
later	later

2. When two bodies are connected by inextensible tight string or rope then their motion will be interdependent. If we connect two blocks as shown in diagram and pull block B towards right the block A must cover same distance as B to keep string tight.

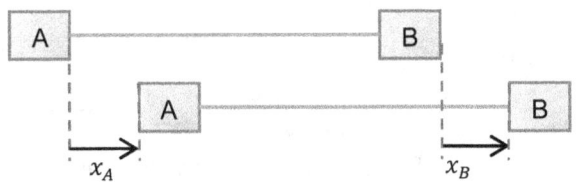

i.e., $\quad x_A = x_B$

Differentiating above equation w.r.t. time, we get

velocity $\quad v_A = v_B$

Further differentiating w.r.t. time, we get

acceleration $a_A = a_B$

Although if we push B towards left there is no constraint relation as string will slack. If A & B are connected by rod then A will have to move as B is moving in both the above cases. Along the string $v_A = v_B$.

EXAMPLE 41. FIGURE 1 shows a hemisphere and a supported rod. Hemisphere is moving in right direction with a uniform velocity v_2 and the end of rod which is in contact with the ground is moving in left direction with a velocity v_1. Find the rate at which the angle θ is changing in terms of $v_1, v_2, R,$ and θ.

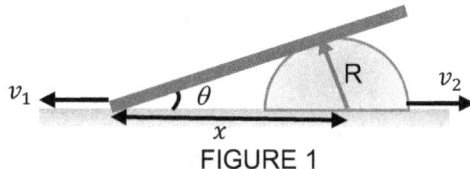

FIGURE 1

APPROACH Let x is the separation between centre of hemisphere and the end of the rod at any time t. Rate of change of x can be taken as the relative velocity of end of rod and hemisphere centre i.e., $(v_1 + v_2)$. We are required to find the rate of change of θ and rate of change of x. So, we have to develop a relation between x and θ.

SOLUTION From above figure, we have $x = R\, cosec\,\theta$

$\Rightarrow \frac{dx}{dt} = -R\, cosec\,\theta \cot\theta \frac{d\theta}{dt}$

or $\frac{d\theta}{dt} = \frac{(v_1+v_2)\sin^2\theta}{R\cos\theta}$

EXAMPLE 42. In adjoining figure, find the relation between velocity and acceleration of the rod and wedge.

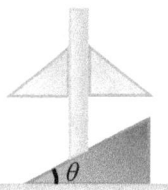

SOLUTION. Let's imagine what happens when wedge is shifted towards right. We make a superimposing diagram on the initial diagram. In this case, three constraints are involved-
(i) The wedge can move only horizontally
(ii) the rod can move only vertically
(iii) the rod and wedge will always remain in contact with each other, i.e., the velocity components of wedge and rod perpendicular to the inclined surface of wedge will always be equal. Thus, we have to relate their motions geometrically.

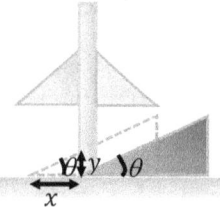

From adjoining figure, it is clear that-
when wedge moves a distance x along horizontal direction, the rod lowers by vertical distance y.

$\therefore \quad \tan\theta = \frac{y}{x} \Rightarrow y = x\tan\theta$

Differentiating both sides w. r. t., time t, we get
$v_R = v_w \tan\theta$,
Again differentiating,
$a_R = a_w \tan\theta$
here $v_R, a_R \to$ velocity and acceleration of rod in vertical direction, $v_W, a_W \to$ velocity and acceleration of wedge in horizontal direction.

EXAMPLE 43. If velocity of A is $2\,ms^{-1}$ downwards what is the velocity of B in all three cases?

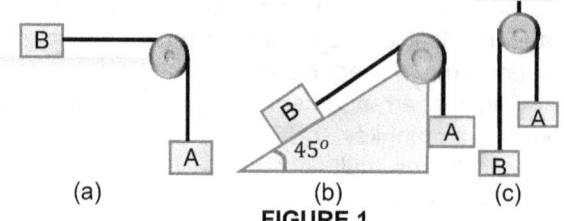

FIGURE 1

SOLUTION Since in all the given figure, the blocks are connected by inextensible strings, therefore the speeds will be same $(2\,ms^{-1})$ in all the three cases.

27. MATHEMATICAL ANALYSIS OF CONSTRAINED MOTION

Sometimes the motions of particles are interrelated because of the constraints imposed by interconnecting members. In such cases it is necessary to account for these constraints in order to determine the respective motions of the particles.

27.1. One Degree of Freedom

Now, consider first a very simple system of two interconnected particles A and B as shown in FIGURE 1(a).

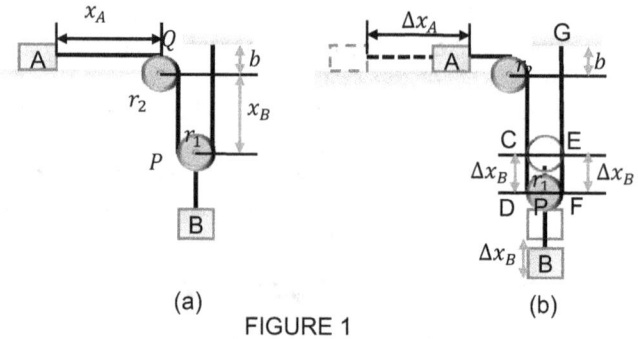

FIGURE 1

Clearly, the motion of B is the same as that of the centre of its pulley, so we establish position coordinates x_A and x_B measured from a convenient fixed reference. The total length of the cable is

$L = x_A + \frac{\pi r_2}{2} + 2x_B + \pi r_1 + b$; here $L, r_2, r_1,$ and b, all are constants, therefore

$x_A + 2x_B = $ constant

Since only one variable, either x_A or x_B, is needed to specify the positions of all parts of the

> Number. of degree of freedom = Minimum Number of arbitrary coordinates needed to specify the positions of all parts of the system

system i.e., only one of the two coordinates x_A and x_B can be chosen arbitrarily, therefore, we say that the system shown in FIGURE 1(a) has *one degree of freedom*.

The first and second time derivatives of above equation give
$\dot{x}_A + 2\dot{x}_B = 0$ or $v_A + 2v_B = 0$
$\ddot{x}_A + 2\ddot{x}_B = 0$ or $a_A + 2a_B = 0$
The velocity and acceleration constraint equations indicate that, for the selected coordinates, the velocity of A must have a sign which is opposite to that of the velocity of B, and similarly for the accelerations. The constraint equations are valid for the motion of the system in either direction. We emphasize that $v_A = \dot{x}_A$ is positive to the left and that $v_B = \dot{x}_B$ is positive down. Because the results do not depend on the lengths or pulley radii, we should be able to analyze the motion without considering them. From the relation between the position coordinates x_A and x_B, it follows that if x_B is given an increment Δx_B—that is, if block B is lowered by an amount Δx_B—the coordinate x_A receives an increment $\Delta x_A = -2\Delta x_B$. In other words, block A moves by two times of the same amount. Here negative sign shows that sense of displacement of the blocks from fixed pulley Q is opposite to each other, i.e., if one block is moving towards pulley, then other will move away from the pulley.

Constraint Relations by inspection
Suppose, the displacement of A corresponding to displacement Δx_B in B is Δx_A (FIGURE 1(b)), The inextensible string connected with block A is fixed at G. If the block B displaced downward by distance Δx_B, then the displacement in the pulley connected with B, will also be Δx_B. Therefore, the extra down length of the string passing over the pulley P will be $CD + EF = \Delta x_B + \Delta x_B = 2\Delta x_B$.
As the string is fixed at G, so this down length cannot come from right side of the pulley P. The extra down length comes only from the displacement of A. Therefore, to keep string tight, we have
$\Delta x_A = 2\Delta x_B$
Similarly, $v_A = 2v_B$ and $a_A = 2a_B$
These equations show the relation between magnitudes only. The directions can be easily determined by the diagram.

☞ The first method of analysis is used for more complex situations where the results cannot be easily reached by inspection.

In FIGURE 2(a), the position of block B depends upon the position of block A. Because the rope ACDEFG is of constant length, and because the lengths of the portions of rope CD and EF wrapped around the pulleys remain constant, it follows that the sum of the lengths of the segments AC, DE, and FG is constant. Observing that the length of the segment AC differs from x_A only by a constant and that, similarly, the lengths of the segments DE and FG differ from x_B only by a constant, we have

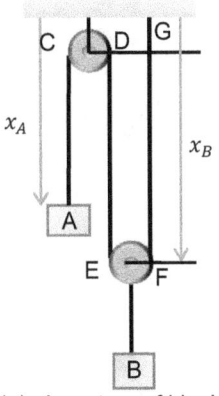
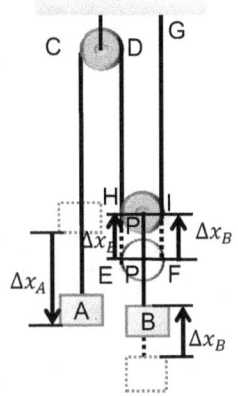

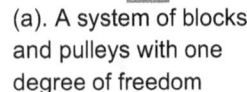

(a). A system of blocks and pulleys with one degree of freedom

(b). A system of blocks and pulleys with one degree of freedom

FIGURE 2.

$x_A + 2x_B$ = constant
Differentiating above relation with respect to time 't', we get
$v_A + 2v_B = 0$
Again, differentiating it, we get
$a_A + 2a_B = 0$
Because only one of the two coordinates x_A and x_B can be chosen arbitrarily, we say that the system shown in FIGURE 3 also has *one degree of freedom*. From the relation between the position coordinates x_A and x_B, it follows that if x_A is given an increment Δx_A—that is, if block A is lowered by an amount Δx_A—the coordinate x_B receives an increment $\Delta x_B = -\frac{1}{2}\Delta x_A$. Here $-ve$ sign shows that, the block B rises by half the same amount.

Constraint Relations by Inspection
In FIGURE 2(b), the block B is connected to the axel of lower pulley P. If the block B moves in upward direction by a distance Δx_B, then, the displacement in the lower pulley in upward direction will also be Δx_B. So, the rope passing over the lower pulley, gets slacked by amount $EH + FI = \Delta x_B + \Delta x_B = 2\Delta x_B$. As this slacked length cannot shift towards fixed end G, so it will get transferred towards block A and block A moves downward by
$\Delta x_A = 2\Delta x_B$
Therefore, the velocities will be related by, $v_A = 2v_B$ and accelerations of A and B will be related by, $a_A = 2a_B$

27.2. Two Degrees of Freedom

In FIGURE 3 (a), the lengths of the cables attached to cylinders A and B can be written, respectively, as

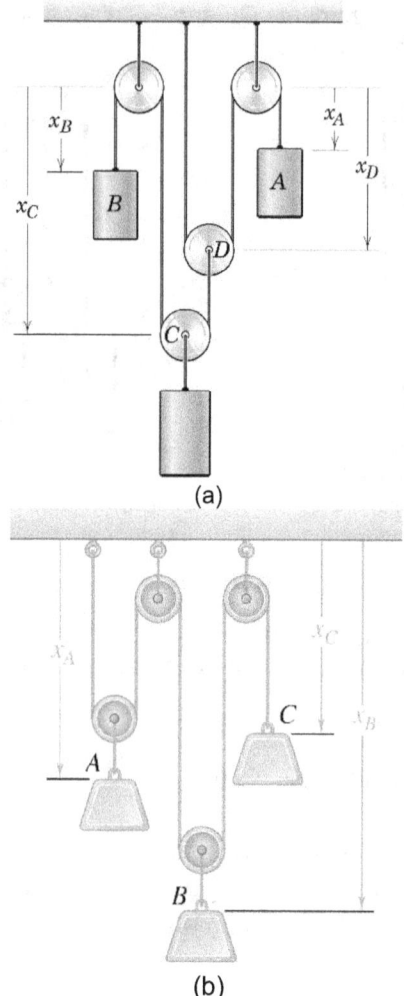

FIGURE 3. A system of blocks and pulleys with two degrees of freedom.

$$L_A = x_A + 2x_D + \text{constant} \quad \ldots (1)$$
$$L_B = x_B + x_C + (x_C - x_D) + \text{constant}$$
or $\quad L_B = x_B + 2x_C - x_D + \text{constant} \quad \ldots (2)$

Eliminating x_D from (1) and (2), we get
$$L_A + 2L_B = x_A + 2x_B + 4x_C + \text{constant}$$
or $\quad x_A + 2x_B + 4x_C = \text{constant} \quad \ldots (3)$
$\quad [\because L_A \text{ and } L_B \text{ are constants}]$

Because two of the coordinates can be chosen arbitrarily, we say that the system shown in FIGURE 5(a) has two degrees of freedom.

Differentiating (1), and (2), with respect to time t, we get
$$0 = \dot{x}_A + 2\dot{x}_D \quad \text{and} \quad 0 = \dot{x}_B + 2\dot{x}_C - \dot{x}_D$$
$$0 = \ddot{x}_A + 2\ddot{x}_D \quad \text{and} \quad 0 = \ddot{x}_B + 2\ddot{x}_C - \ddot{x}_D$$

Eliminating the terms in $\dot{x}_D$ and $\ddot{x}_D$ gives
$$\dot{x}_A + 2\dot{x}_B + 4\dot{x}_C = 0 \quad \text{or} \quad v_A + 2v_B + 4v_C = 0$$
$$\ddot{x}_A + 2\ddot{x}_B + 4\ddot{x}_C = 0 \quad \text{or} \quad a_A + 2a_B + 4a_C = 0$$

It is clearly impossible for the signs of all three terms to be positive simultaneously. So, for example, if both A and B have downward (positive) velocities, then C will have an upward (negative) velocity.

These results can also be found by inspection of the motions of the two pulleys at C and D. For an increment Δx_A (with x_B held fixed), the center of D moves up an amount $\Delta x_A/2$, which causes an upward movement $\Delta x_A/4$ of the center of C. For an increment Δx_B (with x_A held fixed), the center of C moves up a distance $\Delta x_B/2$. A combination of the two movements gives an upward movement

$$-\Delta x_C = \frac{\Delta x_A}{4} + \frac{\Delta x_B}{2}$$

so that $-v_C = \frac{v_A}{4} + \frac{v_B}{2}$ as before. Visualization of the actual geometry of the motion is an important ability.

In the case of the three blocks of FIGURE. 3 (b), we can again observe that the length of the rope that passes over the pulleys is constant. Thus, the following relation must be satisfied by the position coordinates of the three blocks:

$$2x_A + 2x_B + x_C = \text{constant}$$

Because two of the coordinates can be chosen arbitrarily, we say that the system shown in FIGURE 3(b) has two degrees of freedom.

When the relation existing between the position coordinates of several particles is *linear*, a similar relation holds between the velocities and between the accelerations of the particles. In the case of the blocks of FIGURE 3(b), for instance, we can differentiate the position equation twice and obtain

$$2\frac{dx_A}{dt} + 2\frac{dx_B}{dt} + \frac{dx_C}{dt} = 0 \quad \text{or} \quad 2v_A + 2v_B + v_C = 0$$
$$2\frac{dv_A}{dt} + 2\frac{dv_B}{dt} + \frac{dv_C}{dt} = 0 \quad \text{or} \quad 2a_A + 2a_B + a_C = 0$$

27.3. CONSTRAINT WHERE THE DIRECTION OF THE CONNECTING MEMBER CHANGES WITH THE MOTION

A second type of constraint where the direction of the connecting member changes with the motion is illustrated in the following example.

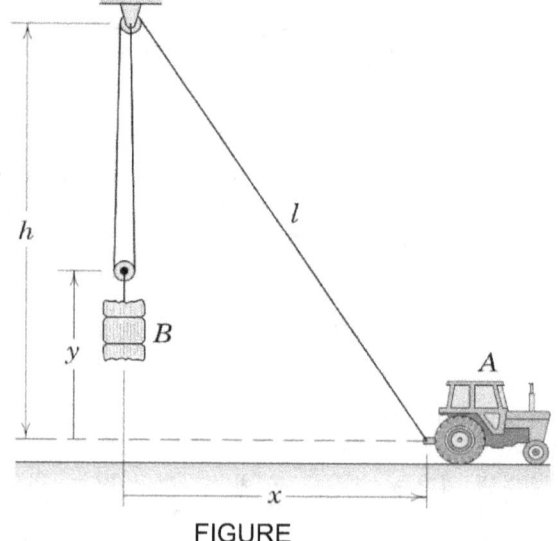

FIGURE

EXAMPLE 44. The tractor A is used to hoist the bale B with the pulley arrangement shown. If A has a forward velocity v_A, determine an expression for the upward velocity v_B of the bale in terms of x.

SOLUTION We designate the position of the tractor by the coordinate x and the position of the bale by the coordinate y, both measured from a fixed reference. The total constant length of the cable is

$$L = 2(h - y) + l = 2(h - y) + \sqrt{h^2 + x^2}$$

Differentiating with respect to time 't', we get

$$0 = -2\dot{y} + \frac{x\dot{x}}{\sqrt{h^2+x^2}}$$

Substituting $v_A = \dot{x}$ and $v_B = \dot{y}$ gives

$$v_B = \frac{1}{2}\frac{xv_A}{\sqrt{h^2+x^2}}$$

Helpful Hint
Differentiation of the relation for a right triangle occurs frequently in mechanics.

27.4. IMPORTANT RESULTS FOR PULLEY CONSTRAINTS

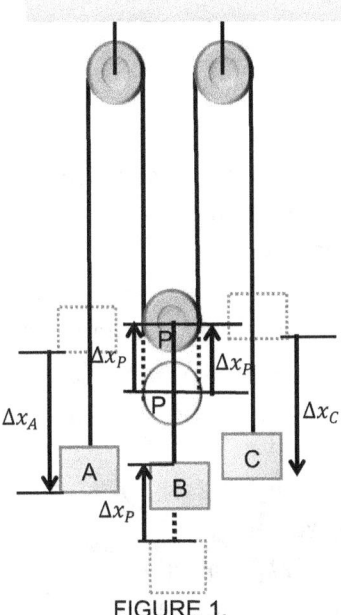

FIGURE 1.

In adjoining FIGURE 1, suppose pulley P moves upward by distance Δx_P, then slacked length of rope will be $2\Delta x_P$. Now assume that blocks A and C moves downward by distances Δx_A and Δx_C respectively, then to keep rope tight

$$2\Delta x_P = \Delta x_A + \Delta x_C \quad \text{or} \quad \Delta x_P = \frac{\Delta x_A + \Delta x_C}{2}$$

⇒ Displacement of pulley = Average of displacements of A and C
⇒ Displacement of pulley = Average of displacements of left and right side of rope passing over pulley

Similarly, the velocity of pulley P, $v_P = \frac{v_A + v_C}{2}$

and acceleration, $a_P = \frac{a_A + a_C}{2}$

These constraints are very useful in problems containing more than one pulleys.

EXAMPLE 45. In the pulley configuration shown, cylinder A has a downward velocity of $0.3\, m/s$. Determine the velocity of B. Solve in two ways.

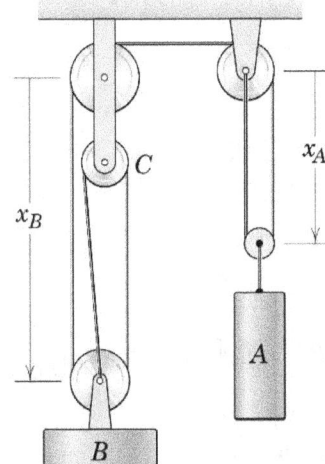

SOLUTION (I). The centers of the pulleys at A and B are located by the coordinates x_A and x_B measured from fixed positions. The total constant length of cable in the pulley system is

$$L = 3x_B + 2x_A + \text{constants}$$

where the constants account for the fixed lengths of cable in contact with the circumferences of the pulleys and the constant vertical separation between the two upper left-hand pulleys. Differentiation with time gives

$$0 = 3\dot{x}_B + 2\dot{x}_A$$

Substituting $v_A = 2\dot{x}_A = 0.3$ m/s and $v_B = \dot{x}_B$ gives

$$0 = 3(v_B) + 2(0.3) \text{ or } v_B = -0.2 \text{ m/s}$$

(II). An enlarged diagram of the pulleys at A, B, and C is shown. During a differential movement ds_A of the center of pulley A, the left end of its horizontal diameter has no motion since it is attached to the fixed part of the cable. Therefore, the right-hand end has a movement of $2ds_A$ as shown. This movement is transmitted to the left-hand end of the horizontal diameter of the pulley at B. Further, from pulley C with its fixed center, we see that the displacements on each side are equal and opposite. Thus, for pulley B, the righthand end of the diameter has a downward displacement equal to the upward displacement ds_B of its center. By inspection of the geometry, we conclude that

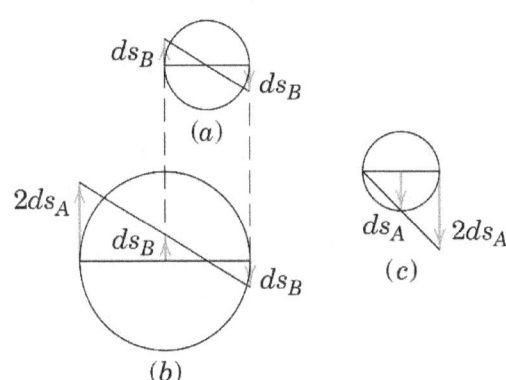

$$2ds_A = 3ds_B \quad \text{or} \quad ds_B = \tfrac{2}{3}ds_A$$

Dividing by dt gives

$$|v_B| = \tfrac{2}{3}v_A = \tfrac{2}{3}(0.3) = 0.2 \text{ m/s (upward)}$$

> **Helpful Hints**
> 1. We neglect the small angularity of the cables between B and C.
> 2. The negative sign indicates that the velocity of B is upward.

EXAMPLE 46. In adjoining figure, if $v_2 = 2$ m/s upwards; $v_B = 1$ m/s upwards. Find the velocity of block 1 and block 3?

SOLUTION. Method 1:
Relative velocity of m_2 with respect to pulley B in upward direction = Relative velocity of m_3 with respect to pulley in downward direction

$$v_2 - v_B = v_3 + v_B$$
or $\quad 2 - 1 = v_3 + 1$
or $\quad v_3 = 0$
$\Rightarrow v_1 = -v_B = 1 \text{ m/s} \downarrow$

Method 2: $\quad v_B = \dfrac{v_2 + v_3}{2}$ (Note)

$$1 = \dfrac{2 + v_3}{2} \Rightarrow v_3 = 0$$

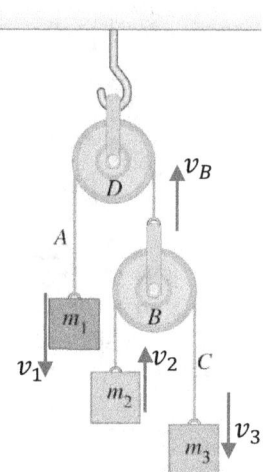

EXAMPLE 47. Let $m_1 = 3\ kg$, $m_2 = 2\ kg$ and $m_3 = 1\ kg$ in FIGURE 1. Find the accelerations of m_1, m_2 and m_3. The string from the upper pulley to m_3, is 20 cm when the system is released from rest. How long will it take before m_3 strikes the pulley?

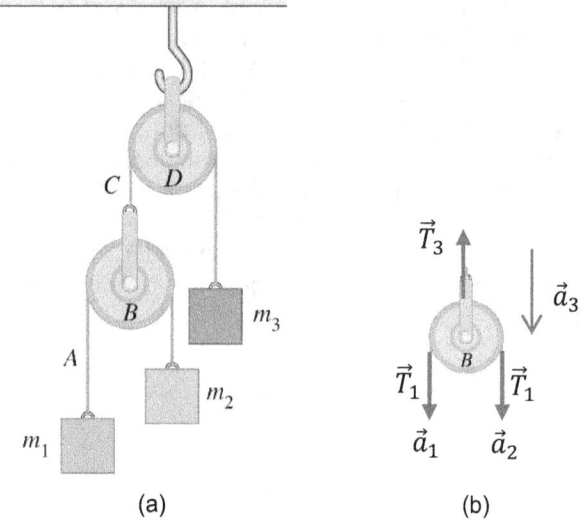

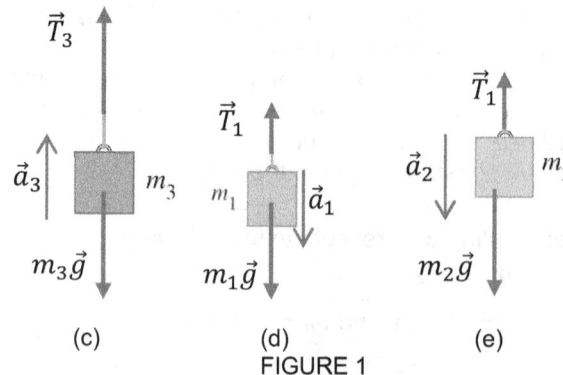

(c) (d) (e)
FIGURE 1

SOLUTION Suppose, the acceleration of m_3 in upward direction is a_3, acceleration of m_1 in downward direction is a_1 and that of m_2 in downward direction is a_2.
The acceleration of pulley B in downward direction $a_B = a_3$.

Constraint Relation for pulley B (Figure 1b)

$$a_B = \dfrac{a_1 + a_2}{2},$$

or $\quad a_3 = \dfrac{a_1 + a_2}{2} \qquad \ldots (1)$

$$2T_1 - T_3 = (0)a_3$$

or $\quad 2T_1 = T_3 \qquad \ldots (2)$

Suppose, the tension in the string connected with pulley B and block of mass m_3 is T_3, therefore from the FBD of block m_3, we have

$$T_3 - m_3 g = m_3 a_3 \qquad \ldots (3)$$

Suppose the tension in the string connecting m_1, and m_2 is T_1, therefore for block m_1

$$m_1 g - T_1 = m_1 a_1 \qquad \ldots (4)$$

and for block m_2, we have

$$m_2 g - T_1 = m_2 a_2 \qquad \ldots (5)$$

Adding (4) and (5), we get

$$m_1 g + m_2 g - 2T_1 = m_1 a_1 + m_2 a_2$$

or $\quad m_1 g + m_2 g - T_3 = m_1 a_1 + m_2 a_2 \qquad \ldots (6)$

(Using equation (2))

Subtracting, equation (5) in (4), we get

$$(m_1 - m_2)g = m_1 a_1 - m_2 a_2 \qquad \ldots (7)$$

$\because \quad m_1 = 3\ kg, \quad m_2 = 2\ kg \quad \text{and} \quad m_3 = 1\ kg,$ therefore

$$(3 - 2)g = 3a_1 - 2a_2$$

or $\quad g = 3a_1 - 2a_2 \qquad \ldots (8)$

Now, adding equations (3) and (6), we get

$$m_1 g + m_2 g - m_3 g = m_1 a_1 + m_2 a_2 + m_3 a_3$$

Substituting, the value of a_3, from equation (1), in above equation, we get

$$m_1 g + m_2 g - m_3 g = m_1 a_1 + m_2 a_2 + m_3 \left(\dfrac{a_1 + a_2}{2}\right)$$

$$2(m_1 + m_2 - m_3)g = 2m_1 a_1 + 2m_2 a_2 + m_3(a_1 + a_2)$$

or $2(m_1 + m_2 - m_3)g = 2m_1a_1 + 2m_2a_2 + m_3(a_1 + a_2)$

or $2(m_1 + m_2 - m_3)g = (2m_1 + m_3)a_1 + (2m_2 + m_3)a_2$

or $2(m_1 + m_2 - m_3)g = (2m_1 + m_3)a_1$
$\qquad\qquad + (2m_2 + m_3)a_2 \qquad\qquad$ ... (9)

Substituting, the values of m_1, m_2 and m_3, we get

$2(m_1 + m_2 - m_3)g = (2m_1 + m_3)a_1 + (2m_2 + m_3)a_2$

or $\quad 2(3 + 2 - 1)g = (6 + 1)a_1 + (4 + 1)a_2$

or $\quad 8g = 7a_1 + 5a_2 \qquad\qquad$ ... (10)

To find a_1 and a_2, multiply equation (8) by 5 and (10), by 2 and then adding (10) in (8), we get

$$21g = 29a_1$$

or $\quad a_1 = \frac{21}{29}g$ (downward)

Now, substituting this value of a_1, in (10), we get

$$8g = 7\left(\frac{21}{29}g\right) + 5a_2$$

or $\quad 5a_2 = \left(8 - \frac{147}{29}\right)g$

or $\quad 5a_2 = \left(\frac{232-147}{29}\right)g = \frac{85}{29}g$

or $\quad a_2 = \frac{17}{29}g$ (downward)

From (1), $a_3 = \frac{a_1+a_2}{2} = \frac{\frac{21}{29}g+\frac{17}{29}g}{2} = \frac{38}{58}g = \frac{19}{29}g$

Time taken by m_2 to reach the pulley D

$$s = ut + \frac{1}{2}a_3t^2$$

$\Rightarrow \quad 0.20 = (0)t + \frac{1}{2}\left(\frac{19}{29}g\right)t^2$

or $\quad \frac{0.20 \times 2 \times 29}{19g} = t^2$

or $\quad t^2 = \frac{0.40 \times 29}{19 \times 9.8}$

or $\quad t^2 = \frac{0.40 \times 29}{19 \times 9.8} = 0.06229 \quad$ or $\quad t = 0.25$ s

EXAMPLE 48. Figure shows a rod of length l resting on a smooth wall and the smooth floor. Its lower end A is pulled towards right with a constant velocity v_A. Find the velocity of the other end B downward when the rod makes an angle θ with the horizontal.

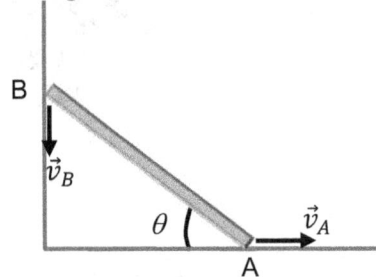

SOLUTION As length of the rod is constant, therefore the relative velocity of ends of rod along its length should be zero. It means the velocities of ends of rod along its length must be equal.

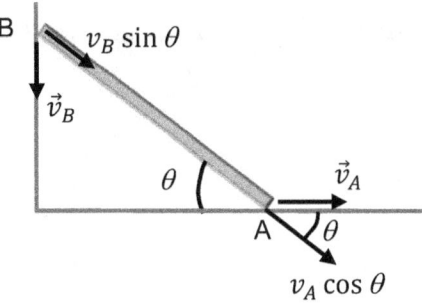

$v_A \cos\theta = v_B \sin\theta$

$\Rightarrow v_B = v_A \frac{\cos\theta}{\sin\theta} \Rightarrow v_B = v_A \cot\theta$

EXAMPLE 49. In the FIGURE, a ball of mass m_1, and a block of mass m_2 are joined together with an inextensible string. The ball can slide on a smooth horizontal surface. If v_1 and v_2 are the respective speeds of the ball and the block, then determine the constraint relation between the two.

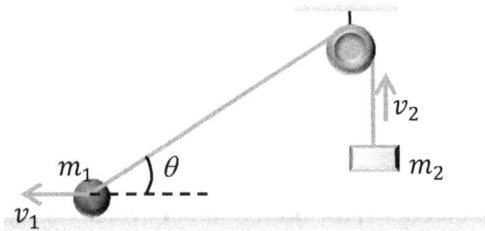

SOLUTION As length of the string is constant, therefore the relative velocity of ends of string along its length should be zero. It means the velocities of ends of string along its length must be equal.

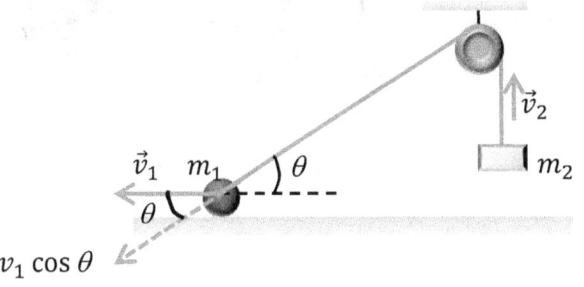

$v_1 \cos\theta = v_2$

EXAMPLE 50. A ring A which can slide on a smooth wire is connected to one end of a string as shown in FIGURE 1. Other end of the string is connected to a hanging mass B. Find the speed of the ring when the string makes an angle θ with the wire and mass B is going down with a velocity v.

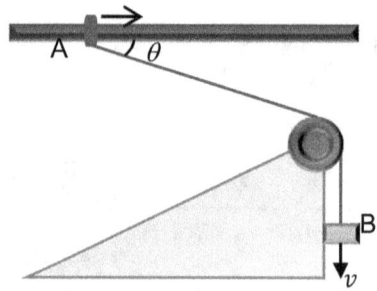

SOLUTION. As length of the string is constant, therefore the relative velocity of ends of string along its length should be zero. It means the velocities of ends of string along its length must be equal.

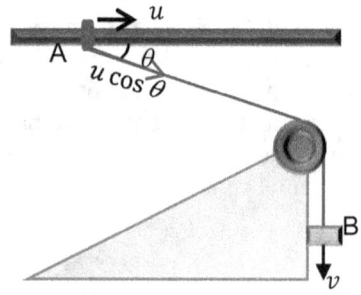

$u \cos \theta = v \Rightarrow u = v \sec \theta$

EXAMPLE 51 The ring m_2 and block m_1 are held in the position shown in figure and the system is released. If $m_2 > m_1$, the ring m_2 slides down along the smooth fixed vertical rod. Find $\frac{v}{u}$.

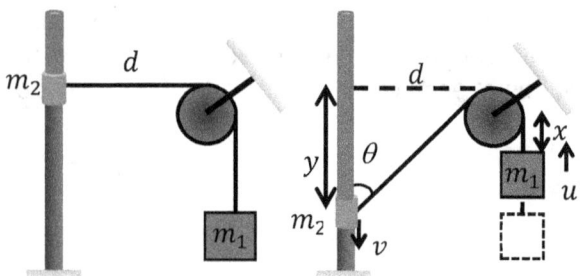

SOLUTION. The length of string is constant,

$x + c + \sqrt{y^2 + d^2} = $ constant

$\Rightarrow \frac{dx}{dt} + 0 + \frac{1}{2\sqrt{y^2+d^2}} \times 2y \frac{dy}{dx} + 0 = 0$

As $\frac{dx}{dt} = u$ and $\frac{dy}{dt} = v$, $u + \cos \theta \cdot v = 0$

$\Rightarrow u + v \cos \theta = 0 \qquad \Rightarrow v = \left| -\frac{u}{\cos \theta} \right|$

$\Rightarrow v = \frac{u}{\cos \theta} \qquad \Rightarrow \frac{v}{u} = \frac{1}{\cos \theta}$

27.5. WEDGE CONSTRAINTS

EXAMPLE 52. *Find the acceleration of rod A and wedge B in the arrangement shown in Figure if the mass of rod equals that of the wedge and the friction between all contact surfaces is negligible. Take angle of wedge as 45°.*

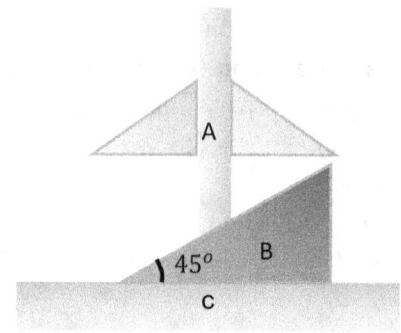

APPROACH Here, the rod is confined to move in vertical direction, whereas wedge is confined to move in horizontal direction only. Since, the rod always remains in contact with the wedge, therefore the relative displacement, relative velocity or relative acceleration of the rod perpendicular to the inclined plane of the wedge with respect to the wedge, will always be zero.

SOLUTION. If, with respect to ground, the displacement of rod in downward direction is ds_A and that of wedge in right side is ds_B (Fig. 1a), then

(i) the component of displacement of rod along the plane of the wedge $= ds_A \sin 45° = ds_A/\sqrt{2}$ (Fig. 1b),

(ii) the component of displacement of rod perpendicular to the plane of the wedge $= ds_A \cos 45° = ds_A/\sqrt{2}$ (Fig. 1b)

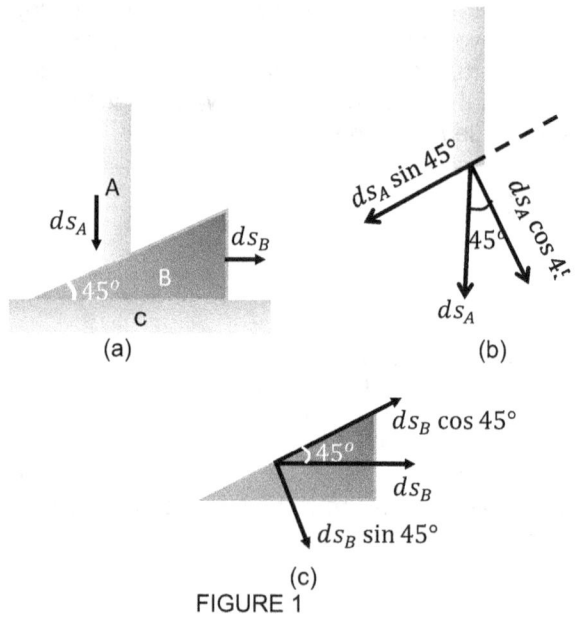

FIGURE 1

Now,

(iii) the component of displacement of wedge along the plane of the wedge $= ds_B \cos 45° = ds_B/\sqrt{2}$ (Fig. 1c),

(iv) the component of displacement of wedge perpendicular to its inclined plane $= ds_B \sin 45° = ds_B/\sqrt{2}$ (Fig. 1c)

The displacement of rod perpendicular to the plane with respect to the wedge is $\dfrac{ds_A}{\sqrt{2}} - \dfrac{ds_B}{\sqrt{2}}$.

Since, the rod is always in contact with the wedge, therefore,

$$\dfrac{ds_A}{\sqrt{2}} - \dfrac{ds_B}{\sqrt{2}} = 0,$$

i.e., $\dfrac{ds_B}{\sqrt{2}} = \dfrac{ds_A}{\sqrt{2}} \Rightarrow ds_B = ds_A$

Differentiating both sides of above equation gives
$$a_B = a_A \qquad (= a, \text{ say})$$
If N is the normal reaction between the wedge and rod, then from Fig. 2a,

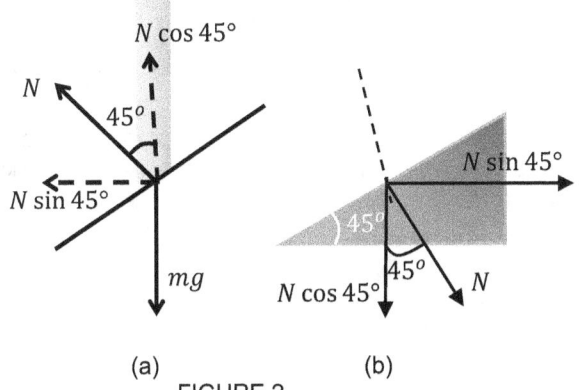

(a) (b)
FIGURE 2

(i) vertical component of normal reaction on rod
 $= N \cos 45°$ (upward)
(ii) horizontal component of normal reaction on rod
 $= N \sin 45°$ (leftward)
(iii) vertical component of normal reaction on wedge
 $= N \cos 45°$ (downward)
(iv) horizontal component of normal reaction on wedge
 $= N \sin 45°$ (rightward)

Now, for vertical motion of the rod, we have
$$\Sigma F_{rod} = ma_A$$
$$mg - \dfrac{N}{\sqrt{2}} = ma \qquad (\because a_A = a) \qquad \ldots (1)$$

For horizontal motion of the wedge,
$$\Sigma F_{wedge} = ma_B$$
or $\quad N \sin 45° = ma_B$
or $\quad \dfrac{N}{\sqrt{2}} = ma \qquad (\because a_B = a) \qquad \ldots (2)$

On adding (1) and (2), we get
$$\Rightarrow mg = 2ma \qquad \ldots (3)$$
$$\Rightarrow a = \dfrac{g}{2}$$

Alternate method for finding the relation between acceleration of rod and wedge

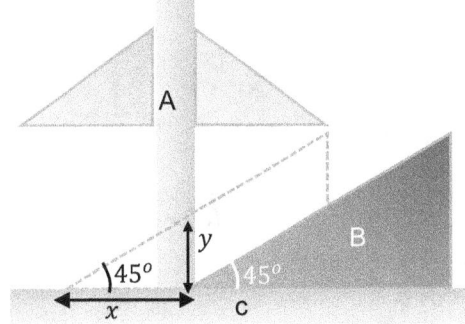

FIGURE 3

If x and y are the displacements of the rod and wedge respectively, then from Fig. 3, we have

$$\tan 45° = \dfrac{y}{x} \Rightarrow y = x \tan 45°$$

$$\therefore \qquad \dfrac{d^2y}{dt^2} = (1)\dfrac{d^2x}{dt^2}$$

therefore $a_{rod} = a_{wedge}$

EXAMPLE 53. There is no friction at any contact. Wedge is free to move. Find force acting on wedge due to block. Also find acceleration of wedge.

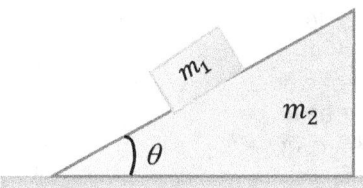

APPROACH Don't reach directly to conclusion that answer is $m_1 g \cos \theta$. As, the horizontal component of normal contact force applied by block on wedge will accelerate the wedge, therefore, the reference frame attached to wedge is non-inertial reference frame. Always start such problems by letting the acceleration with respect to the wedge.

SOLUTION If $\vec{a}_w$ is the acceleration of wedge and $\vec{a}_{b/w}$ is the acceleration of block with respect to wedge, then the acceleration of block with respect to ground,

$$\vec{a}_b = \vec{a}_{b/w} + \vec{a}_w \qquad \ldots (1)$$

*An easy way to remember the setup of these equations is to note the "cancellation" of the subscript w between the two terms, e.g. $\vec{a}_b = \vec{a}_{b/w} + \vec{a}_w$

From acceleration diagram of block, we have

$$\vec{a}_{b/w} = (-a_{b/w} \cos \theta)\hat{\imath} + (-a_{b/w} \sin \theta)\hat{\jmath}$$
and $\vec{a}_w = a_w \hat{\imath}$
Substituting these values in (1), we get
$$\vec{a}_b = (-a_{b/w} \cos \theta)\hat{\imath} + (-a_{b/w} \sin \theta)\hat{\jmath} + a_w \hat{\imath}$$
$$= (a_w - a_{b/w} \cos \theta)\hat{\imath} - a_{b/w} \sin \theta\, \hat{\jmath}$$

or $\vec{a}_b = a_{bx}\hat{i} + a_{by}\hat{j}$ ... (2)
here, $a_{bx} = a_w - a_{b/w}\cos\theta$, $a_{by} = -a_{b/w}\sin\theta$

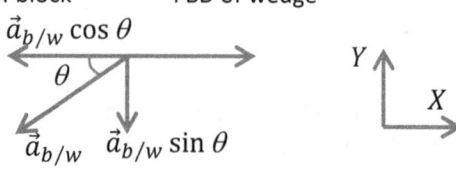

FBD of block   FBD of wedge

Acceleration diagram of block
FIGURE 1

Now, from the FBD of block, we have
Along X axis,
$\quad -N\sin\theta = m_1 a_{bx}$
or $\quad -N\sin\theta = m_1(a_w - a_{b/w}\cos\theta)$
or $\quad N\sin\theta = m_1(a_{b/w}\cos\theta - a_w)$ ... (3)
Along, Y axis, we have-
$\quad N\cos\theta - m_1 g = m_1 a_y$
or $\quad N\cos\theta - m_1 g = m_1(-a_{b/w}\sin\theta)$
or $\quad N\cos\theta - m_1 g = -m_1 a_{b/w}\sin\theta$
or $\quad N\cos\theta = m_1 g - m_1 a_{b/w}\sin\theta$ ... (4)

From the F.B.D of wedge,
along X axis, we have
$\quad N\sin\theta = m_2 a_w$ ... (5)
and along Y axis,
$\quad N_1 - N\cos\theta - m_2 g = 0$ ... (6)

On solving above equation, we get
$$N = \frac{m_1 m_2 g\cos\theta}{(m_2 + m_1\sin^2\theta)}, \quad a_w = \frac{m_1 g\cos\theta\sin\theta}{(m_2 + m_1\sin^2\theta)},$$
$$a_{b/w} = \frac{(m_1 + m_2)g\sin\theta}{(m_1\sin^2\theta + m_2)}$$

EXAMPLE 54. Find velocity vector of m_1 if m_2 is pulled with constant velocity $v_2 = 2$ m/s

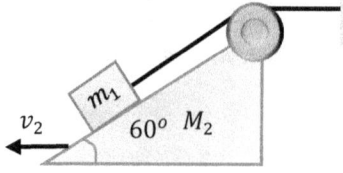

SOLUTION This problem involves two constraints. One involves the rope and other involves m_1 and m_2 remaining in contact with each other.
These two constraints can be understood easily if we shift our reference frame to the wedge. By doing this we will be able to simplify motion of block m_1 and thus solve constraint of rope easily.

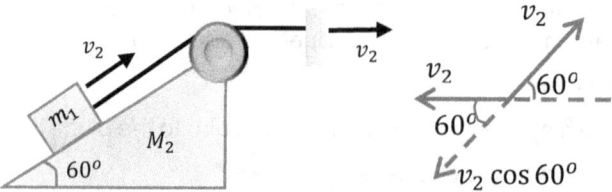

In reference frame attached to the wedge, the wall will move horizontally towards right with speed v_2 as shown.
Thus, it can be easily deduced that m_1 will move with velocity v_2 upwards. But this is velocity of m_1 with respect to the frame attached to m_2.
Now, by the concept of relative motion, we have-
$$\vec{v}_{B/W} = \vec{v}_B - \vec{v}_W$$
here, $v_{B/W}$ = velocity of block m_1, with respect to wedge m_2.
v_B = velocity of block m_1 with respect to ground
v_W = velocity of wedge with respect to ground
$$\vec{v}_B = \vec{v}_{B/W} + \vec{v}_W$$
or $\quad v_B = v_2 - v_2\cos 60 = \frac{v_2}{2}$
(along the plane of the wedge)

28. CHECKPOINT 7

1. •• If the velocity $\dot{x}$ of block A up the incline is increasing at the rate of 0.044 m/s each second, determine the acceleration of B.

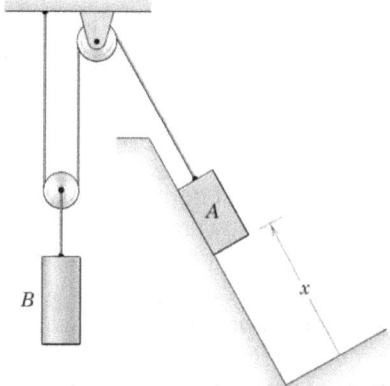

2. •• At the instant represented, $\vec{v}_{B/A} = 3.5\hat{j}$ m/s. Determine the velocity of each body at this instant. Assume that the upper surface of A remains horizontal.

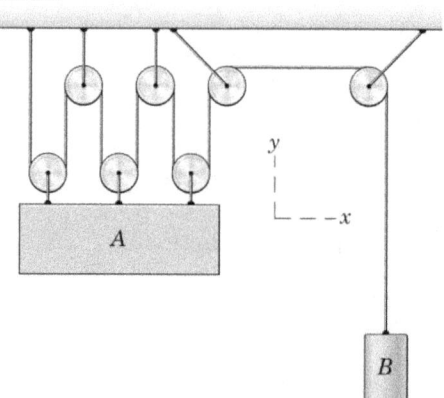

3. •• At a certain instant, the velocity of cylinder B is 1.2 m/s down and its acceleration is 2 m/s² up. Determine the corresponding velocity and acceleration of block A.

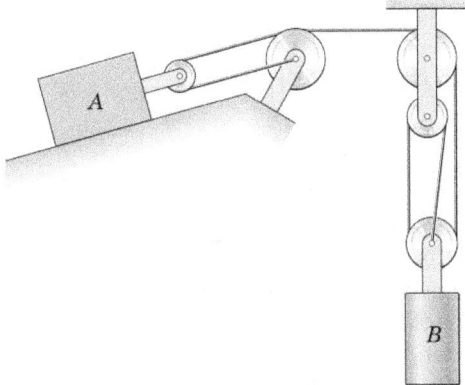

4. •• Determine the velocity of cart A if cylinder B has a downward velocity of 2 ft/sec at the instant illustrated. The two pulleys at C are pivoted independently.

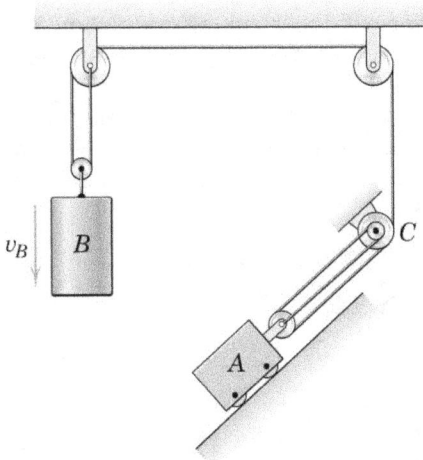

5. •• A hemisphere of mass M and radius R rests on a smooth horizontal table. A vertical rod of mass m is held between two smooth guide walls supported on the sphere as shown. There is no friction between the rod and the sphere. A horizontal string tied to the sphere keeps the system at rest. (a) Find tension in the string. (b) Find the acceleration of the hemisphere immediately after the string is cut.

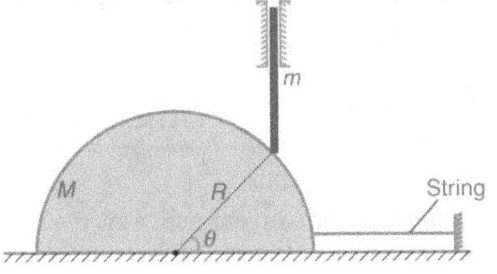

6. ••An electric motor M is used to reel in cable and hoist a bicycle into the ceiling space of a garage. Pulleys are fastened to the bicycle frame with hooks at locations A and B, and the motor can reel in cable at a steady rate of 12 in./sec. At this rate, how long will it take to hoist the bicycle 5 feet into the air? Assume that the bicycle remains level.

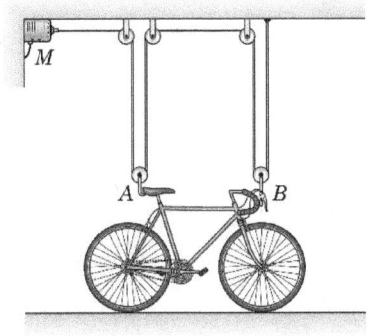

7. •• Determine the relation which governs the accelerations of A, B, and C, all measured positive down. Identify the number of degrees of freedom.

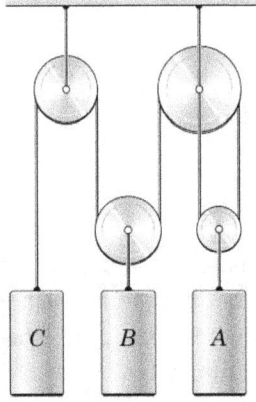

8. ••• Determine an expression for the velocity v_A of the cart A down the incline in terms of the upward velocity v_B of cylinder B.

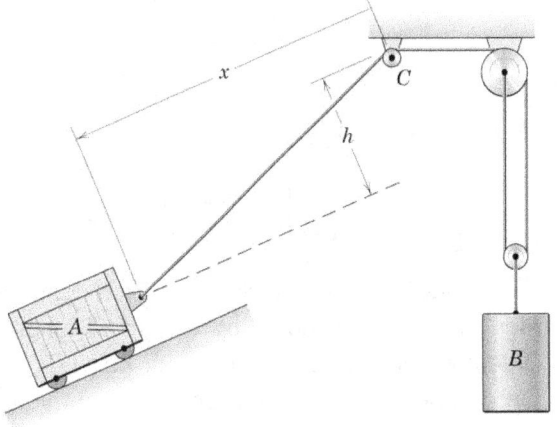

9. ••• Neglect the diameters of the small pulleys and establish the relationship between the velocity of A and the velocity of B for a given value of y.

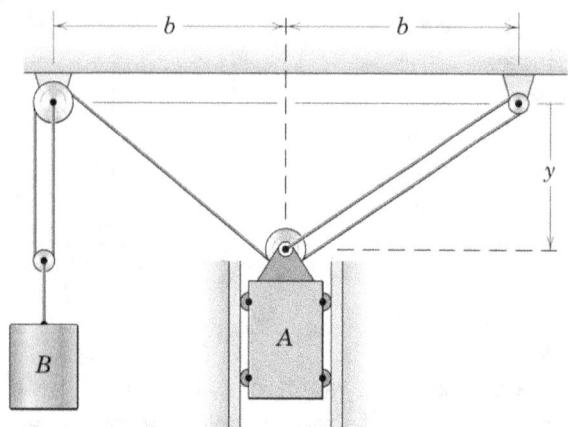

10. ••• Under the action of force P, the constant acceleration of block B is 6 ft/sec² up the incline. For the instant when the velocity of B is 3 ft/sec up the incline, determine the velocity of B relative to A, the acceleration of B relative to A, and the absolute velocity of point C of the cable.

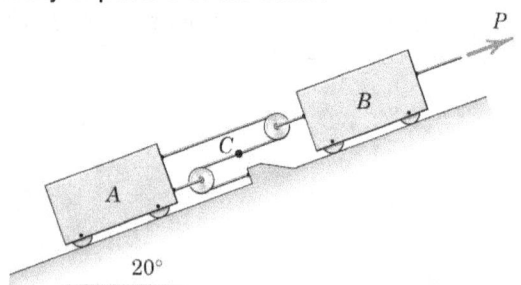

11. ••• Determine the relationship which governs the velocities of the four cylinders. Express all velocities as positive down. How many degrees of freedom are there?

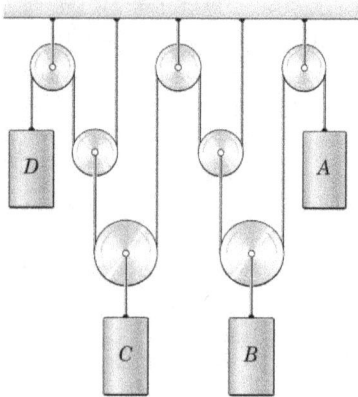

12. •• Collars A and B slide along the fixed right-angle rods and are connected by a cord of length L. Determine the acceleration a_x of collar B as a function of y if collar A is given a constant upward velocity v_A.

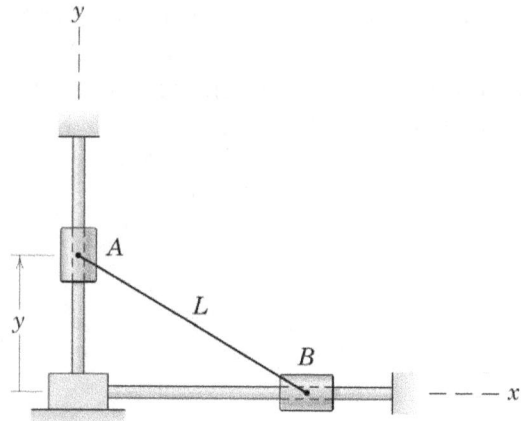

13. •• In the system shown in the figure all surfaces are smooth, pulley and string are massless. The string between the two pulleys and between pulley and block of mass 5 m is parallel to the incline surface of the block of mass 4 m. The system is released from rest. Find the acceleration of the block of mass 4 m.

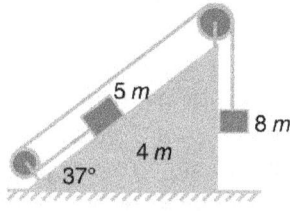

14. •• For a given value of y, determine the upward velocity of A in terms of the downward velocity of B. Neglect the diameters of the pulleys.

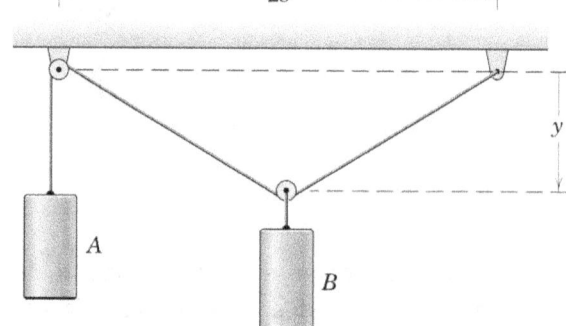

15. ••• Cart A has a leftward velocity v_A and acceleration a_A at the instant represented. Determine the expressions for the velocity and acceleration of cart B in terms of the position x_A of cart A. Neglect the diameters of the pulleys and assume that there is no mechanical interference. The two pulleys at C are pivoted independently.

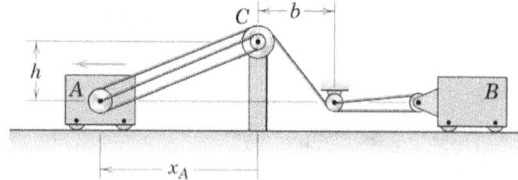

16. ••• Determine the vertical rise h of the load W during 10 seconds if the hoisting drum draws in cable at the constant rate of 180 mm/s.

LAWS OF MOTION AND FRICTION 61

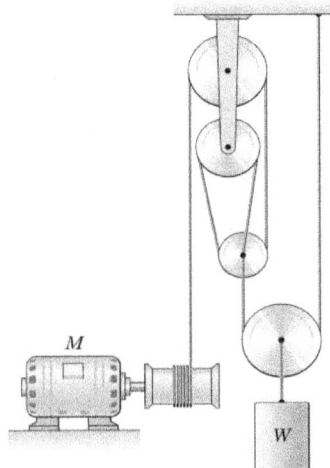

17. ••• The hoisting system shown is used to easily raise boats for overhead storage. Determine expressions for the upward velocity and acceleration of the boat at any height y if the winch M reels in cable at a constant rate $\dot{l}$. Assume that the boat remains level.

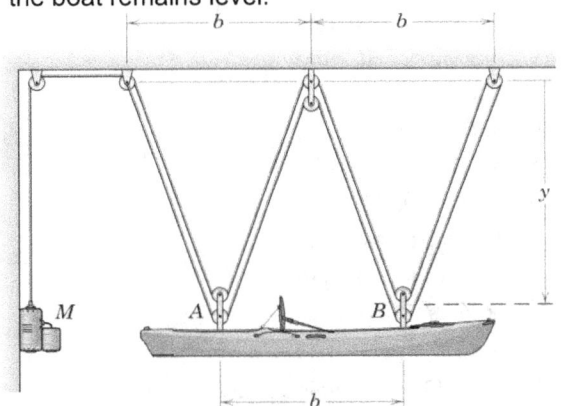

18. ••• Develop an expression for the upward velocity of cylinder B in terms of the downward velocity of cylinder A. The cylinders are connected by a series of n cables and pulleys in a repeating fashion as shown.

19. ••• In the system shown in figure, the two springs S_1 and S_2 have force constant k each. Pulley, springs and strings are all massless. Initially, the system is in equilibrium with spring S_1 stretched and S_2 relaxed. The end A of the string is pulled down slowly through a distance L. By what distance does the block of mass M move?

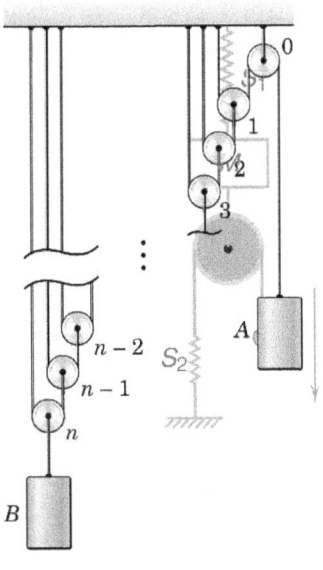

20. ••• The rod of the fixed hydraulic cylinder is moving to the left with a constant speed $v_A = 25$ mm/s. Determine the corresponding velocity of slider B when $s_A = 425$ mm. The length of the cord is 1050 mm, and the effects of the radius of the small pulley A may be neglected.

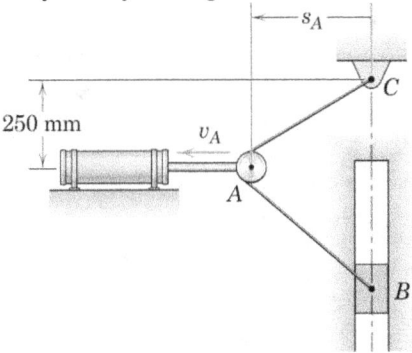

21. •• The 50-kg block A is released from rest. Determine the velocity of the 15-kg block B in 2 s.

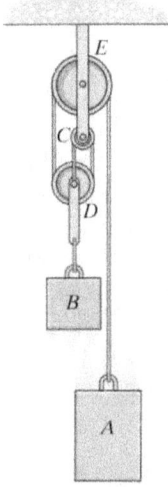

22. •• If the supplied force $F = 150\ N$, determine the velocity of the 50-kg block A when it has risen $3\ m$, starting from rest.

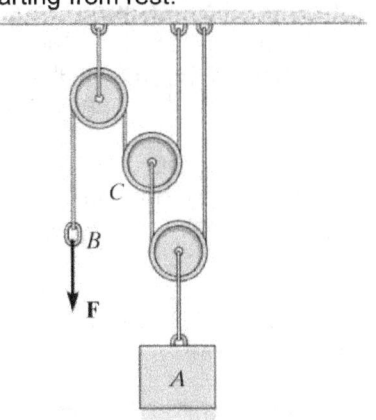

23. ••• An athlete pulls handle A to the left with a constant velocity of 0.5 m/s. Determine (a) the velocity of the weight B, (b) the relative velocity of weight B with respect to the handle A.

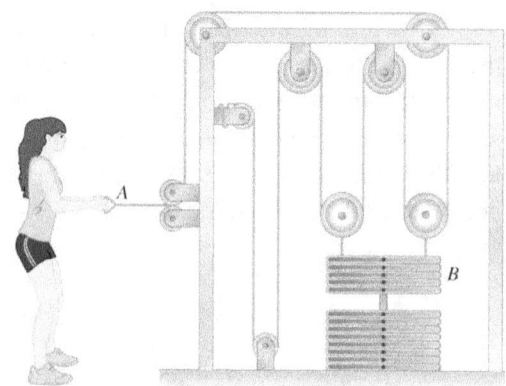

24. ••• The elevator E shown in the figure moves downward with a constant velocity of 4 m/s. Determine (a) the velocity of the cable C, (b) the velocity of the counterweight W, (c) the relative velocity of the cable C with respect to the elevator, (d) the relative velocity of the counterweight W with respect to the elevator.

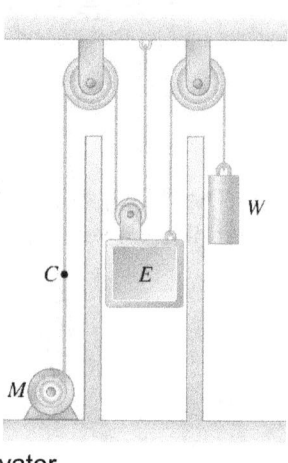

25. ••• In the position shown, collar B moves to the left with a constant velocity of 300 mm/s. Determine (a) the velocity of collar A, (b) the velocity of portion C of the cable, (c) the relative velocity of portion C of the cable with respect to collar B.

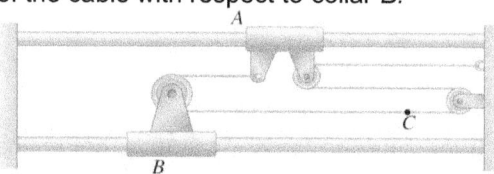

26. ••• Collar A starts from rest and moves to the right with a constant acceleration. Knowing that after 8 s the relative velocity of collar B with respect to collar A is 610 mm/s, determine (a) the accelerations of A and B, (b) the velocity and the change in position of B after 6 s.

27. •• Determine the initial acceleration of the 10-kg smooth collar. The spring has an unstretched length of 1 m.

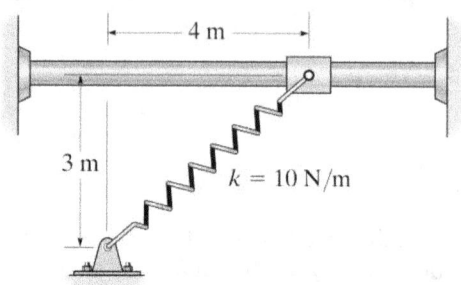

28. ••• Block B moves downward with a constant velocity of 20 mm/s. At $t = 0$, block A is moving upward with a constant acceleration, and its velocity is 30 mm/s. Knowing that at $t = 3$ s slider block C has moved 57 mm to the right, determine (a) the velocity of slider block C at $t = 0$, (b) the accelerations of A and C, (c) the change in position of block A after 5 s.

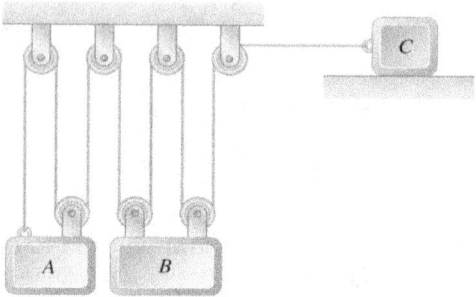

29. ••• The system shown starts from rest, and each component moves with a constant acceleration. If the relative acceleration of block C with respect to collar B is 60 mm/s^2 upward and the relative acceleration of block D with respect to block A is 110 mm/s^2 downward, determine (a) the velocity of block C after 3 s, (b) the change in position of block D after 5 s.

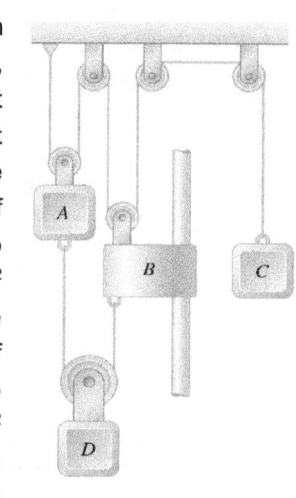

30. ••• The motor M reels in the cable at a constant rate of 100 mm/s. Determine (a) the velocity of load L, (b) the velocity of pulley B with respect to load L.

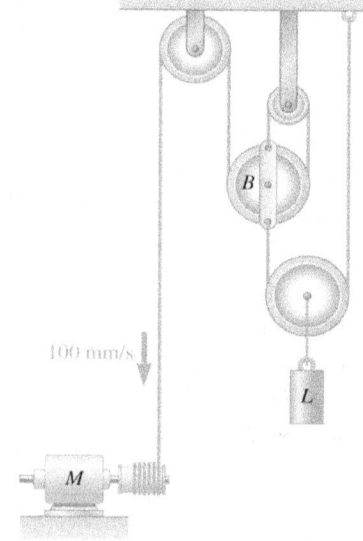

31. •• A man pulls himself up the 15° incline by the method shown. If the combined mass of the man and cart is 100 kg, determine the acceleration of the cart if the man exerts a pull of 175 N on the rope. Neglect all friction and the mass of the rope, pulleys, and wheels.

32. ••• Three identical smooth balls are placed between two vertical walls as shown in adjoining figure. Mass of each ball is m and radius is $r = \frac{5R}{9}$, where $2R$ is separation between the walls. (a) Force between which two contact surfaces is maximum? Find its value. (b) Force between which two contact surfaces is minimum and what is its value?

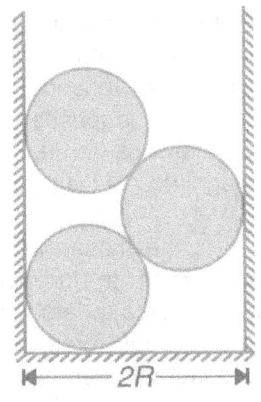

33. •• A wedge is placed on the smooth surface of a fixed incline having inclination θ with the horizontal. The vertical wall of the wedge has height h and there is a small block A on the edge of the horizontal surface of the wedge. Mass of the wedge and the small block are M and m respectively. (a) Find the acceleration of the wedge if friction between block A and the wedge is large enough to prevent slipping between the two. (b) Find friction force between the block and the wedge in the above case. Also find the normal force between the two. (c) Assuming there is no friction between the block and the wedge, calculate the time in which the block will hit the incline.

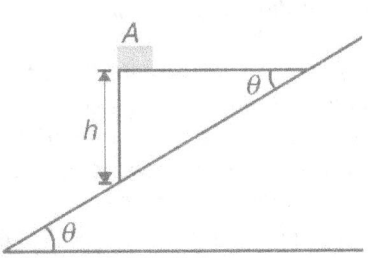

34. ••• A smooth rod is fixed at an angle α to the horizontal. A small ring of mass m can slide along the rod. A thread carrying a small sphere of mass M is attached to the ring. To keep the system in equilibrium, another thread is attached to the ring which carries a load of mass m_0 at its end (see adjoining figure). The thread runs parallel to the rod between the ring and the pulley. All threads and pulley are massless. (a) Find m_0 so that system is in equilibrium. (b) Find acceleration of the sphere M immediately after the thread supporting m_0 is cut.

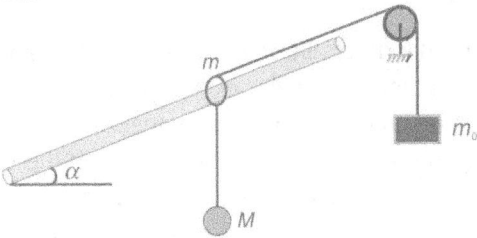

35. •• In the system shown in figure all surfaces are smooth and string and pulleys are light. Angle of wedge $\theta = \sin^{-1}\left(\frac{3}{5}\right)$. When released from rest it was found that the wedge of mass m_0 does not move. Find $\frac{M}{m}$.

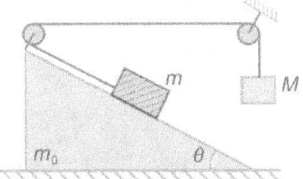

36. •• In the above problem take $M = m$ and $m_0 = 2m$ and calculate the acceleration of the wedge.

37. ••• The system shown in figure is in equilibrium. Surface PQ of wedge A, having mass M, is horizontal. Block B, having mass $2M$, rests on wedge A and is supported by a vertical spring. The spring balance S is showing a reading of $\sqrt{2}\,Mg$. There is no friction anywhere and the thread QS is parallel to the incline surface. The thread QS is cut. Find the acceleration of A and the normal contact force between A and B immediately after the thread is cut.

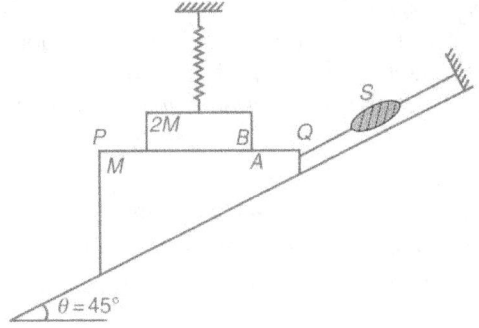

38. ••• In the system shown in the figure all surfaces are smooth and both the pulleys are mass less. Block on the incline surface of wedge A has mass m. Mass of A and B are $M = 4m$ and $M_0 = 2m$ respectively. Find the acceleration of wedge A when the system is released from rest

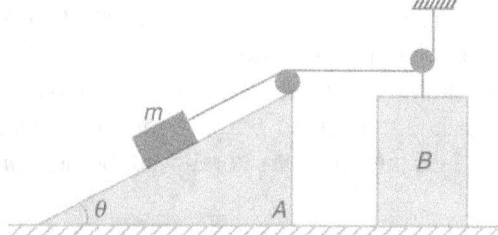

39. ••• A uniform chain of mass $M = 4.8$ kg hangs in vertical plane as shown in the fig. (a) Show that horizontal component of tension is same throughout the chain. (b) Find tension in the chain at point P where the chain makes an angle $\theta = 15°$ with horizontal. (c) Find mass of segment AP of the chain. [Take $g = 10$ m/s²; $\cos 15° = 0.96$, $\sin 15° = 0.25$]

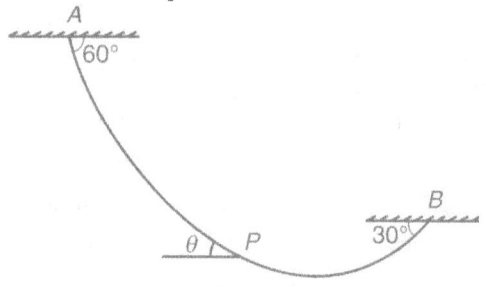

29. FICTITIOUS (OR PSEUDO) FORCE

29.1. MOTION IN ACCELERATED FRAMES:

Till now we have restricted ourselves to apply Newton's laws of motion, only to describe observations that are made in an inertial frame of reference. What can we do to obtain the correct equations of motion from observations in a non-inertial system? The answer lies in the relation

$$\vec{F}_{apparent} = \vec{F}_{true} - m\vec{a}$$

We can treat the last term $(-m\vec{a})$ like an additional force. Because it is not really a force—no interaction is involved—we shall refer to it as a *fictitious force*. We then have

$$\vec{F}_{apparent} = \vec{F}_{true} + \vec{F}_{fictitious}$$

where $\vec{F}_{fictitious} = -m\vec{a}$. Here m is the mass of the particle and $\vec{a}$ is the acceleration of the non-inertial system with respect to any inertial system.

i.e., $\vec{F}_{fictitious} = -$ mass of the object × acceleration of the frame $(\vec{a})$

$-ve$ sign shows that $\vec{F}_{fictitious}$ is opposite to acceleration $\vec{a}$ of the frame of reference.

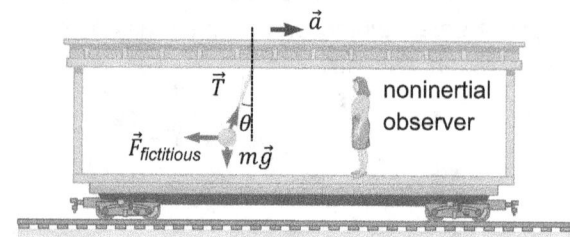

For example, consider a ball suspended from the celling of an accelerated train. Due to acceleration of the train, the ball gets deflected by an angle θ to the left as shown in the following figure. Now, suppose an observer is standing in this noninertial frame (train). To apply Newton's second law in this accelerated train (i.e., noninertial frame of reference), we need to apply a fictitious (or pseudo) force, acting in backward direction, i.e., opposite to the acceleration of non-inertial reference frame.

The magnitude of fictitious (or pseudo) force

$$F_{fictitious} = ma,$$

here m is the mass of the object (ball) and $\vec{a}$ is the acceleration of the frame of reference (train).

Fictitious force appears to act on an object in the same way as a real force, but real forces are always interactions between two objects. On the other hand, there is no second object for a fictitious force.

Direction of Fictitious (or pseudo) force: Direction of fictitious (or pseudo) force is always opposite to the direction of acceleration of the frame of reference.

EXAMPLE 55. *A small ball of mass m hangs by a cord from the ceiling of a compartment of a train that is accelerating to the right as shown in FIGURE 1(a). Analyse the situations for two observers A & B. Observer in the compartment, cannot see the car's motion so that he is not aware of its acceleration. Because he does not know of this acceleration, he will say that Newton's second law is not valid as the object has net horizontal force (the horizontal component of tension) but no horizontal acceleration.*

SOLUTION. The observer A on the ground, is inertial Frame. He sees the compartment is accelerating and knows that the deviation of the cord provides the ball, required horizontal force. The non-inertial observer on the compartment, cannot see the car's motion so that he is not aware of its acceleration. Because he does not know of this acceleration, he will say that Newton's second law is not valid as the object has net horizontal force (the horizontal component of tension) but no horizontal acceleration.

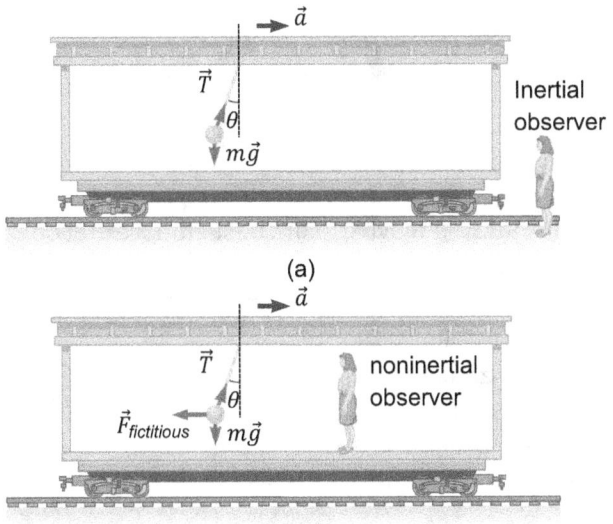

FIGURE 1

For the inertial observer, ball has a net force in the horizontal direction and is in equilibrium in the vertical direction. For the non-inertial observer, we apply fictitious force towards left and consider it to be in equilibrium.

According to the inertial observer A, the ball experience two forces, T exerted by the cord and the weight.
Apply Newton's second law in vertical and horizontal direction we get
Inertial observer
$$T \cos \theta - mg = 0 \qquad \ldots (1)$$
(in vertical direction)
$$T \sin \theta = ma \qquad \ldots (2)$$
(in horizontal direction)
From, equations (1) and (2), we have
$$\tan \theta = \frac{a}{g}$$
or $\theta = \tan^{-1}\left(\frac{a}{g}\right)$

According to the non-inertial observer B riding in the car (**FIGURE** 1b), the ball is always at rest and so its acceleration is zero. The non-inertial observer applies a fictitious force in the horizontal direction of magnitude ma towards left. This fictious force balances the horizontal component of T and thus the net force on the ball is zero.

Apply Newton's second law in vertical and horizontal direction we get

$T \cos \theta = mg$ (in vertical direction)

$T \sin \theta = ma$ (in horizontal direction)

These expressions are equivalent to Equations (1) and (2).

The non-inertial observer B obtains the same equations as the inertial observer. The physical explanation of the cord's deflection, however, differs in the two frames of reference.

EXAMPLE 56 A block of mass m is placed on an inclined plane. With what acceleration and direction, should the system move on a horizontal surface so that m does not slide on the surface of inclined plane? Assume all surfaces are smooth.

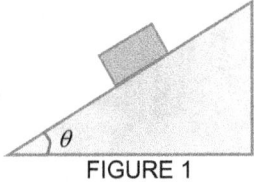

FIGURE 1

SOLUTION Let us consider the block on smooth inclined surface of the wedge. Suppose, initially the wedge is fixed (FIGURE 2). Now, the forces acting on the block are,

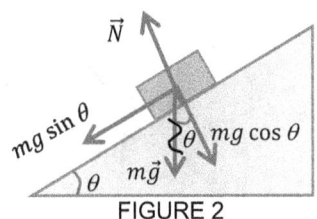

FIGURE 2

1. Gravitational force $m\vec{g}$ in downward direction

2. Normal reaction $\vec{N}$ perpendicular to the surface of the wedge

The force component $mg \sin \theta$, will act along the inclined surface of the wedge in downward direction. Due to this force, the block will accelerate in downward direction along the inclined surface. *To prevent its sliding motion, we have to accelerate the system horizontally leftward.* Now, if acceleration is $\vec{a}$, then the pseudo force on the block will be $m\vec{a}$ in horizontally right ward direction (FIGURE 3).

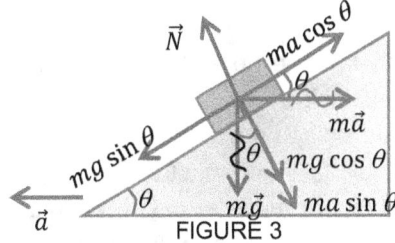

FIGURE 3

From the FBD of the block, in equilibrium,
$$ma \cos \theta = mg \sin \theta$$
or $a = g \tan \theta$

EXAMPLE 57: A pendulum of mass m is hanging from the ceiling of a car having an acceleration a with respect to the road in the direction shown. Find the angle made by the string with the vertical

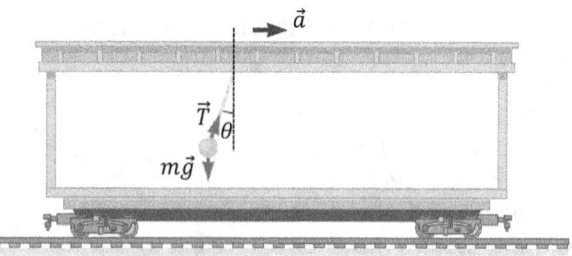

SOLUTION: Since bob of the pendulum is stationary relative to car, Hence,

$T \sin \theta = ma$ (pseudo force)   ... (1)

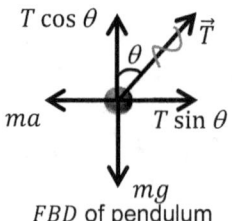

FBD of pendulum

$T \cos \theta = mg$ ... (2)

Dividing, (1), by (2), we get

$\tan \theta = \dfrac{a}{g}$ or $\theta = \tan^{-1} \dfrac{a}{g}$

30. CHECKPOINT 8

1. • We have seen that an observer in an inertial reference frame can use Newton's laws, while observers in a noninertial frame must include fictitious forces. Identify the following reference frames as either inertial or noninertial and explain your answers.
   (A) A car traveling on a level road at a constant velocity
   (B) A car traveling up a steep mountain road at a constant velocity
   (C) A child sitting on a rotating merry-go-round
   (D) An apple falling from a tree

2. •• An object of mass $m = 5.00$ kg, attached to a spring scale, rests on a frictionless, horizontal surface as shown in adjoining Figure. The spring scale, attached to the front end of a boxcar, reads zero when the car is at rest. (a) Determine the acceleration of the car if the spring scale has a constant reading of 18.0 N when the car is in motion. (b) What constant reading will the spring scale show if the car moves with constant velocity? Describe the forces on the object as observed (c) by someone in the car and (d) by someone at rest outside the car.

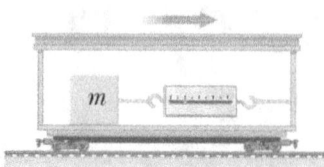

3. •• A truck is moving with constant acceleration a up a hill that makes an angle ϕ with the horizontal as in FIGURE 1. A small sphere of mass m is suspended from the ceiling of the truck by a light cord. If the pendulum makes a constant angle θ with the perpendicular to the ceiling, what is a?

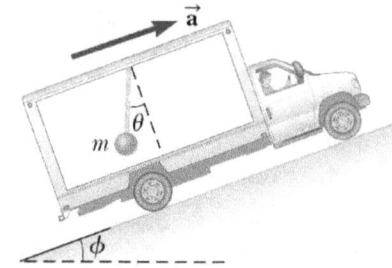

4. ••• The smooth block B of negligible size has a mass m and rests on the horizontal plane. If the board AC pushes on the block at an angle θ with a constant acceleration a_0, determine the velocity of the block along the board and the distance s the block moves along the board as a function of time t. The block starts from rest when $s = 0$, $t = 0$.

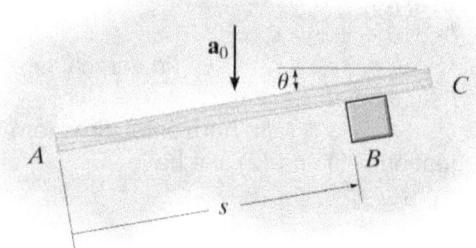

5. •• A wedge of mass m is placed on a horizontal smooth table. A block of mass m is placed at the midpoint of the smooth inclined surface having length L along its line of greatest slope. Inclination of the inclined surface is $\theta = 45°$. The block is released and simultaneously a constant horizontal force F is applied on the wedge as shown. (a) What is value of F if the block does not slide on the wedge? (b) In how much time the block will come out of the incline surface if applied force is 1.5 times that found in part (a)

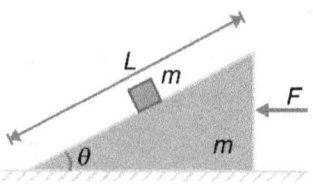

6. •• A wedge of mass m is kept on a smooth table and its inclined surface is also smooth. A small

block of mass *m* is projected from the bottom along the incline surface with velocity u. Assume that the block remains on the incline and take $\theta = 45°$, $g = 10\ m/s^2$.

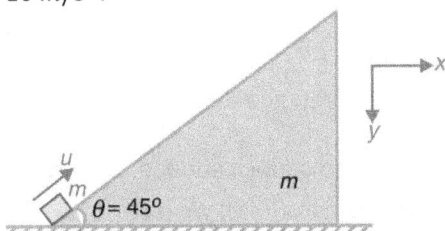

7. ••• A triangular wedge *W* having mass *M* is placed on an incline plane with its face *AB* horizontal. Inclination of the incline is θ. On the flat horizontal surface of the wedge there lies an infinite tower of rectangular blocks. Blocks 1, 2, 3, 4, ... have masses $M, \frac{M}{2}, \frac{M}{4}, \frac{M}{8}$, ... respectively. All surfaces are smooth. Find the contact force between the block 1 and 2 after the system is released from rest. Also find the acceleration of the wedge.

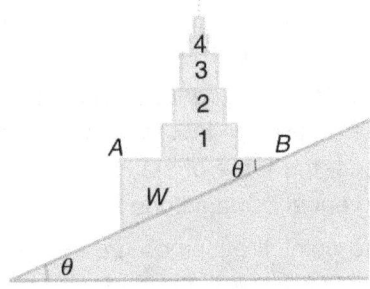

31. ANALYSIS OF FRICTION

Friction is a force that resists the movement of two contacting surfaces that slide relative to one another. This force always acts tangent to the surface at the points of contact and is directed so as to oppose the possible or existing motion between the surfaces.

31.1. TYPES OF FRICTION

(a) Dry Friction: **Dry friction occurs when the unlubricated surfaces of two solids are in contact under a condition of sliding or a tendency to slide.**

(b) Fluid Friction: **Fluid friction occurs when adjacent layers in a fluid (liquid or gas) are moving at different velocities.**

(c) Internal Friction: **Internal friction occurs in all solid materials which are subjected to cyclical loading.**

In this chapter, we will study the effects of dry friction, which is sometimes called Coulomb friction since its characteristics were studied extensively by C. A. Coulomb in 1781. Dry friction occurs between the contacting surfaces of bodies when there is no lubricating fluid.

Three Experiments. Here we deal with the frictional forces that exist between dry solid surfaces, either stationary relative to each other or moving across each other at slow speeds. Consider three simple thought experiments:

1. Send a book sliding across a long horizontal counter. As expected, the book slows and then stops. This means the book must have an acceleration parallel to the counter surface, in the direction opposite the book's velocity. From Newton's second law, then, a force must act on the book parallel to the counter surface, in the direction opposite its velocity. That force is a frictional force.

2. Push horizontally on the book to make it travel at constant velocity along the counter. Can the force from you be the only horizontal force on the book? No, because then the book would accelerate $(a = F/m)$. From Newton's second law, there must be a second force, directed opposite to your force but with the same magnitude, so that the two forces balance. That second force is a frictional force, directed parallel to the counter.

3. Push horizontally on a heavy crate. The crate does not move. From Newton's second law, a second force must also be acting on the crate to counteract your force. Moreover, this second force must be directed opposite your force and have the same magnitude as your force, so that the two forces balance. That second force is a frictional force. Push even harder. The crate still does not move. Apparently, the frictional force can change in magnitude so that the two forces still balance. Now push with all your strength. The crate begins to slide. Evidently, there is a maximum magnitude of the frictional force. When you exceed that maximum magnitude, the crate slides.

31.2. TWO TYPES OF DRY FRICTION

When you try to slide a heavy box of books across the floor, the box doesn't move at all unless you push with a certain minimum force. Then the box starts moving, and you can usually keep it moving with less force than you needed to get it started. If you take some of the books out, you need less force than before to get it started or keep it moving. What general statements can we make about this behaviour?

First, when a body rests or slides on a surface, we can think of the surface as exerting a single contact force on the body, with force components perpendicular and parallel to the surface (FIGURE 1). The perpendicular component vector is the normal force, denoted by $\vec{N}$. The component vector parallel to the surface (and perpendicular to $\vec{N}$) is the **friction force**, denoted by $\vec{f}$. If the surface is frictionless, then $\vec{f}$ is zero but there is still a normal force. (Frictionless surfaces are an unattainable idealisation, like a massless rope. But we can approximate a surface as frictionless if the effects of friction are negligibly small.) The direction of the friction force is always such as to oppose relative motion of the two surfaces.

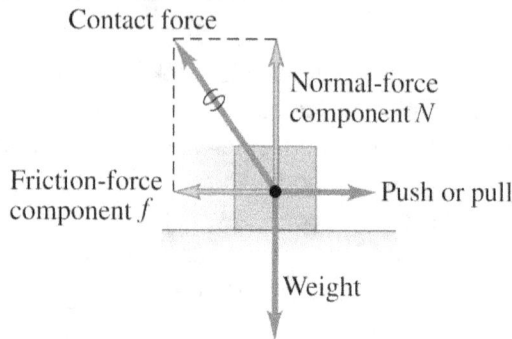

FIGURE1 When a block is pushed or pulled over a surface, the surface exerts a contact force on it.
The friction and normal forces are really components of a single contact force.

31.2.1. KINETIC (SLIDING) DRY FRICTION

The kind of friction that acts when a body slides over a surface is called a **kinetic friction force** $\vec{f}_k$. The *magnitude* of the kinetic friction force usually increases when the normal force increases. This is why it takes more force to slide a box full of books across the floor than to slide the same box when it is empty. In many cases the magnitude of the kinetic friction force f_k is found experimentally to be approximately *proportional* to the magnitude N of the normal force. i.e.,

$$f_k \propto N$$

or $\quad f_k = \mu_k N \quad \ldots (1)$

(magnitude of kinetic friction force)

here μ_k is a proportionality constant called the **coefficient of kinetic friction**. The more slippery the surface, the smaller the coefficient of friction. Because it is a quotient of two force magnitudes, μ_k is a pure number, without units.

☞ For any two parallel vectors $\vec{A}$ and $\vec{B}$, the vector relationship between $\vec{A}$ and $\vec{B}$ can be represented as $\vec{B} = k\vec{A}$. Here k is a scalar constant. But in Eq. (1), the force of kinetic friction $\vec{f}_k$ is perpendicular to nomal reaction $\vec{N}$, therefore Eq. (1) cannot be a vector relationship. Rather, it is a scalar relationship between the magnitudes of the two forces.

As for a given body and contact surface, μ_k and N, both are constant, therefore from (1), we have

$$f_k = \mu_k N = \text{constant}.$$

31.2.2. STATIC DRY FRICTION

Friction forces may also act when there is *no* relative motion. If you try to slide a box across the floor, the box may not move at all because the floor exerts an equal and opposite friction force on the box. This is called a **static friction force** $\vec{f}_s$. In FIGURE 2a, the box is at rest, in equilibrium, under the action of its weight $\vec{W}$ and the upward normal force $\vec{N}$. The normal force is equal in magnitude to the weight ($N = W$) and is exerted on the box by the floor. Now we tie a rope to the box (FIGURE 2b) and gradually increase the tension T in the rope. At first the box remains at rest because, as T increases, the force of static friction f_s also increases (staying equal in magnitude to T).

At some point T becomes greater than the maximum static friction force f_s the surface can exert. Then the box 'breaks loose' (the tension T is able to break the bonds between molecules in the surfaces of the box and floor) and starts to slide. FIGURE 2c shows the forces when T is at this critical value. If T exceeds this value, the box is no longer in equilibrium. For a given pair of surfaces the maximum value of f_s depends on the normal force. Experiment shows that in many cases this maximum value, called $f_{s,max}$, is approximately *proportional* to N; i.e.,

$$f_{s,max} \propto N$$

or $\quad f_{s,max} = \mu_s N \quad \ldots (2a)$

(condition of limiting equilibrium)

we call the proportionality factor μ_s the **coefficient of static friction**. Since, the maximum force of friction is applied at the verge of slipping, therefore, this condition is called the condition of limiting equilibrium. Table 1 lists some representative values of μ_s. In a particular situation, the actual force of static friction can have any magnitude between zero (when there is no other force parallel to the surface) and a maximum value given by $\mu_s N$ In symbols,

$$f_s \leq \mu_s N \quad \ldots (2b)$$

(magnitude of static friction force)

Like Eq. (1), this is a relationship between magnitudes, *not* a vector relationship. The equality sign holds only when the applied force T has reached the critical value at which motion is about to start (FIGURE 2c). When T is less than this value (FIGURE 2b), the inequality sign holds. In that case we have to use the equilibrium conditions ($\sum \vec{F} = 0$) to find f_s. If there is no applied force ($T = 0$) as in FIGURE 2a, then there is no static friction force either ($f_s = 0$).

As soon as the box starts to slide (FIGURE 2d), the friction force usually *decreases*; it's easier to keep the box moving than to start it moving. *Hence the coefficient of kinetic friction is usually less than the coefficient of static friction for any given pair of surfaces*, as Table 1 shows. If we start with no applied force ($T = 0$) and gradually increase the force, the friction force varies somewhat, as shown in FIGURE 2e.

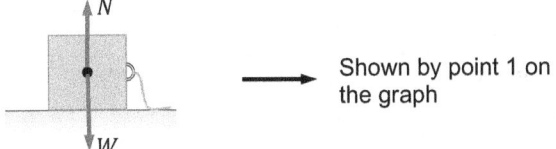

No applied force, box at rest.
No friction: $f_s = 0$

(a)

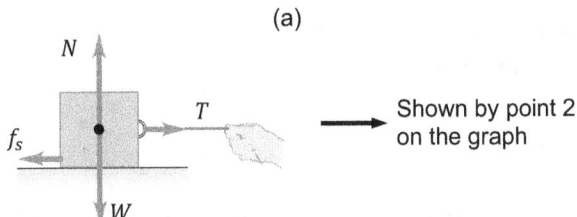

Weak applied force, box remains at rest.
Static friction: $f_s = T < \mu_s N$

(b)

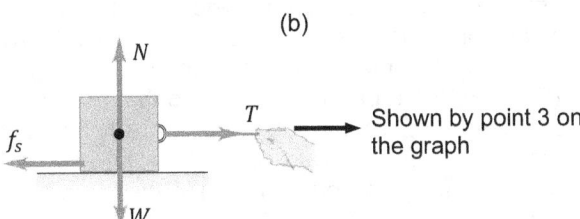

Stronger applied force, box just about to slide.
Static friction: $f_s = T = \mu_s N$

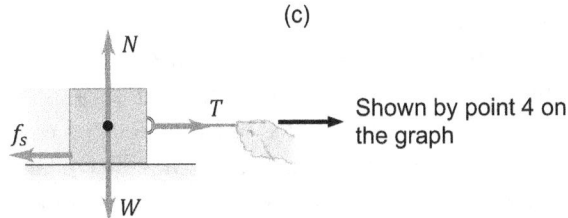

Box sliding at constant speed.
Kinetic friction: $f_s = \mu_k N$

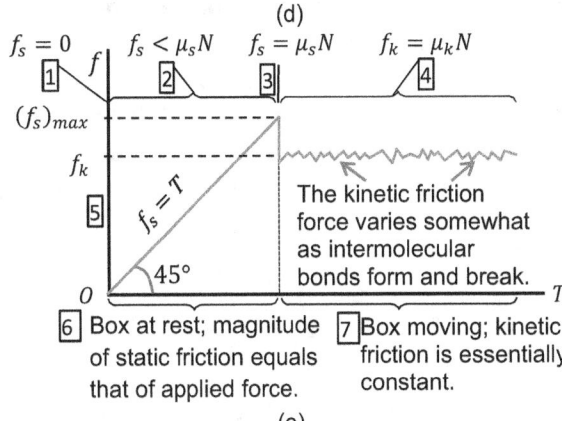

(e)

FIGURE 2. When there is no relative motion, the magnitude of the static friction force f_s is less than or equal to $\mu_s N$. When there is relative motion, the magnitude of the kinetic friction force f_k equals $\mu_k N$.

When a body slides on a layer of gas, friction can be made very small. In the linear air track used in physics laboratories, the gliders are supported on a layer of air. The frictional force is velocity dependent, but at typical speeds the effective coefficient of friction is of the order of 0.001.

☞ Like kinetic force of friction ($\vec{f}_k$), the static force of friction ($\vec{f}_s$) is also perpendicular to normal reaction $\vec{N}$, therefore Eq. (2) cannot be a vector relationship. Rather, it is also a scalar relationship between the magnitudes of the two forces.

31.3. *MICROSCOPIC ORIGIN OF FRICTION*

Equation (1) is only an approximate representation of a complex phenomenon. The microscopic details of the force of friction are still not properly understood. We believe that the two objects in contact make microscopic connections at various points on their surfaces. Even highly polished surfaces are rough and ridged microscopically. Friction and normal forces result from the intermolecular forces (fundamentally electromagnetic in nature) between two rough surfaces at high points where they come into contact (FIGURE 3). If the two objects are pushed together harder, the surfaces deform a little more, enabling more "high points" to bond. That is why the force of kinetic friction and the maximum force of static friction are proportional to the normal force. As a box slides over the floor, bonds between the two surfaces form and break, and the total number of such bonds varies; hence the kinetic friction force is not perfectly constant. Smoothing or polishing the surfaces can actually increase friction, since more molecules are able to interact and bond; bringing two smooth surfaces of the same metal together can cause a 'cold weld.' Lubricating oils work because an oil film between two surfaces (such as the pistons and cylinder walls in a car engine) prevents them from coming into actual contact.

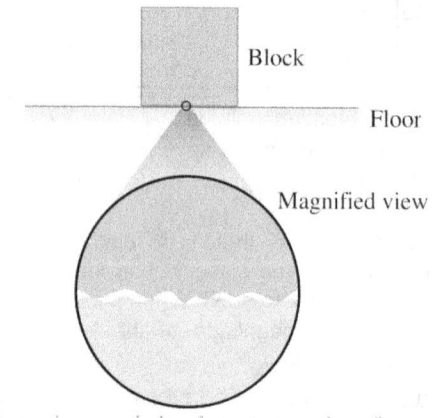

FIGURE 3 The normal and friction forces arise from interactions between molecules at high points on the surfaces of the block and the floor.

In static friction, when these molecular bonds are stretched, they pull back harder. The bonds have to be broken before sliding can begin. Once sliding begins, molecular bonds are continually made and broken as "high points" come together in a hit-or-miss fashion. These bonds are generally not as strong as those formed in the absence of sliding, which is why $\mu_s > \mu_k$.

Table 1 lists some representative values of μ_k. Although these values are given with two significant FIGUREs, they are only approximate, since friction forces can also depend on the speed of the body relative to the surface. For now, we'll ignore this effect and assume that μ_k and f_k are independent of speed, in order to concentrate on the simplest cases. Table 1 also lists coefficients of static friction μ_s.

☞ Notice that, in FIGURE 1, we have shown that the point of action of normal reaction and weight are same, which is incorrect in presence of pulling or pushing force. In this case the line of action of normal reaction shifts towards the pulling or pushing force This shifting, is necessary in order to balance the "tipping effect" caused by force of pull/push. This topic will be discussed in detail in the chapter 'dynamics of rigid body'. For now, it is sufficient to assume that $\vec{N}$ and $\vec{W}$ have same point of action.

Table 1 Approximate Coefficients of Friction

Materials	Coefficient of Static Friction, μ_s	Coefficient of Kinetic Friction, μ_k
Steel on steel	0.74	0.57
Aluminium on steel	0.61	0.47
Copper on steel	0.53	0.36
Brass on steel	0.51	0.44
Zinc on cast iron	0.85	0.21
Copper on cast iron	1.05	0.29
Glass on glass	0.94	0.40
Copper on glass	0.68	0.53
Teflon on Teflon	0.04	0.04
Teflon on steel	0.04	0.04
Rubber on concrete (dry)	1.0	0.8
Rubber on concrete (wet)	0.30	0.25

Experiments have also shown that the magnitude of the *apparent* contact area between the two surfaces does not significantly affect the magnitude of static friction force because the *effective* contact area depends only on the normal force. For example, when the largest side of a rectangular block rests on a table, there are many small regions of contact between the block and the table (Fig. 4a). If the block is rotated so that one of its smallest sides rests on the table, there are fewer, but larger regions of contact (Fig. 4b). The total effective contact areas (and the normal force) in the two cases are equal (Fig. 4c), so friction between the block and the table is the same, too.

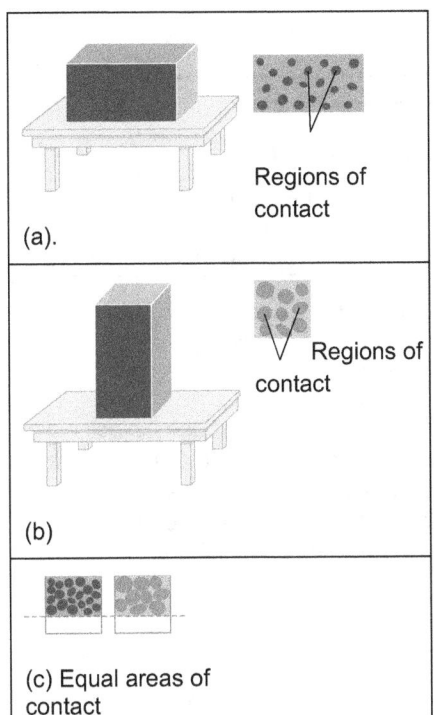

FIGURE 4. (a) When the large side touches the table, there are many small regions of contact. (b) When the small side touches the table, there are fewer, but larger areas of contact. (c) Total effective areas of contact are equal.

32. LAWS OF STATIC FRICTION

Static Friction, acting between the surfaces in contact, (not in relative motion) opposes the tendency of relative motion between the surfaces.

It is independent of the area of contacting surfaces.

Now, $f_{s,max} \propto N$

$f_{s,max} = \mu_s N$

(maximum or limiting static friction)

here, μ_s = coefficient of static friction.

N = normal reaction on the block from the surface.

$0 \leq f_s \leq \mu_s N$

When applied force 'F' exceeds $f_{s,max}$, block starts moving and frictional force decreases to a constant value f_k. f_k is called kinetic friction and it has unique value which is given by

$f_k = \mu_k N$

Here μ_k = co-efficient of kinetic friction.

N = normal reaction.

➤ If the component of applied force, on the block of mass m, along the surface is F and μ_s, μ_k are coefficients of static and kinetic friction respectively, then force of friction is given by

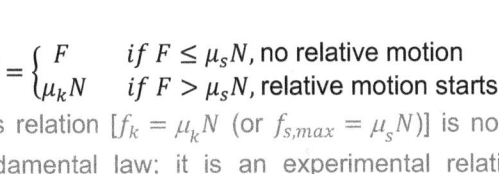

$$f = \begin{cases} F & if\ F \leq \mu_s N, \text{no relative motion} \\ \mu_k N & if\ F > \mu_s N, \text{relative motion starts} \end{cases}$$

➤ This relation [$f_k = \mu_k N$ (or $f_{s,max} = \mu_s N$)] is not a fundamental law; it is an experimental relation between the magnitude of the friction force f, which acts parallel to the two surfaces, and the magnitude of the normal force N, which acts perpendicular to the surfaces. It is *not* a vector equation since the two forces have different directions, perpendicular to one another.

➤ If there are several forces acting on the block along the surface, then calculate net force on it (don't consider force of friction here). The force of friction will act opposite to the net force.

➤ Force of friction is independent of the area of contact between the surfaces

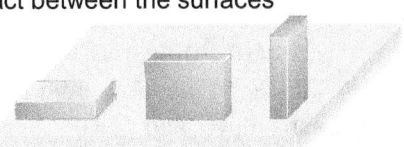

FIGURE 1 The maximum static frictional force $\vec{f}_{s,max}$ would be the same, no matter which side of the block is in contact with the table

➤ It is parallel to the surface of contact, and in the direction that opposes relative motion.

EXAMPLE 58. A 10.0-kg box rests on a horizontal floor. The coefficient of static friction is $\mu_s = 0.40$ and the coefficient of kinetic friction is $\mu_k = 0.30$. Determine the force of friction, acting on the box if a horizontal external applied force is exerted on it of magnitude: (a) 0, (b) 10 N, (c) 20 N, (d) 38 N, and (e) 40. N

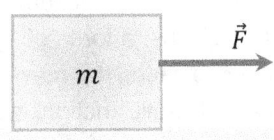

FIGURE 1

APPROACH The force of friction is given by-

$$f = \begin{cases} F & if\ F \leq \mu_s N, \text{no relative motion} \\ \mu_k N & if\ F > \mu_s N, \text{relative motion starts} \end{cases} \quad \ldots (1)$$

So, first of all, find the maximum possible force of static friction $\mu_s N$ and compare it with the given force in each part by using Eq. (1).

SOLUTION The forces on the box are gravity $m\vec{g}$, the normal force exerted by the floor $\vec{N}$, the horizontal applied force $\vec{F}$ and the friction force f as shown in Fig. 2.

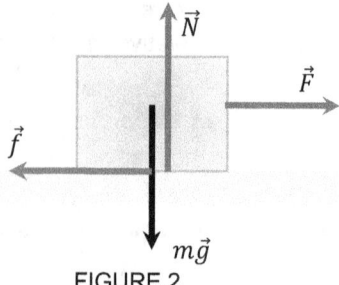

FIGURE 2

The free-body diagram of the box is shown in Fig. 2. In the vertical direction there is no motion, so Newton's second law in the vertical direction gives

$\Sigma F_y = ma_y = 0$

which tells us $N - mg = 0$. Hence the normal force is

$N = mg = (10.0\ kg)(9.80\ m/s^2) = 98.0\ N$

The force of static friction will oppose any applied force up to a maximum of

$\mu_s N = (0.40)(98.0\ N) = 39.2\ N.$

(a) Because, $F = 0 < \mu_s N$, therefore from Eq. (1), it is clear, the box doesn't move, and $f = 0$

(b) In this case also, the applied force is $F = 10$ newton $< \mu_s N$, therefore the box will not move and force of friction on the box, $f = F = 10\ N$.

(c) As applied force $F = 20$ newton $< \mu_s N$, therefore the force is not sufficient to move the box and $f = F = 20\ N$.

(d) The applied force $F = 38$ newton $< \mu_s N$, therefore this force is still not quite large enough to move the box; so the force of friction $f = F = 38\ N$, i.e., the force of friction has now increased to 38 N to keep the box at rest.

(e) In this case, the given force, $F = 40$ newton $> \mu_s N$, therefore the force of friction $f = \mu_k N = (0.30)(98.0 N) = 29.4\ N$. Thus, a force of 40 N will start the box moving since it exceeds the maximum force of static friction, Instead of static friction, we now have kinetic friction, and its magnitude is 29.4 N.

There is now a net (horizontal) force on the box of magnitude $F = 40\ N - 29.4\ N = 10.6\ N$.

So, the box will accelerate at a rate

$a_x = \dfrac{\Sigma F}{m} = \dfrac{10.6\ N}{10.0\ kg} = 1.06\ m/s^2$

as long as the applied force is 40 N. Figure 3 shows a graph that summarizes this Example.

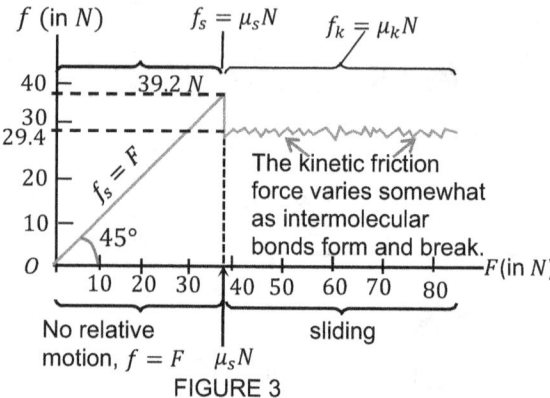

FIGURE 3

EXAMPLE 59. In the Fig. 1, suppose a block weighs 2 kg, and the force T can be increased to 8N before the block starts to slide. A force of 4 N will keep the block moving at constant speed once it has been set in motion. Find the co-efficient of static and kinetic friction.

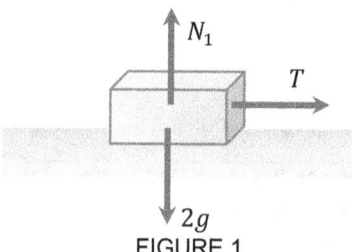

FIGURE 1

CALCULATION OF COEFFICIENT OF STATIC FRICTION

APPROACH Since, the applied force can be increased up to 8 N before the block starts to slide, therefore for 8 N force, the block is in limiting equilibrium and in this case, the force of static friction will be maximum.

For limiting equilibrium in horizontal direction, we have

$\Sigma F_x = T - f_{s,max} = T - \mu_s N_1 = 0$

Here, N_1 is the normal reaction of on the block and it can be obtained by applying the equilibrium condition in vertical direction, and $T = 8\ N$.

By substituting the value of normal reaction N_1 and T in above relation, you can obtain the required coefficient of static friction.

SOLUTION Since, there is no acceleration of the body along y-axis.

Therefore $\Sigma F_y = N_1 - 2g = 0$

$\Rightarrow N_1 = 20N$

$\Sigma F_x = T - f_{s,max} = 8N - \mu_s N_1 = 0$

$\Rightarrow \mu_s N_1 = 8N$

$\therefore \mu_s = \dfrac{8N}{N_1} = \dfrac{8N}{20\ N} = \dfrac{2}{5} = 0.40$

CALCULATION OF COEFFICIENT OF KINETIC FRICTION

APPROACH Since, *a force of 4 N keeps the block moving at constant speed once it has been set in motion*, therefore, the force of friction will be kinetic in nature and the net force on the block along the direction of motion is zero, i.e.,

$$\Sigma F_x = T - f_k = 4N - \mu_k N_1 = 0$$
or $\quad \mu_k N_1 = 4\ N$
or $\quad \mu_k = \frac{4\ N}{N_1}$

Now, substitute the value of N_1 and solve for μ_k.

SOLUTION $\mu_k = \frac{4N}{N_1} = \frac{4N}{20\ N} = 0.20$

33. HOLDING A BOX AGAINST A ROUGH WALL

You can hold a box against a rough wall (Fig. 1) and prevent it from slipping down by pressing hard horizontally. This won't work well if the wall is slippery. You need friction. Even then, if you don't press hard enough, the box will slip. The horizontal force you apply produces a normal force on the box exerted by the wall (net force horizontally is zero since box doesn't move horizontally.) The force of gravity $m\vec{g}$, acting downward on the box, can now be balanced by an upward static friction force whose maximum magnitude is proportional to the normal force. The harder you push, the greater $\vec{N}$ is and the greater f_s can be. If you don't press hard enough, then $mg > \mu_s N$ and the box begins to slide down.

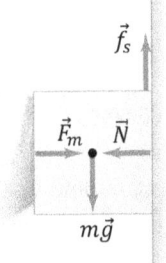

FIGURE 1

CALCULATION OF MINIMUM FORCE REQUIRED TO HOLD THE BOX AGAINST A ROUGH VERTICAL WALL

Suppose, F_m is the minimum required force to hold the box against the rough vertical wall and μ_s is the coefficient of static friction, then
For horizontal equilibrium, we have
$$\Sigma F_x = 0$$
or $\quad F_m - N = 0 \Rightarrow F_m = N \quad \ldots (1)$
And for vertical equilibrium
$$\Sigma F_y = 0$$
or $\quad f_s - mg = 0$
or $\quad f_s = mg$
or $\quad \mu_s N = mg \quad \ldots (2)$
Substituting the value of N from Eq. (1), in Eq. (2), we get
$$\mu_s F_m = mg$$
or $\quad F_m = mg/\mu_s \quad \ldots (3)$

EXAMPLE 60. If $\mu_s = 0.40$ and $mg = 20\ N$, Find the minimum required force F to prevent the box from falling.

APPROACH Apply the relation (3) obtained in above article.

SOLUTION From Eq. (3), we have
$$F = mg/\mu_s = 20\ N/0.40 = 50\ N$$

34. PUSHING VERSUS PULLING

Pushing, a body on a rough surface, is always harder than pulling. The mathematical explanation is given below-

EXPLANATION: The maximum possible force of static friction is given by-
$$f_{s,max} = \mu_s N$$
It depends on normal reaction N. So, larger the value of N means the larger value of limiting frictional force and requires larger value of external force to slide a block on the rough surface.

Now, consider following two cases-

CASE 1: PUSHING THE BLOCK: Suppose, a pushing force $\vec{F}$, is acting on a block of mass m at an angle θ from the horizontal. If N_1 is the normal reaction of the floor on the block, then Newton's second law in vertical direction gives-

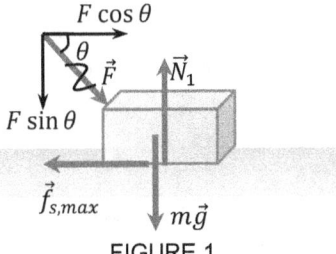

FIGURE 1

$$\Sigma F = N_1 - F \sin\theta - mg = 0$$
or $\quad N_1 = mg + F \sin\theta \quad \ldots (1)$

CASE 2. PULLING THE BLOCK: Now, consider the same block pulled by force $\vec{F}$ at the same angle θ, as shown in following FBD. If in this case, the normal reaction of the floor on the block is N_2, then, on applying Newton's second law in vertical direction, we get-

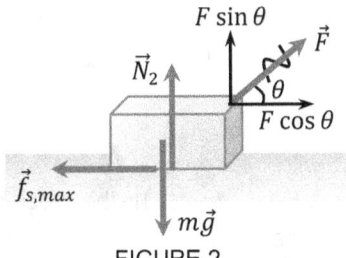

FIGURE 2

$$\Sigma F = N_2 + F \sin\theta - mg = 0$$
or $\quad N_2 = mg - F \sin\theta \quad \ldots (2)$

On comparing, Eq. (1) and (2), we get-
$$N_2 > N_1$$

Therefore, $f_{s,max} = \mu_s N$ will be greater for pushing as compared to pulling. Thus, pushing is much difficult and requires more force as compared to pulling of the block.

EXAMPLE 61. A block placed on a horizontal surface is being pushed by a force $\vec{F}$ making an angle θ with the vertical as shown in Fig. 1. The coefficient of friction between block and surface is μ. (a) Find the force required to slide the block with uniform velocity on the floor.

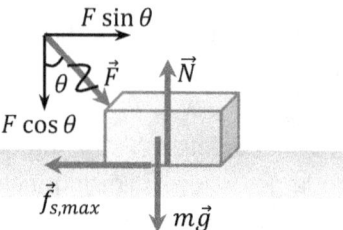

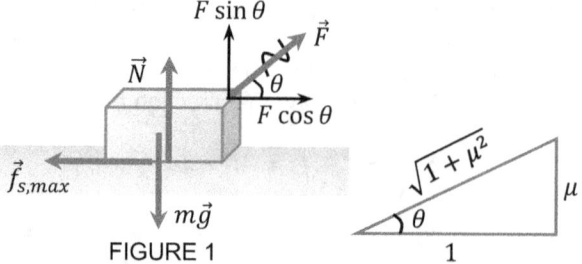

FIGURE 1

(b) Show that if θ is smaller than a certain angle θ_0, the block cannot be made to slide across the floor, no matter how great the force be.

(a) APPROACH The block is in translational equilibrium in vertical direction whereas moving with uniform velocity along the floor i.e., in horizontal direction, therefore apply Newton's 2nd law in vertical and horizontal directions and solve for force F.

SOLUTION (a) Forces acting on the block are shown in Fig. 1. For vertical equilibrium of the block,
$$N = F\cos\theta + mg \qquad \ldots (1)$$
While for horizontal motion,
$$F\sin\theta - \mu N = ma$$
or $\quad F\sin\theta = \mu N \quad$ [as v = constant, $a = 0$]
Substituting R from Eq. (1) in the above, we get
$$F\sin\theta = \mu(F\cos\theta + mg)$$
or $\quad F = \dfrac{\mu mg}{(\sin\theta - \mu\cos\theta)} \qquad \ldots (2)$

(b) APPROACH In presence of friction, a block can slide only if the magnitude of net force on it is > 0. In Eq. (2), μ, m and g are already positive. Therefore for +ve value of F, the denominator of RHS should be +ve, i.e., $\quad \sin\theta - \mu\cos\theta > 0 \qquad \ldots (3)$
From this condition, calculate the required value of θ.

SOLUTION From condition (3), we have
$\sin\theta - \mu\cos\theta > 0$
i.e., $\tan\theta > \mu$ or $\theta > \tan^{-1}(\mu)$
or $\quad \theta > \theta_0 \quad$ with $\quad \theta_0 = \tan^{-1}\mu$
So, for angle $\theta < \theta_0 [= \tan^{-1}(\mu)]$, the block cannot be made to slide across the floor, no matter how great the force be.

EXAMPLE 62. A body of mass m rests on a horizontal floor with which it has a coefficient of static friction μ and is desired to make the body move by applying the minimum possible force $\vec{F}$. Find the magnitude of $\vec{F}$ and the direction in which it has to be applied.

APPROACH Since, pulling requires less force as compared to pushing, therefore, for minimum required force, we apply it to pull the block not as a pushing force. Now, find the expression for force by applying Newton's laws of motion in vertical and horizontal direction and apply the condition of minima for the obtained value of pulling force.

SOLUTION Let the force $\vec{F}$ be applied at an angle θ with the horizontal as shown in Fig.1. For vertical equilibrium,
$N + F\sin\theta = mg$
i.e., $N = mg - F\sin\theta \qquad \ldots (1)$
while for horizontal motion
$F\cos\theta \geq f_L$
& $\quad F\cos\theta \geq \mu N \quad$ [as $f_L = \mu N$] $\qquad \ldots (2)$
Substituting value of N from Eq. (1) in (2), we get
$F\cos\theta \geq \mu(mg - F\sin\theta)$
or $\quad F \geq \dfrac{\mu mg}{(\cos\theta + \mu\sin\theta)} \qquad \ldots (3)$
For the force F to be minimum $(\cos\theta + \mu\sin\theta)$ must be maximum, i.e.
$$\dfrac{d}{d\theta}(\cos\theta + \mu\sin\theta) = 0$$
or $-\sin\theta + \mu\cos\theta = 0 \qquad \ldots (4)$
i.e., $\qquad \tan\theta = \mu$
or $\qquad \sin\theta = \dfrac{\mu}{\sqrt{1+\mu^2}}$ and $\cos\theta = \dfrac{1}{\sqrt{1+\mu^2}}$

On substituting these values in Eqn. (3), we get
$$F \geq \dfrac{\mu mg}{\dfrac{1}{\sqrt{1+\mu^2}} + \dfrac{\mu^2}{\sqrt{1+\mu^2}}} \quad \text{i.e., } F \geq \dfrac{\mu mg}{\sqrt{1+\mu^2}}$$

so that, $F_{min} = \dfrac{\mu mg}{\sqrt{1+\mu^2}}$ with $\theta = \tan^{-1}(\mu)$

EXAMPLE 63. A 10.0-kg box is pulled along a horizontal surface by a force of $F_P = 40.0\, N$ applied at a 30.0° angle above horizontal. We assume a coefficient of kinetic friction of 0.30. Calculate the acceleration.

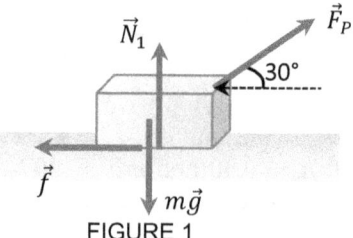

FIGURE 1

APPROACH The free-body diagram is shown in Fig. 1. But with one more force, that of friction.

SOLUTION The calculation for the vertical (y) direction:
$$mg = (10.0\ kg)(9.80\ ms^{-2}) = 98.0\ N$$
and $F_{Py} = (40.0\ N)(\sin 30.0°) = 20.0\ N$.
With y positive upward and $a_y = 0$, we have
$$N_1 - mg + F_{Py} = ma_y$$
or $N_1 - 98.0\ N + 20.0\ N = 0$,
So, the normal force is $N_1 = 78.0\ N$. Now we apply Newton's second law for the horizontal (x) direction (positive to the right), and include the friction force:
$$F_{Px} - f = ma_x$$
The friction force is kinetic as long as $f = \mu_k N_1$ is less than $F_{Px} = (40.0\ N)\cos 30.0° = 34.6\ N$, which it is:
$$f = \mu_k N_1 = (0.30)(78.0\ N) = 23.4\ N.$$
Hence the box does accelerate:
$$a_x = \frac{F_{Px}-f}{m} = \frac{34.6N - 23.4N}{10.0 kg} = 1.1\ m/s^2$$

NOTE Our final answer has only two significant figures because our least significant input value $\mu_k = 0.30$ has two.

EXAMPLE 64. At the moment $t = 0$ the force $F = ct$ is applied to a small body of mass m resting on a smooth horizontal plane (c is a constant). Find:
(a) the velocity of the body at the moment of its breaking off the plane;
(b) the distance traversed by the body up to this moment.

APPROACH First of all calculate acceleration of the block by applying, Newton's second law in vertical and horizontal direction. Since, at the time of break off the surface, the normal reaction, $N = 0$, therefore using this condition, we can find the time ($t = t_0$) when the body breakoff the contact with the plane. Now replace the acceleration a by dv/dt and integrate it from $t = 0$ to $t = t_0$ and solve for velocity v.
Again replacing v with ds/dt and integrating it from $t = 0$ to $t = t_0$ gives the distance traveled by the body up to the breaking off time.

SOLUTION The FBD is shown in following figure.

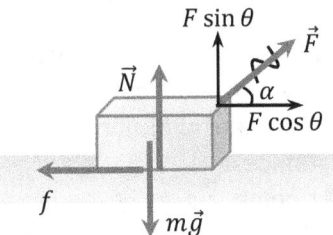

Along vertical direction: By Newton's 2nd law,
$$\Sigma F_y = 0$$
or $\quad N + F\sin\alpha - mg = 0$
or $\quad N + F\sin\alpha = mg \qquad \ldots (1)$
Along horizontal direction, by Newton law, we have

$$F\cos\alpha = ma \qquad \ldots (2)$$
Given that, $F = ct$, therefore, substituting this value in above Eq. (2), we get
$$ct\cos\alpha = ma$$
$$\therefore \quad a = \frac{ct\cos\alpha}{m} \qquad \ldots (2)$$
Let after time t_0, the body is breaking the surface. i.e., at $t = t_0$, $N = 0$. Therefore from (1), we get-
$$0 + (ct_0)\sin\alpha = mg$$
or $\quad t_0 = \frac{mg}{c\sin\alpha} \qquad \ldots (3)$
Now, from Eq. (2), we have
$$a = \frac{ct\cos\alpha}{m}$$
or $\quad a = \frac{dv}{dt} = \frac{at\cos\alpha}{m}$
or $\quad \int_0^v dv = \int_0^t \frac{ct\cos\alpha}{m} dt$
or $\quad v = \frac{ct^2\cos\alpha}{2m} \qquad \ldots (4)$
At, $t = t_0$, $v = v_0$, therefore, $v_0 = \frac{ct_0^2\cos\alpha}{2m}$
(a) at the time of break off, $t = t_0$, $v = v_0$
$$\therefore \quad v_0 = \frac{ct_0^2\cos\alpha}{2m}$$
Putting the value of t_0, from eq. (3) we get,
$$v_0 = \frac{mg^2\cos\alpha}{2c\sin^2\alpha}$$
Now, from Eq. (4), we have
$$v = \frac{ds}{dt} = \frac{ct^2\cos\alpha}{2m}$$
or $\quad \int_0^s ds = \int_0^{t_0} \frac{ct^2\cos\alpha}{2m} dt$
$\therefore \quad s = \frac{ct_0^3\cos\alpha}{6m} = \frac{m^2g^3\cos\alpha}{6c^2\sin^3\alpha} \quad \left(\because t_0 = \frac{mg}{c\sin\alpha}\right)$

EXAMPLE 65. A 60 kg block is pushed up an inclined plane by means of a horizontal push P as shown in the following figure. The coefficients of static and kinetic friction between incline and block are $\mu_s = 0.6$ and $\mu_k = 0.4$ respectively and the ramp makes an angle of 30° with the horizontal.
(a) What value of P is required to move the block at a constant speed of $0.20\ m/s$ along the incline?
(b) If the person pushing should stop for rest and let $P = 0$ does the block slide back on the Incline?

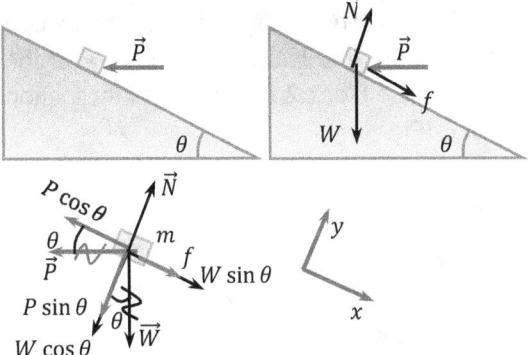

(a) APPROACH Apply Newton's second law along x and y axis and solve for P.

SOLUTION From the force diagrams as shown in figure:

Along y axis, by Newton's 2nd law, we get-
$$N - W\cos\theta - P\sin\theta = 0$$
or $\quad\quad N = W\cos\theta + P\sin\theta \quad\quad ...(1)$

Along x-direction, Newton's 2nd law gives-
$$f + W\sin\theta - P\cos\theta = 0 \quad\quad ...(2)$$

To move the block with constant velocity, we have
$$f = \mu_k N$$
Therefore, Eq. (2), gives-
$$\mu_k N + W\sin\theta - P\cos\theta = 0 \quad\quad ...(3)$$
Substituting the value N From Eqns. (1) and (2), we get
$$\mu_k(W\cos\theta + P\sin\theta) + W\sin\theta - P\cos\theta = 0$$
or $P[\mu_k\sin\theta - \cos\theta] = -W[\mu_k\cos\theta + \sin\theta]$

or $P = \frac{\sin\theta + \mu_k\cos\theta}{\cos\theta - \mu_k\sin\theta}W = 747.24$ N

(b) APPROACH (i) If $W\sin\theta > \mu_s N$, the block slips down.

(ii) If $W\sin\theta \leq \mu_s N$, the block will stay in equilibrium

SOLUTION If force P is removed, the block will have a tendency to slide down due to $W\sin\theta$ (component of weight along incline), Maximum possible force of friction is $\mu_s N$,

(i) If $W\sin\theta > \mu_s N$, the block slips down.

(ii) If $W\sin\theta \leq \mu_s N$, the block will stay in equilibrium in this situation.

As here,
$W\sin\theta = W \times \frac{1}{2} < \mu_s N \quad\quad (\because \theta = 30°)$
$= 0.6 \times W\cos\theta = 0.6W \times \frac{\sqrt{3}}{2}$
[from (2), put $P = 0$]

therefore, the block will stay in equilibrium.

35. CONDITION OF CONTACT OF TWO BLOCKS ON INCLINED PLANE

Suppose, the two blocks in contact with each other are sliding on an inclined board with a common acceleration 'a'. The mass of first block is m_1 and that of second block is m_2. The coefficient of kinetic friction between the block and the inclined plane is μ_1 for the block 1 and μ_2 for the block 2. The angle of inclination of the plane is α (Fig. 1)

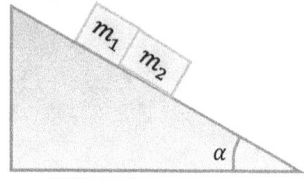

FIGURE 1

Now, consider the FBDs of m_1 and m_2. N represents contact force between m_1 and m_2; N_1, N_2 are normal reactions between blocks and the inclined plane.

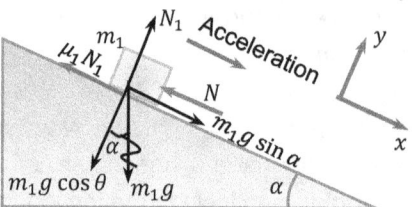

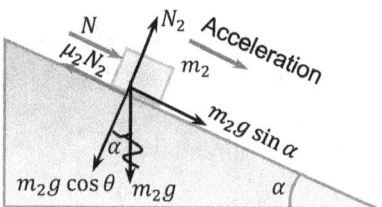

Balancing forces along y-axis:
For block of mass m_1:
$$N_1 = m_1 g\cos\alpha \quad\quad ...(1)$$
For block of mass m_1:
$$N_2 = m_2 g\cos\alpha \quad\quad ...(2)$$

Along x-axis:
For block m_1:
$$m_1 g\sin\alpha - N - \mu_1 N_1 = m_1 a \quad\quad ...(3)$$
For block m_2:
$$m_2 g\sin\alpha + N - \mu_2 N_2 = m_2 a \quad\quad ...(4)$$
On adding Eq. (3) and (4), we get
$$(m_1 + m_2)g\sin\alpha - \mu_1 N_1 - \mu_2 N_2 = (m_1 + m_2)a$$
Now, substituting the values of N_1 and N_2 from (1) and (2), in above equation, we get
$(m_1 + m_2)g\sin\alpha - \mu_1 m_1 g\cos\alpha - \mu_2 m_2 g\cos\alpha$
$= (m_1 + m_2)a$
$$a = \frac{g[(m_1+m_2)\sin\alpha - (\mu_1 m_1 + \mu_2 m_2)\cos\alpha]}{(m_1+m_2)}, g = 9.8\, m/s^2$$

Substitution of this value of a in Eqn. (3) or (4), gives
$$N = (\mu_2 - \mu_1)m_1 m_2 g\cos\alpha/(m_1 + m_2)$$
Necessary condition for contact: $N > 0$
or $\quad (\mu_2 - \mu_1)m_1 m_2 g\cos\alpha/(m_1 + m_2) > 0$
or $\quad \mu_2 > \mu_1$

i.e., for contact between the blocks, the coefficient of friction between the lower block and inclined plane must be greater than the coefficient of friction between upper block and inclined surface.

☞ If $\mu_1 > \mu_2$ then N will come out to be negative, which is not possible, because, the least value of the reaction between two bodies is zero, which implies that the blocks will slide down without touching each other. They will get separated. So, for $\mu_1 > \mu_2$ always assume different acceleration for each block.

36. ANGLE OF FRICTION

In limiting equilibrium, the angle made by the resultant reaction force (force of contact, S) with the vertical (normal reaction) is known as the angle of the friction.

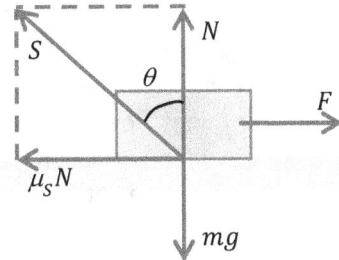

Now, from figure,

or $\tan\theta = \dfrac{f}{N}$

$\because \quad f \leq \mu_s N \quad \therefore \quad \tan\theta \leq \dfrac{\mu_s N}{N}$

or $\quad \tan\theta \leq \mu_s$

or angle of friction $\theta \leq \tan^{-1}\mu_s$

Note: 1. Static friction may be less or more than kinetic friction whereas the limiting static friction is always greater than the kinetic friction.

2. In limiting equilibrium, the resultant of normal reaction and force of friction provides total contact force.

37. ANGLE OF REPOSE

Angle of repose is defined as the angle of an inclined plane, at which a body placed on it just begins to slide.

Place a body of mass m on an inclined plane (Fig. 1). Increase the angle of inclined plane till the body just begins to slide. If α is the angle of inclination at which the body just begins to slide, then α is the angle of repose.

The forces acting on the body are-
1. Weight of the body $(m\vec{g})$ in vertical downward direction.
2. Normal reaction $\vec{N}$, perpendicular to the plane

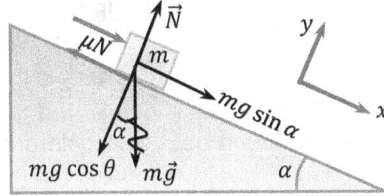

Resolving above forces along and perpendicular to the inclined plane, we get

Along x: $mg\sin\alpha - \mu_s N = 0$

or $\quad mg\sin\alpha = \mu_s N$ ... (1)

Along y: $N - mg\cos\alpha = 0$

or $\quad mg\cos\alpha = N$ ... (2)

Dividing Eq. (1) by (2), we get

$\tan\alpha = \mu_s$ ... (3)

Thus, the coefficient of limiting friction is equal to the tangent of the angle of repose.

RELATION BETWEEN ANGLE OF FRICTION AND ANGLE OF REPOSE

From Eq. (3), the angle of repose

$$\alpha = \tan^{-1}\mu_s$$

And the angle of friction is defined as-

$$\theta = \tan^{-1}\mu_s$$

On comparing above two relations, we find that

$$\alpha = \theta$$

i.e., Angle of repose = Angle of friction

38. MINIMUM AND MAXIMUM VALUES OF CONTACT FORCE

The horizontal component of contact S provides force of friction whereas its normal component provides normal reaction. i.e.,

$S\sin\theta = f_s$ ... (1)

$S\cos\theta = N$ ... (2)

Squaring and adding, we get

$S^2 = f_s^2 + N^2$ ... (3)

Thus, $S = \sqrt{f_s^2 + N^2}$

As, $0 \leq f_s \leq \mu_s N$, therefore

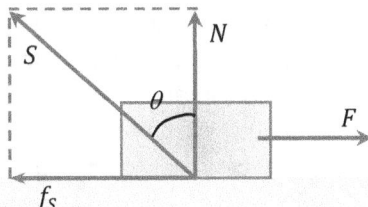

FIGURE 1

$S_{min} = \sqrt{f_{min}^2 + N^2} = \sqrt{0 + N^2} = N = mg$

and $S_{max} = \sqrt{f_{max}^2 + N^2} = \sqrt{(\mu_s N)^2 + N^2} = N\sqrt{\mu_s^2 + 1} = mg\sqrt{\mu_s^2 + 1}$,

therefore, $mg \leq S \leq mg\sqrt{\mu_s^2 + 1}$

4. For a smooth surface, $\mu_s = 0$, therefore $\tan\theta = \mu_s = 0$ or $\theta = 0$

In this case, net contact force $S = N = mg$

39. CHARACTERISTICS OF DRY FRICTION

As a result of *experiments* that pertain to the foregoing discussion, we can state the following rules which apply to bodies subjected to dry friction.

- The frictional force acts *tangent* to the contacting surfaces in a direction *opposed* to the *motion* or tendency for motion of one surface relative to another.
- The maximum static frictional force F_s that can be developed is independent of the area of contact, provided the normal pressure is not very low nor great enough to severely deform or crush the contacting surfaces of the bodies.
- The maximum static frictional force is generally greater than the kinetic frictional force for any two surfaces of contact. However, if one of the bodies is moving with a *very low velocity* over the surface of another, F_k becomes approximately equal to F_s, i.e., $\mu_s = \mu_k$
- When *slipping* at the surface of contact is *about to occur*, the maximum static frictional force is proportional to the normal force, such that $F_s = \mu_s N$.
- When *slipping* at the surface of contact is *occurring*, the kinetic frictional force is proportional to the normal force, such that $F_k = \mu_k N$.

40. ROLLING FRICTION

Undoubtedly the invention of the wheel made life a lot easier. Like objects sliding on a horizontal surface, however, rolling ones are also subject to friction that causes them to slow down and stop.

The friction associated with the rolling motion of one object against a surface is called **rolling friction**, and it is weaker than both kinetic and static friction. That is why it is easier to move a cart with wheels than to move one without wheels.

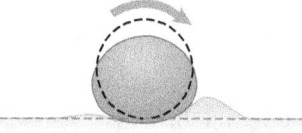

FIGURE 1. A ball rolling on a floor causes the ball to flatten and the floor to deform.

On the microscopic level, rolling friction is similar to both static and kinetic friction. Consider a ball rolling along a horizontal floor. Where the ball is in contact with the floor, molecular bonds form and then immediately break as the ball rolls over the floor. New bonds form and break repeatedly. The number of bonds that forms depends on the real contact area.

The portion of the ball in contact with the pavement at any instant is slightly flattened, and the portion of the floor in contact with the ball is slightly deformed as well (Fig. 1). The more the ball and floor deform, the greater their real contact area. All other things being equal, a greater normal force causes a greater deformation. So, rolling friction depends on the normal force between the two surfaces:

$$f_r = \mu_r N \qquad \ldots (1)$$

Like the other two coefficients of friction, the **coefficient of rolling friction** μ_r is a unitless scalar that is found experimentally. Like kinetic friction, the direction of the rolling friction force is opposite the direction of motion with respect to the surface. Because the object moves with respect to the surface, the term *moving friction* is use to include both kinetic and rolling friction, but not static friction

41. BLOCK OVER BLOCK FRICTION PROBLEMS

Sometimes we have multi block problems in which one block is placed over the other and some externally applied forces are also given. In such cases, the calculation of acceleration corresponding to each block is somewhat tricky. To solve such problems, we should first determine whether there is a relative slipping between the blocks in contact or not. If there is a relative slipping, the kinetic friction come into role otherwise required static friction will be used. To understand the method clearly, we are discussing here some block over block problems.

41.1. TWO BLOCK PROBLEM

EXAMPLE 66. In FIGURE 1, A and B are two blocks of masses m_1 and m_2 respectively. Coefficient of static and kinetic frictions between the blocks A and B are μ_s and μ_k respectively. There is no friction between the ground and lower block B. If a force F is applied to block A, then find out the accelerations and effective force on each block. Discuss all possibilities.

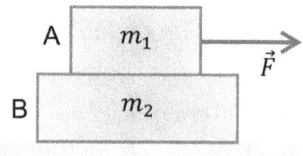

FIGURE 1

SOLUTION To solve such problems, we first calculate the maximum possible static friction between contact surfaces of each block.

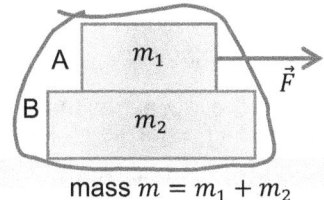

mass $m = m_1 + m_2$

FIGURE 2

Let, N_1 is the normal reaction between block A and B; N_2 is the normal reaction between block B and ground, then from FBD of block A (FIGURE 3), we have
$$N_1 = m_1 g,$$
and $N_2 = N_1 + m_2 g = m_1 g + m_2 g = (m_1 + m_2)g$

∴ the maximum possible static friction between contact surfaces of A and B is
$$f_{1max} = \mu_{s1} N_1, \quad \text{or} \quad f_{max} = \mu_s (m_1 g)$$

Now, if a force F is applied on block A. then, following two cases may be possible.

Case 1: If $F \leq f_{max}$, then the force of friction can easily prevent the relative slipping between the blocks and both blocks will move with a common acceleration. In this case, the nature of force of friction between the two blocks will be static and we can consider both the blocks in a single system of mass $(m_1 + m_2)$ (due to same acceleration) and their common acceleration will be given by [FIGURE 2]
$$\Sigma F = ma \quad \Rightarrow \quad F = (m_1 + m_2)a$$
or $\quad a = \dfrac{F}{m_1 + m_2}$ ... (1)

We can also find the effective or accelerating force on each individual block.

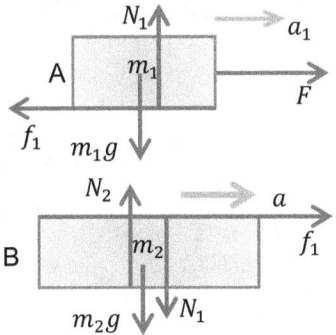

FIGURE 3 (FBDs of A and B)

Calculation of Effective Force on Each Block

Suppose f_1 is the static friction between the blocks, then this force opposes the motion of A and its reactional force of friction on B provides the motion to block B.

From FBD of A in FIGURE 3, the net force on block A is given by
$$\Sigma F_A = F - f_1 = m_1 a \quad \ldots (2)$$

The net force on block B is f_1.

From equation (2), we have
$$f_1 = F - m_1 a$$
or $\quad f_1 = F - m_1 \left(\dfrac{F}{m_1 + m_2}\right) = \left(\dfrac{m_2}{m_1 + m_2}\right) F$

The value of f_1 can also be calculated by FBD of block B shown in FIGURE 3.
$$f_1 = m_2 a = m_2 \left(\dfrac{F}{m_1 + m_2}\right) = \left(\dfrac{m_2}{m_1 + m_2}\right) F$$

Case 2.

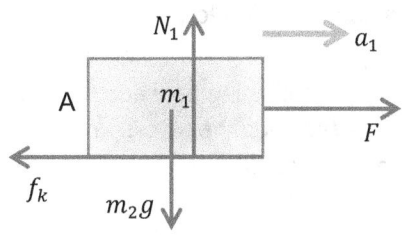

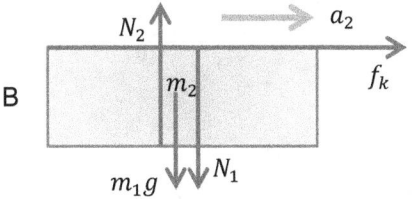

FIGURE 4 (FBDs of A and B)

If $F_1 > f_{max}$, then the force of friction between the two blocks cannot prevent the relative slipping and the friction between the blocks will be kinetic in nature. This kinetic friction always has a constant value, $f_k = \mu_k N_1$, where N_1 is the normal reaction between the blocks. The direction of f_k on block A will be opposite to its relative motion i.e., opposite to F, whereas on block B it will be in forward direction (frictional forces always make action reaction pair)

Kinetic friction between blocks A and B,
$$f_k = \mu_k N_1 = \mu_k (m_1 g)$$
The net force on block A, $\Sigma F_A = F - f_k = F - \mu_k N_1$
By newton's second law we have $\Sigma F_A = m_1 a_1$
$$\Rightarrow a_1 = \dfrac{\Sigma F_A}{m_1} = \dfrac{F - \mu_k N_1}{m_1}$$
From FBD of block B, we have
$$\Sigma F_B = m_2 a_2 \Rightarrow a_2 = \dfrac{\Sigma F_B}{m_2} = \dfrac{f_k}{m_2} = \dfrac{\mu_k N_1}{m_2}$$

EXAMPLE 67. Block B of mass 35 kg is resting on a frictionless floor. Another block A of mass 7 kg is resting on it as shown in the FIGURE 1. The coefficient of static friction between the blocks is 0.5 while kinetic friction is 0.4. If a force F ($= 100$ N) is applied to block A, find the acceleration of each block ($g = 10 \ m/s^2$).

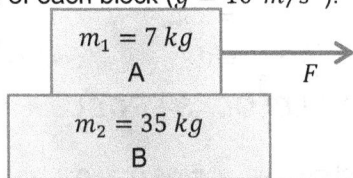

FIGURE 1

SOLUTION. The maximum possible static friction between contact surfaces of A and B is

$$f_{max} = \mu_s N_1,$$

or $\quad f_{max} = \mu_s(m_1 g) = 0.5 \times 7 \times 10 = 35\ N$

The external force applied to block A is $F = 100\ N$

Since, $F > f_{max}$, therefore the friction is unable to prevent the relative slipping between the blocks. Thus, a slipping starts between the blocks and force of kinetic friction comes into play.

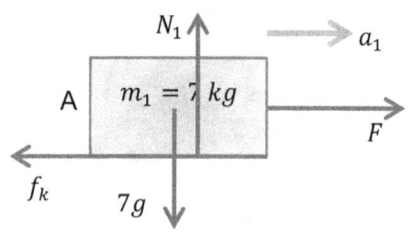

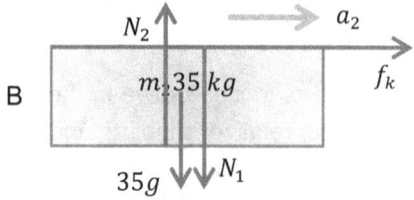

FIGURE 2 (FBDs of A and B)

From, FBD of block A (FIGURE 2), $N_1 = 7g\ N$
From FBD of block B (FIGURE 2), $N_2 = 35g + N_1 = 35g + 7g = 42g\ N$

The kinetic friction between the blocks,

$$f_k = \mu_k N_1 = 0.4 \times 7 \times 10 = 28\ N.$$

The direction of this kinetic friction on block A will be backward whereas on B it is forward.

From the FBD of block A, we have

$$\Sigma F_A = m_1 a_1$$

or $\quad F - f_k = m_1 a_1$

or $\quad a_1 = \left(\dfrac{F - f_k}{m_1}\right)$

or $\quad a_1 = \left(\dfrac{100-28}{7}\right) = \dfrac{72}{7} \approx 10.3\ m/s^2$ (Acceleration of A)

Now, from FBD of B (FIGURE 2), we have

$$\Sigma F_B = m_2 a_2$$

$$\Rightarrow f_k = 35 a_2$$

or $\quad a_2 = \dfrac{f_k}{35} = \dfrac{28}{35} = 0.8\ m/s^2$ (Acceleration of B)

41.2. MULTI BLOCKS PROBLEM

In such problems, we always observe whether there is a relative slipping between the blocks or not. To observe it, always first assume that all blocks are moving with same acceleration. The method will be crystal clear in following example

EXAMPLE 68.. Find the accelerations a_1, a_2, a_3 of the three blocks shown in FIGURE1(a) if a horizontal force of 10 N is applied on (a) 2 kg block, (b) 3 kg block, (c) 7 kg block. Take $g=10\ m/s^2$.

[Note: In adjoining figure, it is not mentioned whether the given coefficients of friction are static or kinetic. So, we will consider them static or kinetic as per requirement]

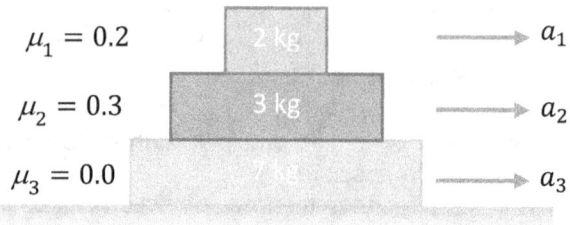

FIGURE 1

SOLUTION. We first calculate the maximum possible force of static friction between each block-

Calculation of maximum possible friction: From FIGURE 2,
the maximum possible friction between 2 kg and 3 kg blocks, $f_{1,max} = \mu_1 N_1 = 0.2 \times 2g = 4\ N$,

the maximum possible friction between 3 kg and 7 kg blocks, $f_{2,max} = \mu_2 N_2 = 0.3 \times 5g = 15\ N$, and

the maximum possible friction between 7 kg block and ground, $f_{3,max} = \mu_3 N_3 = 0.0 \times 12g = 0\ N$

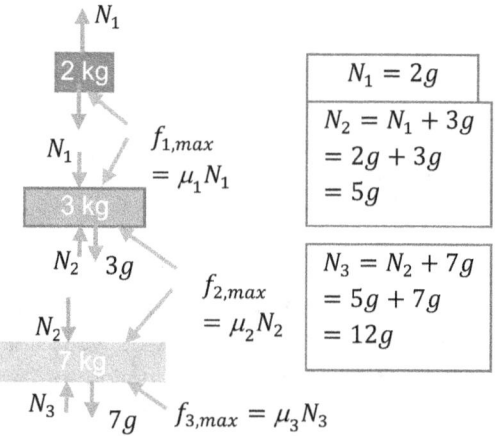

FIGURE 2 FBDs showing directions of normal forces and limiting frictions

Suppose, the applied force is denoted by F, then $F = 10\ N$.

(a) If the horizontal force of 10 N is applied on 2 kg block

If, all the three blocks are moving with same acceleration, then we can consider all the blocks in a single system. In this case common acceleration is given by

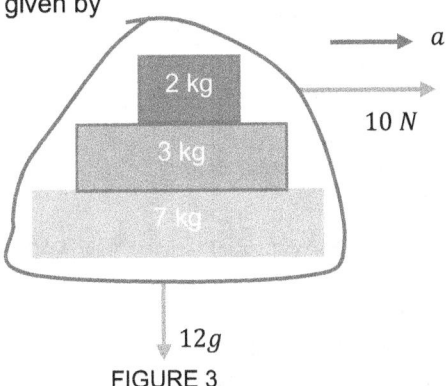

FIGURE 3

$$\Sigma F = ma$$
$$\Rightarrow a = \frac{\Sigma F}{m} = \frac{10\,N}{2\,kg + 3\,kg + 7\,kg} = \frac{10\,N}{12\,kh} = 5/6\ m/s^2$$

Since, applied force on 2 Kg block $F > f_{1,max}$, so the maximum static friction is unable to prevent relative slipping between 2 kg and 3 kg blocks. As a result of which, there will be a relative slipping between 2 kg block and 3 kg block and all the blocks cannot move with the above common acceleration $a = 5/6\ m/s^2$. In this case, between 2 kg block and 3 kg block, the force of friction will be kinetic ($f_1 = \mu_1 N_1$) in nature.

$$f_1 = \mu_1 N_1 = 0.2 \times 2g = 4\,N$$

The direction of f_1 on 2 kg block is backward whereas on 3 kg block it will be forward. Now, from Newton's second law, we have-

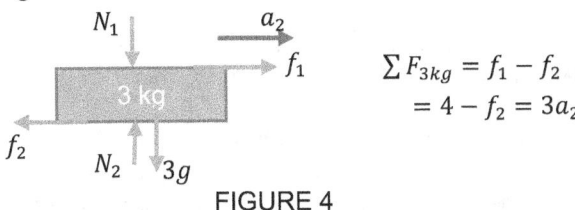

FIGURE 3

$$\Sigma F = ma$$
$$\Rightarrow \Sigma F_{2\,kg} = m_1 a_1$$
$$\Rightarrow 10 - 4 = 2a_1 \quad \Rightarrow a_1 = 3\ m/s^2$$

∴ Acceleration of 2 kg block $= 3\ m/s^2$ and accelerating force on 2 kg block
$$F_1 = \Sigma F_{2\,kg} = 10 - 4 = 6\,N.$$

By Newton's third law, the force of reactional friction of magnitude $f_1 (= 4\,N)$ will act in forward direction on 3 kg block.

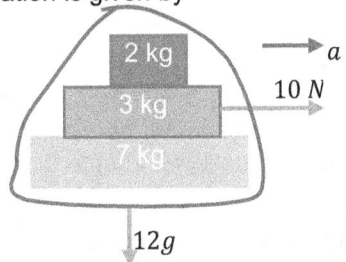

$$\Sigma F_{3kg} = f_1 - f_2$$
$$= 4 - f_2 = 3a_2$$

FIGURE 4

Therefore, the forward force on 3 kg block $f_1 = 4N$

Maximum possible friction between 3 kg and 7 kg block, $f_{2,max} = 15\,N$,

Since, forward force $4N < f_{2,max}$, therefore, the force of friction can easily prevent the relative slipping between the 3 kg and 7 kg blocks. As a result of which, both 3 kg and 7 kg blocks will move with the same acceleration. In this case, we can consider both blocks in a single system. Suppose the common acceleration is a', then from FIGURE 5, we have

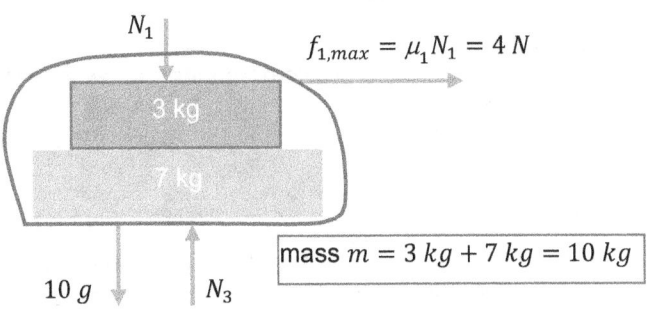

mass $m = 3\,kg + 7\,kg = 10\,kg$

FIGURE 5

$$\Sigma F = ma$$
$$\Rightarrow 4 = 10a'$$
$$\Rightarrow a' = \frac{4}{10} = 0.4\ m/s^2$$

Therefore, the acceleration of 3 kg block = acceleration of 7 kg block $= 0.4\ m/s^2$. i.e.,

$$a_2 = a_3 = 0.4\ m/s^2$$

(b) If the horizontal force of $10\,N$ is applied on $3\ kg$ block

Suppose all the three blocks are moving with a common acceleration a, then we can consider all the blocks in a single system. In this case common acceleration is given by

FIGURE 6

$$\Sigma F = ma$$
$$\Rightarrow a = \frac{\Sigma F}{m} = \frac{10\,N}{2\,kg + 3\,kg + 7\,kg} = \frac{10\,N}{12\,kh} = 5/6\ m/s^2$$

The net accelerating force on 3 kg block
$$F_2 = 3\,kg \times \frac{5}{6} m/s^2 = \frac{5}{2} N = 2.5\,N.$$

If, f_1 and f_2 are force of frictions between 2 kg, 3 kg; and 3kg, 7 kg blocks respectively, then from FBD (FIGURE 7), net accelerating force on 3 kg block is given by-

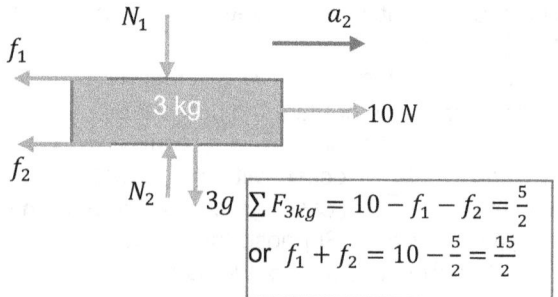

FIGURE 7

$\sum F_{3kg} = 10 - f_1 - f_2$

This is the force which provides required accelerating force to 3 kg block.

Thus, $\sum F_{3kg} = F_2$

or $10 - f_1 - f_2 = \frac{5}{2}$

$\Rightarrow f_1 + f_2 = 10 - \frac{5}{2} = \frac{15}{2}$ ... (1)

The direction of f_1 on 2 kg block will be forward, therefore this will be the accelerating force on 2 kg block.

Now, if 2 kg block also moves with acceleration $a = 5/6 \ m/s^2$, then,

Net accelerating force on 2 kg block

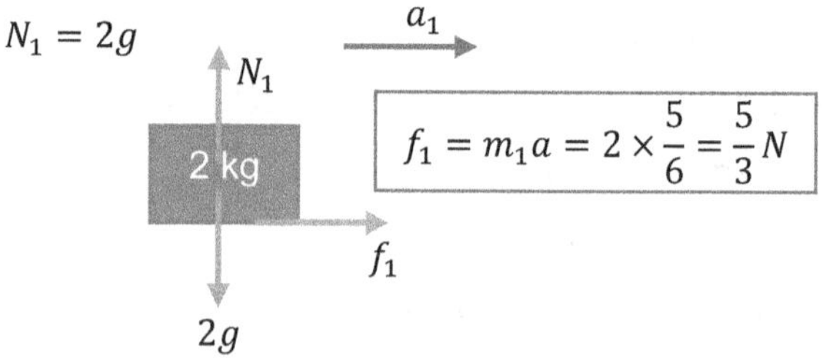

FIGURE 8

$\sum F_{2kg} = f_1 = 2 \ kg \times \frac{5}{6} m/s^2 = \frac{5}{3} N = F_1 \text{(say)}$ ... (2)

and maximum possible friction between 2 kg block and 3 kg block is

$f_{1,max} = 4 \ N$

As $F_1 < f_{1,max}$, therefore the friction is sufficient to prevent the relative slipping between 2 kg and 3 kg blocks. In this case the friction will be static in nature. So, both the blocks will move with same acceleration $a = \frac{5}{6} m/s^2$.

On substituting, the value of f_1 from equation (2), in equation (1), we get-

$10 \ N - \frac{5}{3} N - f_2 = \frac{5}{2} N$

or $f_2 = 10 \ N - \frac{5}{3} N - \frac{5}{2} N = \frac{35}{6} N$

The direction of f_2 on 7 kg block is forward (FIGURE 7).

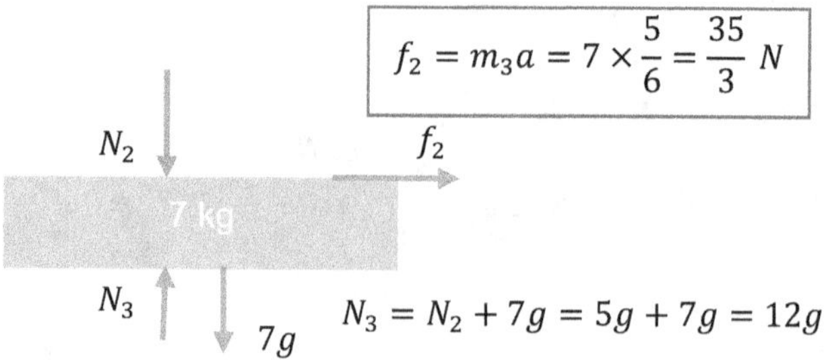

$$f_2 = m_3 a = 7 \times \frac{5}{6} = \frac{35}{3} N$$

$N_3 = N_2 + 7g = 5g + 7g = 12g$

FIGURE 9

Since, there is no friction between ground and 7 kg block, therefore net accelerating force on 7 kg block

$\therefore \quad \sum F_{7kg} = f_2 = \frac{35}{6} N = F_3 \text{ (say)}$ ... (3)

The maximum possible friction between 3 kg block and 7 kg block is $f_{2,max} = 15 \ N$

As, $F_3 < f_{2,max}$, therefore to prevent relative motion between 3 kg and 7 kg blocks, the required force can be easily provided by friction between these two blocks and these two blocks will also move with same acceleration $a = 5/6 \ m/s^2$. In this case the nature of friction between 3 kg block and 7 kg block will be static. We can also find the accelerating force on 7 kg block directly as-

$F_3 = m_3 a$

i.e., $F_3 = 7 \ kg \times \frac{5}{6} m/s^2 = \frac{35}{6} N$ (double check)

Thus, all the three blocks move with same acceleration, i.e., $a_1 = a_2 = a_3 = \frac{5}{6} m/s^2$

(c) If the horizontal force of 10 N is applied on 7 kg block

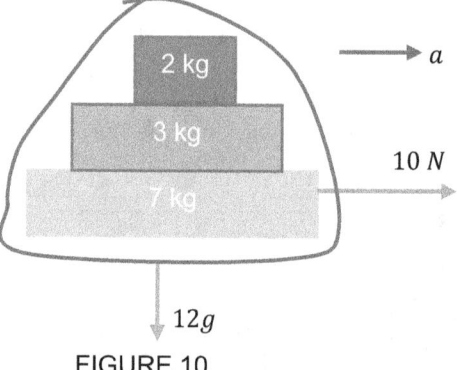

FIGURE 10

Suppose all the three blocks are moving with same acceleration, then we can consider all the blocks in a single system. In this case common acceleration is given by

$$a = \frac{10 \ N}{2 \ kg + 3 \ kg + 7 \ kg} = \frac{10 \ N}{12 \ kh} = 5/6 \ m/s^2$$

Maximum possible frictional force between 7 kg and 3 kg is $f_{2,max} = 15 \ N$

i.e., $F < f_{2,max}$, therefore, the frictional force between 7 kg block and 3 kg block can easily prevent relative slipping between these two blocks and both blocks will move with a common acceleration.

The accelerating force on 7 kg block corresponding to acceleration a is given by,

$$F_3 = m_3 a = 7\,kg \times \frac{5}{6} m/s^2 = \frac{35}{6} N$$

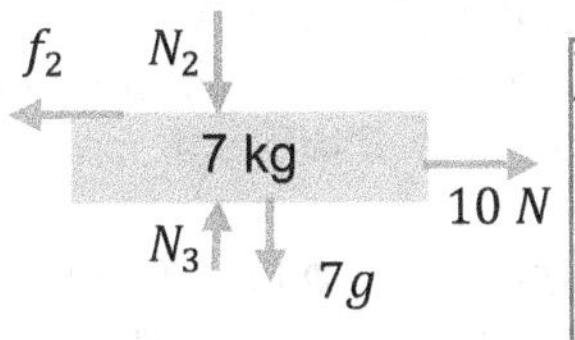

FIGURE 11

$$\Sigma F_{7\,kg} = 10 - f_2 = m_3 a = 7 \times \frac{5}{6} = \frac{35}{6} N$$
$$f_2 = 10 - \frac{35}{6} = \frac{25}{6} N$$

If f_2 is the static friction between 7 kg and 3 kg blocks, then direction of f_2 on 7 kg block will be backward (FIGURE 11) whereas, on 3 kg block it will be forward (FIGURE 12). Now from FBD (FIGURE 11) we have
$\Sigma F_{7\,kg} = m_3 a \Rightarrow 10 - f_2 = \frac{35}{6} N$
$\Rightarrow f_2 = 10 - \frac{35}{6} = \frac{25}{6} N \approx 4.2\,N$

Now, the force of static friction between 7 kg and 3 kg block is $f_2 = \frac{25}{6} N$.

The direction of this frictional force on 7 kg block is in backward direction whereas on 3 kg block it will be in forward direction.

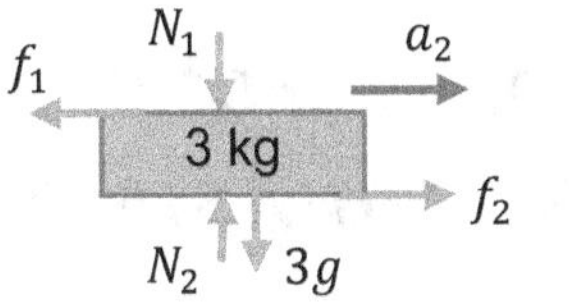

FIGURE 12

$$\Sigma F_{net} = f_2 - f_1 = m_3 a = 3 \times \frac{5}{6}$$
$$\Rightarrow f_2 - f_1 = \frac{5}{2} N$$

Now, net accelerating force on 3 kg block $F_2 = (3\,kg)\left(\frac{5}{6} m/s^2\right) = \frac{5}{2} N = 2.5 N$.

If, f_1 is the force of friction between 2 kg block and 3 kg block, then
$$\Sigma F_{3\,kg} = f_2 - f_1 = F_2$$
i.e., $f_1 = f_2 - F_2 = \frac{25}{6} N - \frac{5}{2} N = \frac{5}{3} N$

As, $f_1 < f_{1,max} (= 4\,N)$, therefore this force can easily prevent the relative slipping between 2 kg block and 3 kg block. As a result, 2 kg block and 3 kg block both will move with same acceleration $a = \frac{5}{6} m/s^2$

So, All three blocks will move with same acceleration $a 5/6\,m/s^2$.
i.e., $a_1 = a_2 = a_3 = 5/6\,m/s^2$

Note that, we can also calculate f_1 for 2 kg block directly by using FBD of 2 kg (FIGURE 13), as follows
$$f_1 = m_1 a = 2 \times \frac{5}{6} = \frac{5}{3} N$$

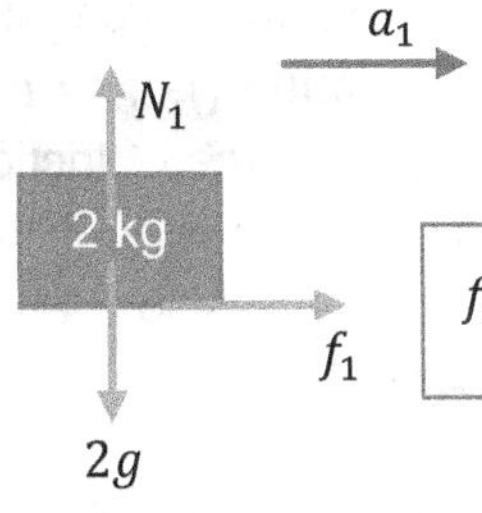

$$f_1 = m_1 a = 2 \times \frac{5}{6} = \frac{5}{3} N$$

FIGURE 13

EXAMPLE 69. A plank of mass m_1 with a bar of mass m_2 placed on it lies on a smooth horizontal plane. A horizontal force growing with time t as $F = at$ (a is constant) is applied to the bar. Find how the accelerations of the plank a_1 and of the bar a_2 depend on t, if the coefficient of friction between the plank and the bar is equal to μ. Draw the approximate plots of these dependences.

SOLUTION Force $F = at$ is time dependent, so we divide the problem into two steps:

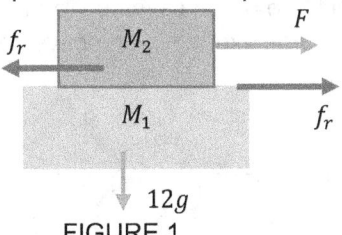

FIGURE 1

Step 1: At $t = 0$, applied force F is zero. When t increases, applied force also increases.
Let at $t = t_0$, applied force is equal to limiting friction between the blocks.
When $t < t_0$, applied force is less than limiting friction
$\Rightarrow$ no relative motion between the bar and plank
$\Rightarrow$ Both move with common acceleration a.
$\therefore \quad a_1 = a_2 = a \qquad$ when $t < t_0$
In this case, both (Plank + bar) can be considered as a single system.
And acceleration is given by, $a = \frac{F}{m_1 + m_2}$

Step 2: When $t > t_0$, applied force is greater than the limiting friction and there will be relative motion between the blocks.
Let accelerations of plank and bar are, a_1 and a_2 respectively
$\therefore$ From FBD of mass m_2, we have
$$F - f_r = m_2 a_2$$
or $\quad F - \mu m_2 g = m_2 a_2$
or $\quad a_2 = \frac{F - \mu m_2 g}{m_2} = \frac{at - \mu m_2 g}{m_2}, \quad t > t_0$

From FBD of plank of mass m_1, we have
$a_1 = \frac{f_r}{m_1} = \frac{\mu m_2 g}{m_1}$

$\therefore$ acceleration of m_2 with respect to m_1
$a_{21} = a_2 - a_1 = \frac{at - \mu m_2 g}{m_2} - \frac{\mu m_2 g}{m_1}$

Calculation of t_0: At $t = t_0$, $a_{21} = 0$,
$\therefore a_{21} = \frac{at_0 - \mu m_2 g}{m_2} - \frac{\mu m_2 g}{m_1} = 0$
or $\quad \frac{at_0}{m_2} - \mu g = \frac{\mu m_2 g}{m_1}$

or $\frac{at_0}{m_2} = \frac{\mu m_2 g}{m_1} + \mu g \quad$ or $\quad t_0 = \frac{\mu g m_2}{m_1 a}(m_1 + m_2)$

EXAMPLE 70 THE HORSE BEFORE THE CART
A horse refuses to pull a cart (FIGURE 1a). The horse reasons, "according to Newton's third law, whatever force I exert on the cart, the cart will exert an equal and opposite force on me, so the net force will be zero and

I will have no chance of accelerating the cart." What is wrong with this reasoning?

APPROACH Because we are interested in the motion of the cart, we draw a simple diagram for it (FIGURE 1b). The force exerted by the horse on the cart is labelled $\vec{F}_{HC}$. (This force is actually exerted on the harness. Because the harness is attached to the cart, we consider it a part of the cart.) Other forces acting on the cart are the gravitational force of Earth on the cart $\vec{F}_{gEC}$, the normal force of the pavement on the cart $\vec{N}_{PC}$ and the frictional force exerted by the pavement on the cart, labelled $\vec{f}_{PC}$.

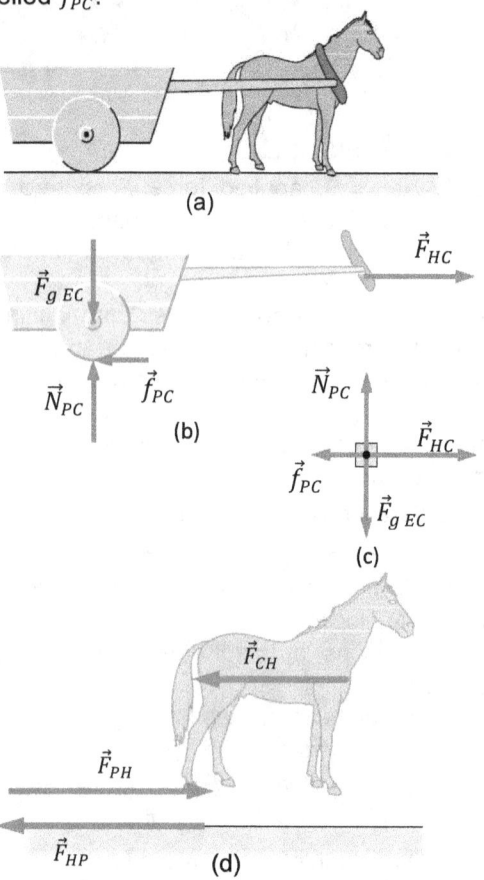

FIGURE 1

SOLUTION 1. Draw a free-body diagram for the cart (see FIGURE 1c). Because the cart does not accelerate vertically, the vertical forces must sum to zero. The horizontal forces are $\vec{F}_{HC}$ to the right and $\vec{f}_{PC}$ to the left. The cart will accelerate to the right if $|\vec{F}_{HC}|$ is greater than $|\vec{f}_{PC}|$.

2. Note that the reaction force to $\vec{F}_{HC}$, which we call $\vec{F}_{CH}$, is exerted on the horse, not on the cart (FIGURE 1d). It has no effect on the motion of the cart, but it does affect the motion of the horse. If the horse is to accelerate to the right, there must be a force $\vec{F}_{PH}$ (to the right) exerted on the horse's hooves by the pavement that is greater in magnitude than $\vec{F}_{CH}$.

Because the reaction force to $\vec{F}_{CH}$ is exerted on the horse, it has no effect on the motion of the cart. This is the mistake in the horse's reasoning.

EXAMPLE 71. A block weighing 2kg rests on a horizontal surface. The coefficient of static friction between the block and surface is 0.40 and kinetic friction is 0.20.
(a) How large is the friction force acting on the block ?
(b) How large will the friction force be if a horizontal force of 5N is applied on the block?
(c) What is the minimum force that will start the block in motion?

SOLUTION: (a) As the block rests on the horizontal surface and no other force parallel to the surface is on the block, the friction force is zero.
(b) With the applied force parallel to the surfaces in contact 5N, opposing friction becomes equal and opposite. Further the limiting friction is $\mu_s N = \mu_s Mg = 8N$
∴ Force of friction is $5N$.
(c) The minimum force that can start motion is the limiting one. $\mu_s N = \mu_s mg = 8N$

42. FORCE OF FRICTION IN CASE OF A BLOCK CONNECTED WITH A STRING

Consider, a block of mass m connected with a massless string on a rough plane surface (Fig.1). Let initially, the tension in the string is zero.

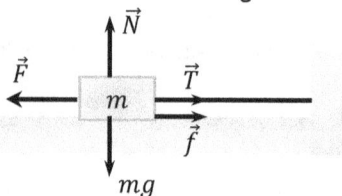

Now, as we apply an external force (F) on the block towards left, a static force of friction (f) develops between the contact surfaces of block and rough plane. On the block, this force opposes the tendency of motion of the block, i.e., it acts opposite to applied force.

Since, the force of static friction is self-adjustable, therefore, as we increase the external force (F), the force of friction (f) also increases. When applied external force reaches to $\mu_s N$, the force of friction attains its maximum possible value ($f_{max} = \mu_s N$). So, up to this limiting value of f, the force of friction (f) is capable to balance the external force and the block remains stationary at its original position, and hence the tension in the string cannot change and remains zero.

Now, when F further increases beyond $f_{max}(=\mu_s N)$, the force of friction alone cannot prevent the block to

move towards applied force and the tension (T) in the string comes into play. In this case, the maximum force of friction (f_{max}) together with tension in string (T), balances the applied external force F, i.e.,
$F = \mu_s N + T$
Therefore, we can summarise the case as

Applied Force (F)	Force of friction (f)	Tension in string (T)
$F \leq \mu_s N$	F	0
$F > \mu_s N$	$\mu_s N$	$F - \mu_s N$ ($\because F = \mu_s N + T$)

To illustrate the above case let us consider following example-

EXAMPLE 72.: Suppose, on a flat and rough horizontal fixed plane, a block of mass $20 kg$ is connected with a massless. string as shown in adjoining figure. The coefficients of static and kinetic friction between the block and the rough plane surface are, $1/4$ and $1/5$ respectively. Now an external force F is applied on the block towards left. Find the force of friction (f) and tension (T) in the string if- (a) $F = 10 N$, (b) $F = 50 N$, (c) $F = 65 N$. Take $g = 10 m/s^2$.

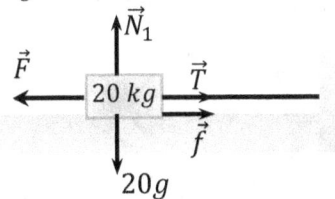

APPROACH First of all, calculate the maximum possible force of Friction which the surfaces can offer on the block, by using the relation $F_{max} = \mu_s N_1$.
After that, compare the applied force of friction with $f_{max} (= \mu_s N_1)$ and use the above table to find the required force of friction (f) and tension (T) in the string.
SOLUTION Maximum possible force of friction between the surfaces,
$f_{max} = \mu_s N_1$,
Here, $N_1 = mg$ is the normal reaction applied by horizontal plane surface on the block, therefore,
$f_{max} = \mu_s N_1 = \mu_s mg = \frac{1}{4} \times 20 kg \times 10 m/s^2 = 50 N$,
So, for $F \leq 50N$, the required force of friction $f = F$ and tension in the string $T = 0$;
and for $F > 50$ newton, the required force of friction $f = f_{max} = 50$ newton and the tension in the string $T = F - \mu_s N_1$, i.e., $T = F - 50 N$.
(a) The given external force, $F = 10$ (for it, the condition $F \leq \mu_s N_1$, satisfies), therefore, force of friction
$f = F = 10 N$
and the tension in the string, $T = 0 N$

(b) In this case, $F = 50 N$ (for it, the condition $F \leq \mu_s N_1$ satisfies), therefore, force of friction $f = F = 50 N$ and the tension in the string, $T = 0$ newton
(c) In this case, $F = 65 N$ (for it, the condition $F > \mu_s N_1$ satisfies), therefore, force of friction
$f = \mu_s N_1 = 50 N$ and the tension in the string, $T = F - \mu_s N_1 = 65 - 50 = 15 N$.

EXAMPLE 73.: The static and kinetic friction coefficient between the two blocks shown in adjoining figure are μ_s and μ_k respectively, but the floor is smooth. (a) What maximum horizontal force F can be applied without disturbing the equilibrium of the system? (b) Suppose the horizontal force applied is double of that found in part (a). Find the accelerations of the two masses.

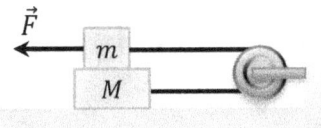

(a) APPROACH Apply, the condition of limiting equilibrium between the two blocks connected with string, and solve for F.

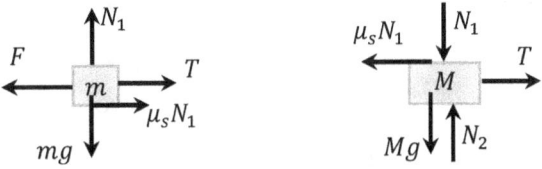

SOLUTION For maximum horizontal force, without disturbing the equilibrium of the system, the required condition is
$\Sigma F = 0$
or $\quad F - \mu_s N_1 - T = 0 \quad \ldots (1)$
Here, N_1 is the normal reaction on upper block by lower block and T is the tension in string in limiting equilibrium.
And for vertical equilibrium of the upper block, we have
$N_1 = mg \quad \ldots (2)$
Substituting this value of N_1, in Eq. (1), we get
$F - \mu_s(mg) - T = 0 \quad \ldots (3)$
Now, for the horizontal equilibrium of lower block, we have
$T - \mu_s N_1 = 0 \quad$ or $\quad T = \mu_s N_1$
or $\quad T = \mu_s mg \quad \ldots (4)$
$[\because$ from (2), $N_1 = mg]$
Substituting, this value of T, in Eq. (3), we get
$F - \mu_s(mg) - \mu_s mg = 0$
or $\quad F = 2\mu_s mg$

(B) APPROACH When the force is increased above the value obtained in above part, then both blocks start moving in opposite directions with same magnitude of acceleration (= a, say). In this case, the force of friction will be kinetic (= $\mu_k N_1$) in nature. Apply, Newtons second law, $\sum F = ma$ for both the blocks and solve for acceleration a.

SOLUTION When, the horizontal force applied is double of that found in above part (a), then from FBD of upper block of mass m, Newton's second law gives-

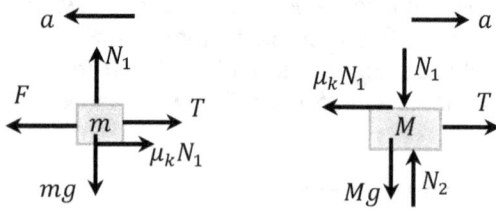

$$\sum F = ma$$

or $\quad 2F - T - \mu_k mg = ma \quad \ldots (5)$

$\because \quad F = 2\mu_s mg$

∴ from Eq. (5), we have

$$4\mu_s mg - T - \mu_k mg = ma \quad \ldots (6)$$

and from the FBD of the lower block, we have-

$$T - \mu_k N_1 = Ma$$

or $\quad T - \mu_k mg = Ma \quad [\because N_1 = mg] \quad \ldots (7)$

On adding Eq. (6) and (7), we get-

$$4\mu_s mg - 2\mu_k mg = (M + m)a$$

or $\quad a = \frac{2(2\mu_s - \mu_k)mg}{(M+m)}$

Both blocks move with this acceleration 'a' in opposite direction.

EXAMPLE 74.: Block A in the figure weighs 0.4 kg and block B weighs 0.5kg. The coefficient of sliding friction between all surfaces is 0.25 (a) find the force F necessary to drag block B to the left at constant speed. (b) find the tension in the string.

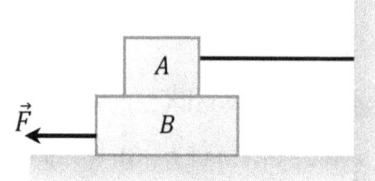

SOLUTION For A as there is no motion along horizontal or vertical. Let m_1 be mass of A

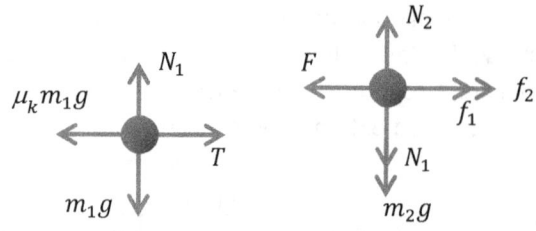

$N_1 - m_1 g = 0 \Rightarrow N_1 = m_1 g$
and $T - \mu_k N_1 = 0 \Rightarrow T = \mu_k N_1$
$\Rightarrow T = \mu_k m_1 g = 0.25 \times 0.4 \times 10 = 1N$

For B as there is no motion along vertical. Let m_2 be mass of B

$N_2 = N_1 + m_2 g \Rightarrow N_2 = (m_1 + m_2)g$

As there is no acceleration along x-axis

$F - f_1 - f_2 = 0$
$\Rightarrow F = f_1 + f_2 = \mu_k N_1 + \mu_k N_2 = \mu_k g(2m_1 + m_2)$
$\quad = 0.25 (0.8 + 0.5)10 = 3.25 \, N$

EXAMPLE 75. A body is moving down along an inclined plane of angle of inclination θ. The coefficient of friction between the body and the plane varies as $\mu = 0.5 \, x$, where x is the distance moved down the plane. The body will have the maximum velocity when it has travelled a distance x given by

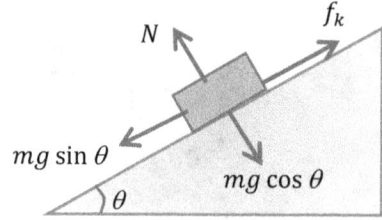

(A) $x = 2 \tan \theta$ \qquad (B) $x = \frac{2}{\tan \theta}$

(C) $x = \sqrt{2} \cot \theta$ \qquad (D) $x = \frac{\sqrt{2}}{\cot \theta}$

SOLUTION (A) $N = mg \cos\theta$, $f_k = \mu_k N$
Applying Newton's second law along the inclined surface, we get
$\sum F = ma$
$\Rightarrow mg \sin \theta - f_k = ma$
$\Rightarrow mg \sin \theta - \mu_k N = ma$
$\Rightarrow mg \sin \theta - 0.5x \, mg \cos \theta = ma$
$\Rightarrow a = g \sin \theta - 0.5 \, xg \cos \theta$
Now $v^2 = u^2 + 2as$

$\Rightarrow v^2 = 0^2 + 2(g \sin \theta - 0.5xg \cos \theta)x$

$\Rightarrow 2v \frac{dv}{dx} = [2g \sin \theta - 0.5g \cos \theta \, 2x]$

For maximum velocity, $\frac{dv}{dx} = 0$

$\Rightarrow 2g \sin \theta - gx \cos \theta = 0$

$\Rightarrow x = 2 \tan \theta$

Alternate Method:

Velocity will be maximum when $f_{k\,max} = \mu N = mg \sin\theta$
$0.5x - mg\cos\theta = mg\sin\theta$
$x = 2\tan\theta$

EXAMPLE 76: A block of mass m is on an inclined plane of angle θ. The coefficient of friction between the block and the plane is μ and $\tan\theta > \mu$. The block is held stationary by applying a force P parallel to the plane. The direction of force pointing up the plane is taken to be positive. As P varied from $P_1 = mg(\sin\theta - \mu\cos\theta)$ to $P_2 = mg(\sin\theta + \mu\cos\theta)$, the frictional force f versus P graph will look like

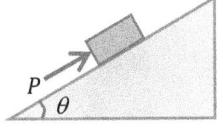

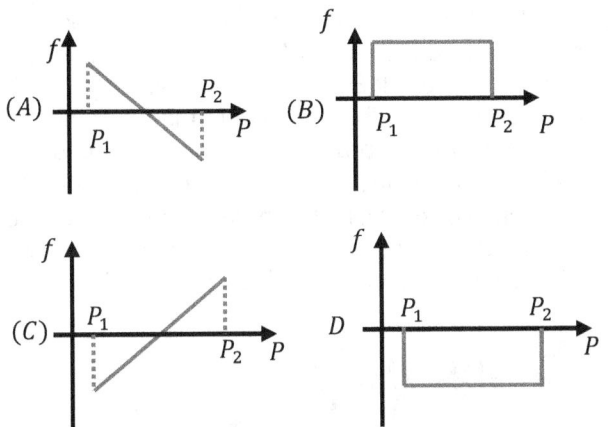

SOLUTION. (A) Initially the frictional force is upwards as P increases frictional force decreases.

43. CHECKPOINT 9

1. • When a horse pulls a cart, the force exerted by the horse on the cart is always equal and opposite to the force exerted by the cart on the horse. Is this statement true? If yes, why does the cart move?

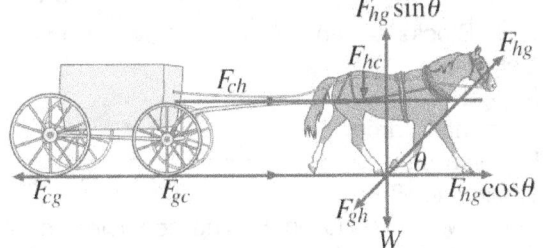

2. •• If motor M exerts a force of $F = (10t^2 + 100)$ N on the cable, where t is in seconds, determine the velocity of the 25-kg crate when $t = 4$ s. The coefficients of static and kinetic friction between the crate and the plane are $\mu_s = 0.3$ and $\mu_k = 0.25$, respectively. The crate is initially at rest.

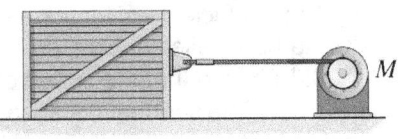

3. •• The motor winds in the cable with a constant acceleration, such that the 20-kg crate moves a distance s = 6 m in 3 s, starting from rest. Determine the tension developed in the cable. The coefficient of kinetic friction between the crate and the plane is $\mu_k = 0.3$

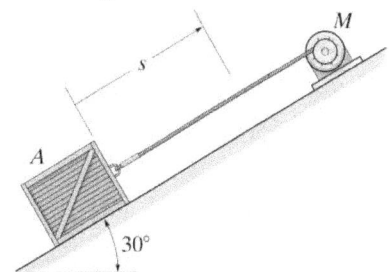

4. •• The conveyor belt is designed to transport packages of various weights. Each 10-kg package has a coefficient of kinetic friction $\mu_k = 0.15$. If the speed of the conveyor is 5 m/s, and then it suddenly stops, determine the distance the package will slide on the belt before coming to rest.

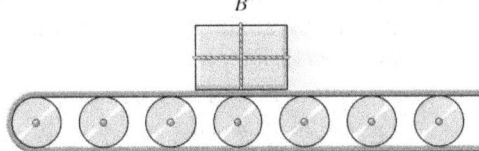

5. •• The 75-kg man pushes on the 150-kg crate with a horizontal force F. If the coefficients of static and kinetic friction between the crate and the surface are $\mu_s = 0.3$ and $\mu_k = 0.2$, and the coefficient of static friction between the man's shoes and the surface is $\mu_s = 0.8$, show that the man is able to move the crate. What is the greatest acceleration the man can give the crate?

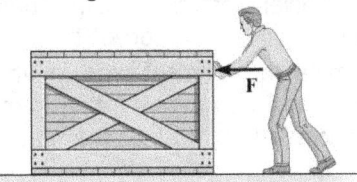

6. ••• The conveyor belt delivers each 12-kg crate to the ramp at A such that the crate's speed is $v_A = 2.5$ m/s, directed down along the ramp. If the coefficient of kinetic friction between each crate and the ramp is $\mu_k = 0.3$, determine the smallest incline θ of the ramp so that the crates will slide off and fall into the cart.

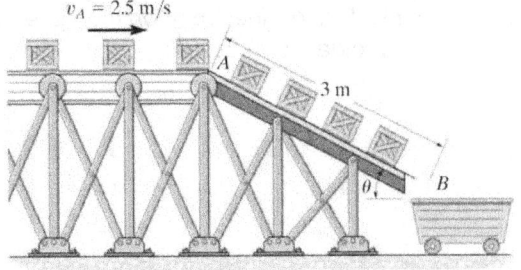

7. ••• A 60-kg suitcase slides from rest 5 m down the smooth ramp. Determine the distance R where it strikes the ground at B. How long does it take to go from A to B?

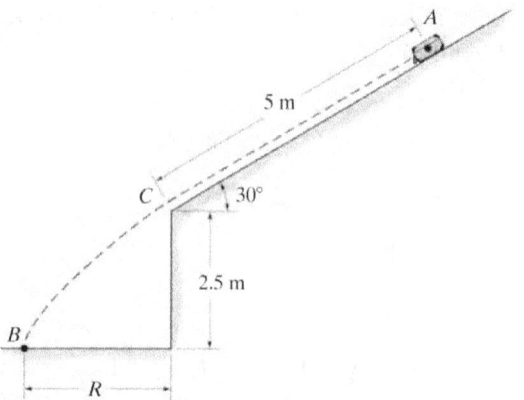

8. ••• Solve above Problem if the suitcase has an initial velocity down the ramp of $v_A = 2$ m/s, and the coefficient of kinetic friction along AC is $\mu_k = 0.2$.

9. •• The conveyor belt is moving downward at 4 m/s. If the coefficient of static friction between the conveyor and the 15-kg package B is $\mu_s = 0.8$, determine the shortest time the belt can stop so that the package does not slide on the belt.

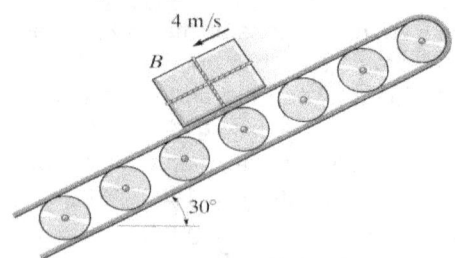

10. •• The coefficient of static friction between the 200-kg crate and the flat bed of the truck is $\mu_s = 0.3$. Determine the shortest time for the truck to reach a speed of 60 km/h, starting from rest with constant acceleration, so that the crate does not slip.

11. ••• The 10-kg block A rests on the 50-kg plate B in the position shown. Neglecting the mass of the rope and pulley, and using the coefficients of kinetic friction indicated, determine the time needed for block A to slide 0.5 m *on the plate* when the system is released from rest.

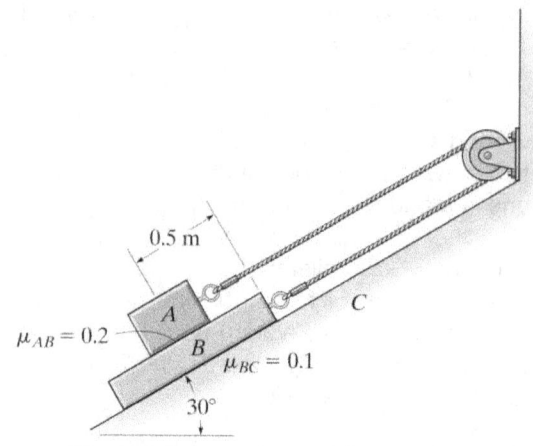

12. ••• Block A has a mass m_A and is attached to a spring having a stiffness k and unstretched length l_0. If another block B, having a mass m_B, is pressed against A so that the spring deforms a distance d, show that for separation to occur it is necessary that $d > 2\mu_k g(m_A + m_B)/k$, where μ_k is the coefficient of kinetic friction between the blocks and the ground. Also, what is the distance the blocks slide on the surface before they separate?

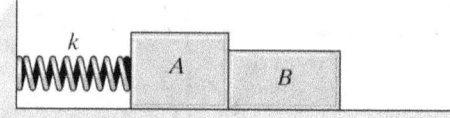

13. •• If the force exerted on cable AB by the motor is $F = (100t^{3/2})$N, where t is in seconds, determine the 50-kg crate's velocity when $t = 5s$. The coefficients of static and kinetic friction between the crate and the ground are $\mu_s = 0.4$ and $\mu_k = 0.3$, respectively. Initially the crate is at rest.

14. ••• Blocks A and B each have a mass m. Determine the largest horizontal force P which can be applied to B so that A will not slip on B. The coefficient of static friction between A and B is μ_s. Neglect any friction between B and C.

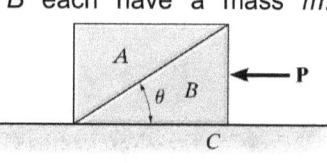

15. •• The coefficients of friction between the load and the flatbed trailer shown are $\mu_s = 0.40$ and $\mu_k = 0.30$. Knowing that the speed of the rig is 72 km/h, determine the shortest distance in which the rig can be brought to a stop if the load is not to shift.

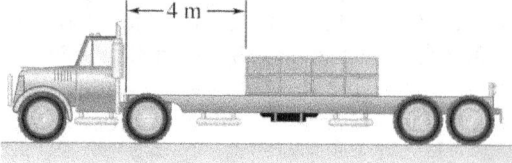

LAWS OF MOTION AND FRICTION 89

16. ••• A small package is deposited by the conveyor belt onto the 30° ramp at A with a velocity of 0.8 m/s. Calculate the distance s on the level surface BC at which the package comes to rest. The coefficient of kinetic friction for the package and supporting surface from A to C is 0.30.

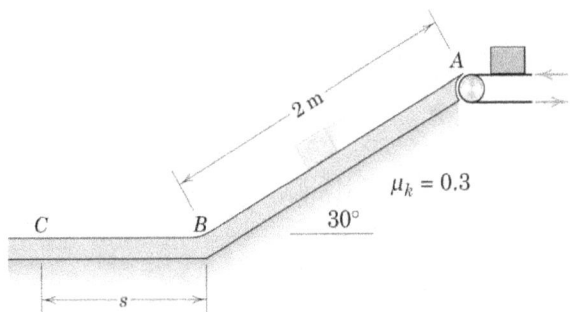

17. ••• Figure shows a cable-car system. built on stilts to keep the passenger compartments level as they go up and down the steep hillsides. When one car ascends, the other descends. The cars use a two-cable arrangement to compensate for friction; one cable passing around a large pulley connects the cars, the second is pulled by a small motor. Suppose the mass of both cars (with passengers) is 1500 kg, the coefficient of rolling friction is 0.020, and the cars move at constant speed. What is the tension in (a) the connecting cable and (b) the cable to the motor? ($\sin 35° = 0.57$)

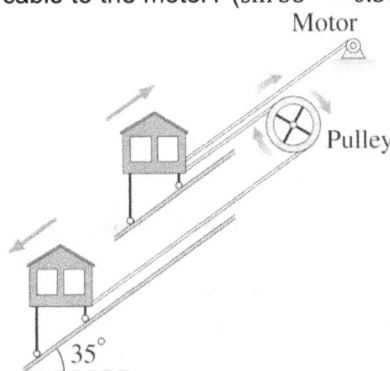

18. ••• In the arrangement shown in figure pulley P can move whereas other two pulleys are fixed. All of them are light. String is light and inextensible. The coefficient of friction between 2 kg and 3 kg block is $\mu_1 = 0.75$ and that between 3 kg block and the table is $\mu_2 = 0.5$. The system is released from rest
(i) Find maximum value of mass M, so that the system does not move. Find friction force between 2 kg and 3 kg blocks in this case.
(ii) If M = 4 kg, find the tension in the string attached to 2 kg block.

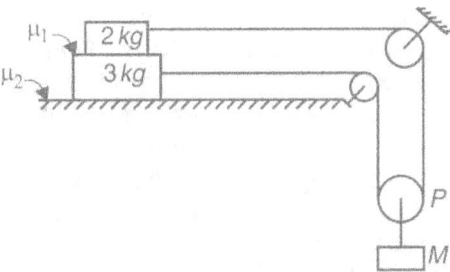

19. •• Block A of mass M is placed on an incline plane, connected to a string, passing over a pulley as shown in the fig. The other end of the string also carries a block B of mass M. The system is held in the position shown such that triangle APQ lies in a vertical plane with horizontal line AQ in the plane of the incline surface. Find the minimum coefficient of friction between the incline surface and block A such that the system remains at rest after it is released. Take $\theta = \alpha = 45°$.

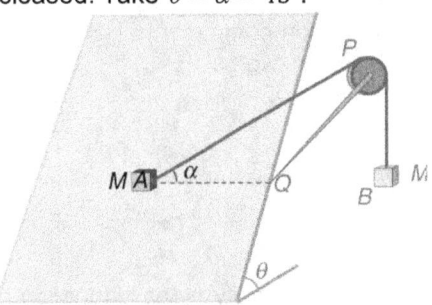

44. DRAG FORCES

44.1. SMALL OBJECTS

When a small object moves at a low speed v through a viscous medium, it experiences a resistive drag force $\vec{F}_D$ that opposes its motion. In such situations, the force has a magnitude given by:
$$F_D = bv \qquad \ldots (1)$$
where b is a proportionality constant that depends on the properties of the medium and on the shape of the object, and b has the units kg/s. Our goal is to find the velocity of the falling object as a function of the time.

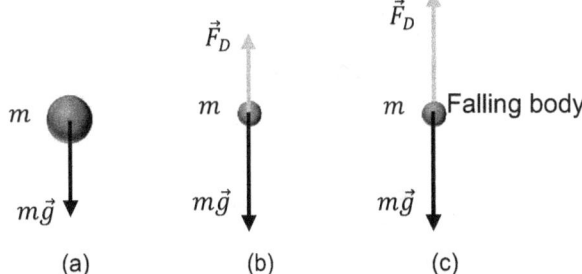

FIGURE 1 Forces acting on a body falling in air. (a) At the instant of release, $v_y = 0$ and there is no drag force.
(b) The drag force increases as the body gains speed.
(c) Eventually the drag force equals the weight; for all later times it remains equal to the weight and the body falls at its constant terminal speed.

Suppose a sphere of mass m and weight $W = mg$ is released from rest in air as in FIGURE 1a. Figure 1 also shows the free-body diagram, which changes with time as the object falls. When the object is released, $F_D = 0$ (because $v_y = 0$). As it falls farther, its speed and hence the drag force F_D increases. As F_D continues to increase, at some point it will equal the weight of the object, and at that point there is no net force acting on the object; its acceleration is zero, and the object falls with a *constant* speed. The speed at which the exact balance between the upward drag force and the downward weight force causes an object to fall without acceleration is called the *terminal speed*. Once an object has reached terminal speed, it will continue falling at that speed until it hits the ground.

We choose the y axis to be vertical and the positive direction to be downward. (The choice of direction is arbitrary, and here it is convenient to work with positive velocity and acceleration components.). By application of Newton's second law, we get

$$mg - F_D = ma_y$$

or $$mg - bv_y = ma_y$$

or $$a_y = g - \frac{bv_y}{m} \qquad \ldots (2)$$

Our goal is to find the velocity v_y as a function of the time. We begin by substituting $a_y = \frac{dv_y}{dt}$ in equation (1), we get

$$\frac{dv_y}{dt} = g - \frac{bv_y}{m}$$

or $$\frac{dv_y}{g - \frac{bv_y}{m}} = dt \qquad \ldots (3)$$

With $v_y = 0$ at time $t = 0$, we seek the velocity v_y at time t. We can therefore integrate the left side of Eq. 3 from velocity 0 to v_y and the right side from time 0 to t, i.e.,

$$\int_0^{v_y} \frac{dv_y}{g - \frac{bv_y}{m}} = \int_0^t dt$$

On integrating above equation, we get

$$-\frac{m}{b}\left[\ln\left(g - \frac{bv_y}{m}\right)\right]_0^{v_y} = [t]_0^t$$

or $$-\frac{m}{b}\left[\ln\left(\frac{mg - bv_y}{m}\right) - \ln g\right] = t$$

or $$-\frac{m}{b}\ln\left(\frac{mg - bv_y}{mg}\right) = t$$

or $$\frac{mg - bv_y}{mg} = e^{-\frac{bt}{m}}$$

or $$mg - bv_y = mge^{-\frac{bt}{m}}$$

or $$v_y = \frac{mg}{b}\left(1 - e^{-\frac{bt}{m}}\right) \qquad \ldots (4)$$

This is the expression for the velocity as a function of time.

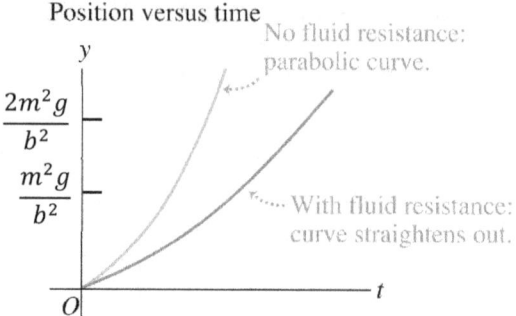

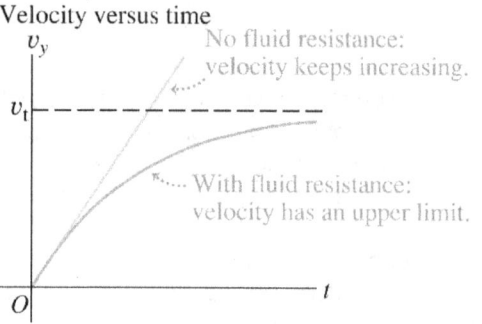

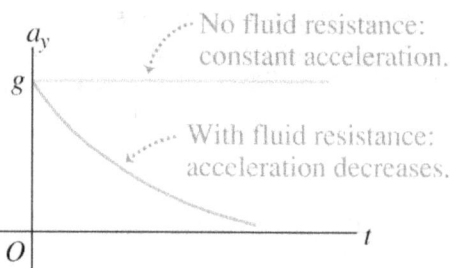

FIGURE 2. Position, velocity, and acceleration for a falling body subject to a drag force. Note that the acceleration starts at g and falls to zero; the velocity starts at zero and approaches v_t. Note also that $y(t)$ becomes nearly linear at large t, as we expect for motion with constant velocity.

It is interesting to examine this result in the two limiting cases of small and large values of t. The velocity starts with $v_y = 0$ at $t = 0$ Just after near the beginning of the projectile's fall, we can find the velocity by approximating the exponential function using $e^{-x} \approx 1 - x$ for small $x (\ll 1)$ This gives

$$v_y = \frac{mg}{b}\left[1 - \left(1 - \frac{bt}{m}\right)\right] = gt \quad \text{(small } t\text{)} \qquad \ldots (5)$$

which agrees with Equation $v_y = v_{0y} + gt$, when $v_{0y} = 0$ (recalling that here we chose the positive y direction to be downward). Early in the motion, when the velocity is small and the drag force has not increased significantly, the object is approximately in free fall.

For large t, the exponential approaches zero ($e^{-x} \to 0$ as $x \to \infty$) and the magnitude of the velocity approaches the *terminal speed* given by

$$v_t = \frac{mg}{b} \qquad \ldots (6)$$

We can also find the terminal speed directly from Eq. 2 —when the speed increases to the point at which the drag force and the weight are equal, $a_y = 0$ and Eq. 2 then gives Eq. 6

We see that, just as we expect, the larger is the drag force coefficient b, the smaller is the terminal speed. The terminal speed of a pebble falling in water is less than that of the same pebble falling in air, because the drag coefficient is much larger in water.

Now that we have an expression for $v_y(t)$, we can differentiate it to find $a_y(t)$ or integrate it to find $y(t)$. Figure 2 shows the time dependence of y, v_y, and a_y. We leave the derivations for you to complete; the results are

$$a_y = g e^{-(b/m)t} \qquad \ldots (7)$$

$$y = v_t \left[t - \frac{m}{b} \left(1 - e^{-(b/m)t} \right) \right] \qquad \ldots (8)$$

A drag force proportional to v is representative of *viscous drag*, which is the force that might be experienced by a small particle falling through a thick fluid. Large objects in air experience *aerodynamic drag*, in which F_D is proportional to v^2. This case is more complicated mathematically, but it also yields a terminal speed (different from the terminal speed calculated for $F_D \propto v$).

Table 1 Some Terminal Speeds in Air

Object	Terminal Speed (m/s)
16-lb (7.263 kg) shot	145
Skydiver (typical)	60
Baseball	42
Tennis ball	31
Basketball	20
Ping-Pong ball	9
Raindrop (radius 1.5 mm)	7
Parachutist (typical)	5

44.1.1. Projectile Motion with Air Resistance

Drag calculations are also important for two-dimensional projectile motion. A baseball, for example, leaves the bat with a speed of about 100 mi/h or 45 m/s. This is already greater than its terminal speed in air when dropped from rest (Table 1). The magnitude of the drag force can be estimated from our previous calculation. From Eq. 6 we see that the constant b is the weight mg of the baseball (about 1.4 N, corresponding to a mass of 0.14 kg) divided by its terminal speed, 42 m/s. Thus $b = 0.033$ N/(m/s). If the ball travels at 45 m/s, it experiences a drag force bv with a magnitude of about 1.5 N, which is greater than its weight and therefore has a substantial effect on its motion.

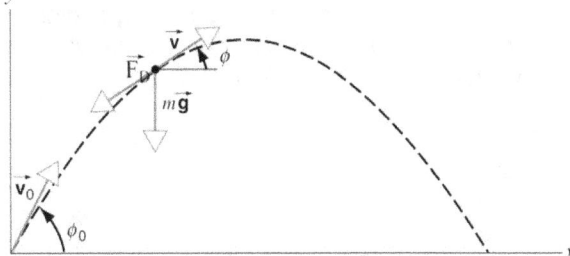

FIGURE 3. A projectile in motion. It is launched with velocity v_0 at an angle ϕ_0 with the horizontal. At a certain time later its velocity is $\vec{v}$ at the angle ϕ. The weight and the drag force (which always points in a direction opposite to $\vec{v}$) are shown at that time.

Figure 3 shows the free-body diagram at a particular point in the baseball's trajectory. Like all frictional forces, $\vec{F}_D$ is in a direction opposite to $\vec{v}$, and we assume no wind is blowing. If we take $\vec{F}_D = -b\vec{v}$ we can use Newton's laws to find an analytic solution for the trajectory, an example of which is illustrated in Fig. 4. When air resistance is taken into account, the range is reduced from 179 m to 72 m and the maximum height from 78 m to 48 m. Note also that the trajectory is no longer symmetric about the maximum; the descending motion is much steeper than the ascending motion. For $\phi_0 = 60°$ the projectile strikes the ground at an angle of $-79°$, while in the absence of drag it would strike the ground at an angle equal to $-\phi_0 = -60°$.

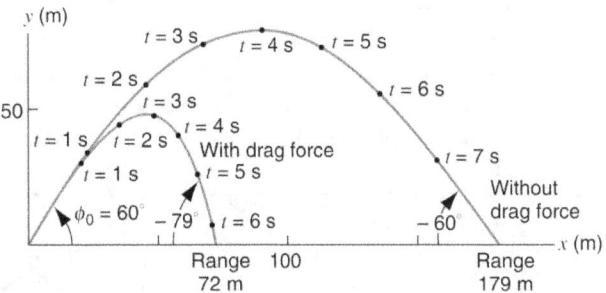

FIGURE 4 Projectile motion with and without a drag force, calculated for $v_0 = 45$ m/s and $\phi_0 = 60°$.

For other (and more realistic) choices for the drag force $\vec{F}_D$ the calculation must be done numerically.

44.2. LARGE OBJECTS

When a large object (such as a baseball, skydiver, or an airplane) moves at a high speed v in a medium (gas or liquid) of density ρ (mass per unit volume), it experiences a drag force ($\vec{F}_D$) that opposes the motion. From experiments, it was found that in these situations the magnitude F_D will be given by:

$$F_D = \tfrac{1}{2} C \rho A v^2 \quad \ldots (1)$$

where C is a dimensionless proportionality constant called the **drag coefficient**, and A is the effective cross sectional area of the object, taken to be perpendicular to its velocity $\vec{v}$. If v varies significantly, C can vary as well, but we ignore such complications.

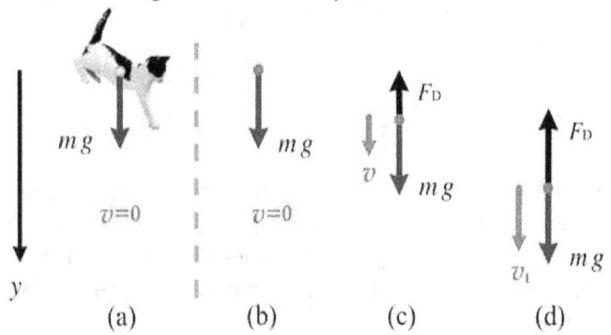

FIGURE 5. Part (a) shows a body (cat) when it has just begun to fall through air and part (b) shows its corresponding free-body diagram. (c) Later, the drag force F_D has developed. (d) F_D has increased until it balances mg and the body falls with constant terminal speed v_t.

If we assume a large body of mass m and weight $W = mg$ falls from rest in air, as shown in FIGURE 5, then application of Newton's second law $\sum \vec{F} = m\vec{a}$ in the vertical direction as in part (c) of the figure gives:

$mg - \tfrac{1}{2} C \rho A v^2 = ma$

$\Rightarrow mg - \tfrac{1}{2} C \rho A v^2 = m \dfrac{dv}{dt} \quad \ldots (2)$

From above equation, we can calculate the velocity of the object at any instant 't'.

By setting $a = \dfrac{dv}{dt} = 0$ in this equation, the terminal speed is given by:

$$v_t = \sqrt{\dfrac{2mg}{C \rho A}} \quad \ldots (3)$$

- ☞ For a small object, the drag force is directly proportional to its speed whereas for a larger object it is proportional to the square of speed.
- ☞ If v is the speed of projection of a body and v_t is its terminal speed, then
  $v = v_t \Rightarrow$ Upthrust = Weight of the body. In this case, the body moves with a constant speed
  $v < v_t \Rightarrow$ Upthrust < Weight of the body. In this case, initially speed of body increases until it becomes equal to v_t. After that it moves with constant speed v_t.
  $v > v_t \Rightarrow$ Upthrust > Weight of the body. In this case speed of the body decreases upto v_t, then body moves with constant speed v_t.

EXAMPLE 77. The terminal speed of a Styrofoam ball is $15\ m/s$. Suppose a Styrofoam ball is shot straight down with an initial speed of $30\ m/s$. Which velocity graph is correct?

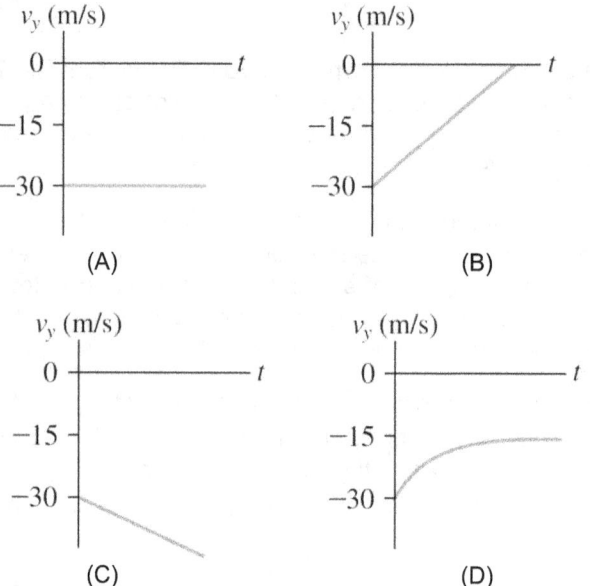

SOLUTION (D) The ball is shot *down* at $30\ m/s$, so $v_{0y} = -30$ m/s. This exceeds the terminal speed, so the upward drag force is *greater* than the downward weight force. Thus the ball *slows down* even though it is "falling." It will slow until $v_y = -15$ m/s, the terminal velocity, then maintain that velocity.

45. CHECKPOINT 10

1. ••A parachutist having a mass m opens his parachute from an at-rest position at a very high altitude. If the atmospheric drag resistance is $F_D = kv^2$, where k is a constant, determine his velocity when he has fallen for a time t. What is his velocity when he lands on the ground? This velocity is referred to as the terminal velocity, which is found by letting the time of fall $t \to \infty$.

2. •The drag coefficient for a spherical raindrop with a radius of 0.415 cm falling at its terminal velocity is $2.43 \times 10^{-5} kg/m$. Calculate the raindrop's terminal velocity in m/s.

3. ••What is the acceleration of a raindrop that has reached half of its terminal velocity? Give your answer in terms of g.
4. ••When a parachute opens, the air exerts a large drag force on it. This upward force is initially greater than the weight of the sky diver and, thus, slows him down. Suppose the weight of the sky diver is 915 N and the drag force has a magnitude of 1027 N. The mass of the sky diver is 93.4 kg. What are the magnitude and direction of his acceleration?

46. CENTRIPETAL & CENTRIFUGAL FORCES

If a body is moving with a constant speed in a circle, as seen from an inertial frame it is continuously accelerated towards the centre of rotation. The magnitude of the centripetal acceleration for a body moving with a tangential velocity v is given by v²/r. If the angular velocity of the body is ω then the centripetal acceleration is $mr\omega^2$.

- *A centripetal force accelerates a body by changing the direction of the body's velocity without changing the body's speed.*

According to Newton's law, force causes acceleration and so, the net centripetal force

$$F = ma = \frac{mv^2}{r} = m\omega^2 r$$

FIGURE 1

An object moving in a circular path may increase or decrease its speed. In such a case, the object has both an acceleration tangential to its path that changes its speed, $\vec{a}_t$, and a centripetal acceleration perpendicular to its path, $\vec{a}_c$, that changes its direction of motion. Such a situation is illustrated in **Figure 1**. The total acceleration, $\vec{a}$, of the object is the vector sum of $\vec{a}_t$ and $\vec{a}_c$.

$$\vec{a} = \vec{a_t} + \vec{a_c}$$

Magnitude of total acceleration is given by

$$a = \sqrt{a_c^2 + a_t^2}$$

CENTRIFUGAL FORCE is described as a force pulling *outward* on an object moving in a circular path. If you are feeling a "centrifugal force" on a rotating carnival ride, what is the other object with which you are interacting? You cannot identify another object because it is a **fictitious** force that occurs when you are in a noninertial reference frame.

47. FICTITIOUS FORCE IN A ROTATING SYSTEM

Suppose a block of mass m lying on a horizontal, frictionless turntable is connected to a string attached to the center of the turntable, as shown in FIGURE 1. According to an inertial observer, if the block rotates uniformly, it undergoes an acceleration of magnitude v^2/r, where v is its linear speed. The inertial observer concludes that this centripetal acceleration is provided by the force T exerted by the string and writes Newton's second law as $T = mv^2/r$.

But, according to a noninertial observer attached to the turntable, the block is at rest and its acceleration is zero. Therefore, he must introduce a fictitious outward force of magnitude mv^2/r to balance the inward force exerted by the string. According to him, the net force on the block is zero, and he writes Newton's second law as $T = mv^2/r$.

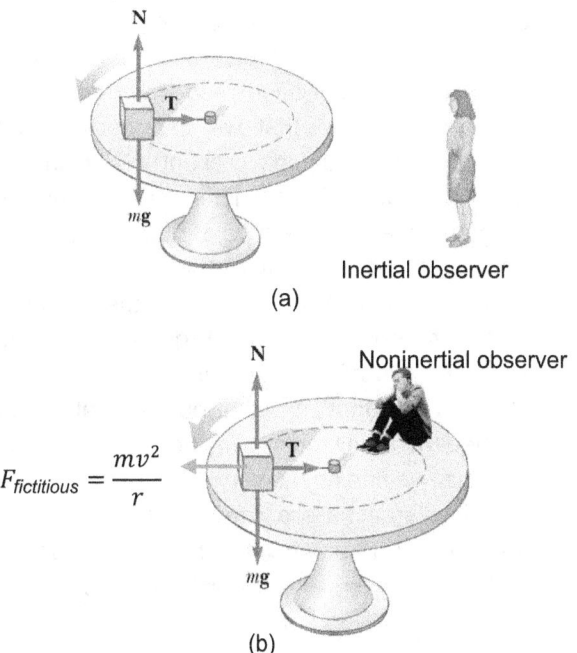

FIGURE 1 A block of mass m connected to a string tied to the center of a rotating turntable. (a) The inertial observer claims that the force causing the circular motion is provided by the force T exerted by the string on the block. (b) The noninertial observer claims that the block is not accelerating, and therefore he introduces a fictitious force of magnitude mv^2/r that acts outward and balances the force T.

EXAMPLE 78. *A 1200 kg automobile rounds a level curve of radius 200m, on an unbanked road with a velocity of 72 km/hr. What is the minimum coefficient of friction between the tyres and road in order that the automobile may not skid (g= 10m/s²)?*

SOLUTION: In an unbanked road the centripetal force is provided by the frictional force.

$$\therefore \quad f_{friction} = \frac{mv^2}{r}$$

But $f_{limiting\ friction} \geq f_{friction}$

or $\quad \mu mg \geq f_{friction}$

or $\quad \mu mg \geq \frac{mv^2}{r}$

$$\therefore \quad \mu_{min} = \frac{v^2}{gr} = \frac{20 \times 20}{10 \times 200} = 0.2$$

48. LIMITATIONS OF NEWTON'S LAWS

Galaxies and clusters of galaxies are often observed to rotate, and by observation we can deduce the speed of rotation. From this we can calculate the amount of matter that must be present in the galaxy or cluster for gravity to supply the centripetal force corresponding to the observed rotation. Yet the amount of matter that we actually observe with telescopes is far less than we expect. Therefore, it has been proposed that there is additional "dark matter" that we can not see with telescopes but that must be present to provide the needed gravitational force. There is as yet no convincing candidate for the type or nature of this dark matter, and so other explanations have been proposed for the apparent inconsistency between the amount of matter actually observed in the galaxies and the amount we think is needed to satisfy Newton's laws. One proposed explanation is that our calculations are incorrect because Newton's laws do not hold in the conditions that we find on the very large scale— that is, when the accelerations are very small (below a few times 10^{-10} m/s²). In particular, it has been proposed that for these very small accelerations, the force is proportional to a^2 instead of a.

Figure 1 shows the results of an experiment testing this supposition. If force depended on the acceleration to some power other than 1, the data would not fall on a straight line. From this extremely precise experiment, we conclude that down to accelerations of about 10^{-10} m/s², force is proportional to acceleration and Newton's second law holds.

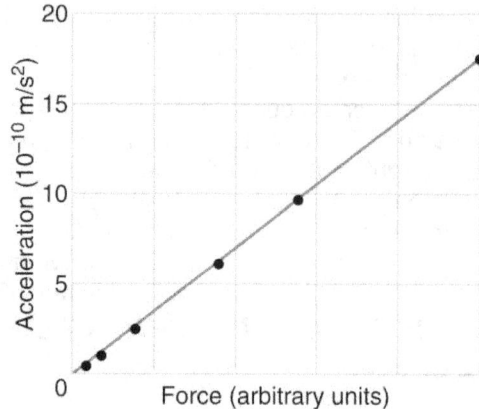

FIGURE 1. Results of an experiment to test whether Newton's second law holds for small accelerations below 109 m/s2. The straight line shows that acceleration is proportional to the applied force down to 1010 m/s2, and so Newton's law remains valid even at such small accelerations.

In the 20th century, we have experienced three other revolutionary developments: Einstein's special theory of relativity (1905), his general theory of relativity (1915), and quantum mechanics (in about 1925). Special relativity teaches us that we cannot extrapolate the use of Newton's laws to particles moving at speeds comparable to the speed of light. General relativity shows that we cannot use Newton's laws in the vicinity of extremely massive objects. Quantum mechanics teaches us that we cannot extrapolate Newton's laws to objects as small as atoms.

Special relativity, which involves a distinctly non-Newtonian view of space and time, can be applied under all circumstances, at both high speeds and low speeds. In the limit of low speeds, it can be shown that the dynamics of special relativity reduces directly to Newton's laws. Similarly, general relativity can be applied to weak as well as strong gravitational forces, but its equations reduce to Newton's laws for weak forces. Quantum mechanics can be applied to individual atoms, where a certain randomness of behavior is predicted, or to ordinary objects containing a huge number of atoms, in which case the randomness averages out to give Newton's laws once again.

49. EXERCISES AND QUESTIONS

49.1. CONCEPTUAL QUESTIONS

1. You can pull a wagon with a rope, but you can't push it with a rope. Is there such a thing as a "negative" tension?

2. While on a very smooth level transcontinental plane flight, your coffee cup sits motionless on your tray. Are there forces acting on the cup? If so, how do they differ from the forces that would be acting on the cup if it sat on your kitchen table at home?
3. When a car accelerates starting from rest, where is the force applied to the car in order to cause its acceleration? What object exerts this force on the car?
4. A block with mass m is supported by a cord C from the ceiling, and a similar cord D) is attached to the bottom of the block. Explain this: If you give a sudden jerk to D, it will break, but if you pull on D steadily, C will break.

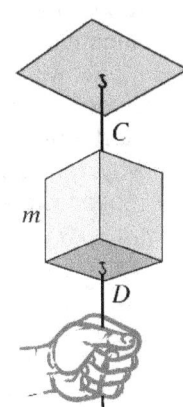

5. When a car stops suddenly, the passengers are "thrown forward" out of their seats (if they are not wearing seat belts). What causes this to happen? Are they really *thrown* forward?
6. Criticize the statement, often made, that the mass of a body is a measure of the "quantity of matter" in it.
7. Can an object exert a force on itself? Argue for your answer.
8. For medical reasons, it is important for astronauts in outer space to determine their body mass at regular intervals. Devise a scheme for measuring body mass in a zero-gravity environment.
9. Can Newton's first law be considered merely the special case $a = 0$ of the second law? If so, is the first law really needed? Discuss.
10. Although in everyday life the earth seems to be an inertial frame of reference, the paths of objects moving freely across the earth's surface, such as air masses, artillery shells, and even thrown baseballs, have a tendency to curve slightly to their right (in the Northern Hemisphere) as seen by an observer on the surface. Can you explain why this is?
11. A passenger in a bus notices that a ball which has been at rest in the aisle suddenly starts to roll toward the front of the bus. What can the passenger conclude about the motion of the bus from this observation?
12. What's the relation— if any— between the force acting on an object and the direction in which the object is moving?
13. Suppose you are in a car in a very dense fog and cannot see the meters on your dashboard or anything outside the car. (Maybe you should not be driving under these conditions!) (a) How can you tell if your car is speeding up or slowing down? (b) Can you tell if your car is at rest or is moving with uniform velocity?
14. It would be much easier to lift a bowling ball on the moon than on the earth. Would it be similarly easier to *catch* the bowling ball on the moon, if someone threw it to you? Explain.
15. If only a single nonzero force acts on an object, does the object accelerate relative to all inertial reference frames? Is it possible for such an object to have zero velocity in some inertial reference frame and not in another? If so, give a specific example.
16. Two students try to break a rope. First they pull against each other and fail. Then they tie one end to a wall and pull together. Is this procedure better than the first? Explain your answer.
17. Comment on whether the following pairs offerees are examples of action-reaction: (a) The Earth attracts a brick; the brick attracts the Earth, (h) A propellered airplane pushes air toward the tail; the air pushes the plane forward, (c) A horse pulls forward on a cart, moving it; the cart pulls backward on the horse, (d) A horse pulls forward on a cart without moving it; the cart pulls back on the horse, (e) A horse pulls forward on a cart without moving it; the Earth exerts an equal and opposite force on the cart. (*f*) The Earth pulls down on the cart; the ground pushes up on the cart with an equal and opposite force.
18. If your hands are wet and no towel is handy, you can remove excess water by shaking them. Use Newton's laws to explain why doing this gets rid of the water.
19. It is possible to play catch with a softball in an airplane in level flight just as though the plane were at rest. Is this still possible when the plane is making a turn?
20. Newton's third law tells us that if you push a box with a 15 N force, it pushes back on you with a 15 N force. How can you ever accelerate this box if it always pushes back with the same force you exert on it?
21. Comment on the following statements about mass and weight taken from examination papers, *(a)* Mass and weight are the same physical quantities expressed in different units. *(b)* Mass is a property of one object alone, whereas weight results from the interaction of two objects, (c) The weight of an object is proportional to its mass, {d) The mass of a body varies with changes in its local weight
22. There is a limit beyond which further polishing of a surface increases rather than decreases frictional resistance. Explain why.
23. A crate, heavier than you are, rests on a rough floor. The coefficient of static friction between the crate and the floor is the same as that between the

soles of your shoes and the floor. Can you push the crate across the floor?

24. What is the purpose of curved surfaces, called spoilers, placed on the rear of sports cars? They are designed so that air flowing past exerts a downward force.
25. A horizontal force acts on a body that is free to move. Can it produce an acceleration if the force is less than the weight of that body?
26. Why does the acceleration of a freely falling object not depend on the weight of the object?
27. Describe several ways in which you could, even briefly, experience weightlessness.
28. Under what circumstances would your weight be zero? Does your answer depend on the choice of a reference system?
29. Two surfaces are in contact but are at rest relative to each other. Nevertheless, each exerts a force of friction on the other. Explain
30. The following statement is true; explain it. Two teams are having a tug of war; the team that pushes harder (horizontally) against the ground wins.
31. The owner's manual of a car suggests that your seat belt should be adjusted "to fit snugly" and that the front seat head rest should *not* be adjusted so that it fits comfortably at the back of your neck but so that "the top of the head rest is level with the top of your ears." How do Newton's laws support these good recommendations?
32. You shoot an arrow into the air and you keep your eye on it as it follows a parabolic flight path to the ground. You note that the arrow turns in flight so that it is always tangent to its flight path. What makes it do that?
33. A massless rope is strung over a frictionless pulley. A monkey holds onto one end of the rope and a mirror, having the same weight as the monkey, is attached to the other end of the rope at the monkey's level. Can the monkey get away from its image seen in the mirror (a) by climbing up the rope, (b) by climbing down the rope, or (c) by releasing the rope?
34. Would a spring scale carried to the Moon give accurate results if the scale had been calibrated on Earth, (a) in pounds, or (b) in kilograms?
35. In a tug of war, three men pull on a rope to the left at A and three men pull to the right at B with forces of equal magnitude. Now a 5-N weight is hung vertically from the center of the rope, (a) Can the men get the rope AB to be horizontal? (b) If not, explain. If so, determine the magnitude of the forces required at A and B to do this.
36. Here is an interesting experiment that you can perform at home: take a wooden block and rest it on the floor or some other flat surface. Attach a rubber band to the block and pull gently on the rubber band in the horizontal direction. Keep your hand moving at constant speed. At some point, the block will start moving, but it will not move smoothly. Instead, it will start moving, stop again, start moving again, stop again, and so on. Explain why the block moves this way. (The start-stop motion is sometimes called "stick-slip" motion.)
[Hint: As the spring is extended, the force exerted by the spring on the block increases. Once that force exceeds the maximum value of the force of static friction, the block will slip. As it does, it will shorten the length of the spring, decreasing the force that the spring exerts. The force of kinetic friction then slows the block to a stop, which starts the cycle over again.]
37. Viewed from an inertial reference frame, an object is seen to be moving in a circle. Which, if any, of the following statements must be true. (a) A nonzero net force acts on the object. (b) The object cannot have a radially outward force acting on it. (c) At least one of the forces acting on the object must point directly toward the center of the circle.
38. The force of gravity on a 2-kg rock is twice as great as that on a 1-kg rock. Why then doesn't the heavier rock fall faster?
39. Your car skids across the centerline on an icy highway. Should you turn the front wheels in the direction of the skid or in the opposite direction (a) when you want to avoid a collision with an oncoming car and (b) when no other car is near but you want to regain control of the steering? Assume rearwheel drive, then front-wheel drive.
40. An elevator is supported by a single cable. There is no counterweight. The elevator receives passengers at the ground floor and takes them to the top floor, where they disembark. New passengers enter and are taken down to the ground floor. During this round trip, when is the tension in the cable equal to the weight of the elevator plus passengers? Greater? Less?
41. You stand on the large platform of a spring scale and note your weight. You then take a step on this platform and note that the scale reads less than your weight at the beginning of the step and more than your weight at the end of the step. Explain.
42. Could you weigh yourself on a scale whose maximum reading is less than your weight? If so, how?

43. A weight is hung by a cord from the ceiling of an elevator. From the following conditions, choose the one in which the tension in the cord will be greatest ... least: (a) elevator at rest; (b) elevator rising with uniform speed; (c) elevator descending with decreasing speed; (d) elevator descending with increasing speed.
44. A woman stands on a spring scale in an elevator. In which of the following cases will the scale record the minimum reading ... the maximum reading: (a) elevator stationary; (b) elevator cable breaks, free fall; (c) elevator accelerating upward; (d) elevator accelerating downward; (e) elevator moving at constant velocity?
45. What conclusion might a physicist draw if unequal masses hung over a pulley inside an elevator remain balanced; that is, there is no tendency for the pulley to turn?
46. Can you think of physical phenomena involving the Earth in which the Earth cannot be treated as a particle?
47. When a golf ball is dropped to the pavement, it bounces back up. (a) Is a force needed to make it bounce back up? (b) If so, what exerts the force?
48. (a) Why do you push down harder on the pedals of a bicycle when first starting out than when moving at constant speed? (b) Why do you need to pedal at all when cycling at constant speed?
49. In a science-fiction movie, the hero must get from one spaceship to another, which is several kilometers away. The hero (wearing the appropriate boots) stands on his ship and runs along its surface toward the other ship. When he reaches the edge of his ship, he jumps toward the other ship. Assuming he is far from any major sources of gravity such as planets and friction is negligible. Does he arrive with a lower speed, higher speed or about the same speed at which he left his ship? Explain.
50. A ball hanging from a light string or rod can be used as an accelerometer (a device that measures acceleration) as shown in adjoining figure. What force causes the deflection of the ball? Is the cart in the lower part of the photo an inertial reference frame? How can the ball's deflection be used to find the cart's acceleration? In which direction is the cart accelerating? Explain your answers

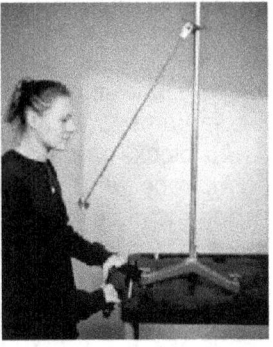

51. Suppose the cart in Figure P5.18 is moving at constant velocity in a strong wind. What force causes the deflection of the ball? Is the cart an inertial reference frame? Explain.
52. You are riding a luxury bus. In front of you is a cup of tea resting on the seat-back tray. Which of the following events may lead to spilled tea in your lap? The bus (a) remains at rest, (b) moves at constant velocity, (c) speeds up or (d). slows down. Don't worry about other circumstances such as a person knocking your cup over. (More than one choice may be correct.) Explain your answers.
53. Here is a story of a horse and a farmer: One day, the farmer attaches a heavy cart to the horse and demands that the horse pull the cart. "Well," says the horse, "I cannot pull the cart, because, according to Newton's third law, if I apply a force to the cart, the cart will apply an equal and opposite force on me. The net result will be that I cannot pull the cart, since all the forces will cancel. Therefore, it is impossible for me to pull this cart." The farmer was very upset! What could he say to persuade the horse to move?
54. (a) A rock and a feather held at the same height above the ground are simultaneously dropped. During the first few milliseconds following release, the drag force on the rock is smaller than the drag force on the feather, but later on during the fall the *opposite* is true. Explain. (b) In light of this result, explain how the rock's acceleration can be so obviously larger than that of the feather. *Hint:* Draw a free-body diagram of each object.

49.2. PROBLEMS

1. •• A certain orthodontist uses a wire brace to align a patient's crooked tooth as in adjoining figure. The tension in the wire is adjusted to have a magnitude of 18.0 N. Find the magnitude of the net force exerted by the wire on the crooked tooth.

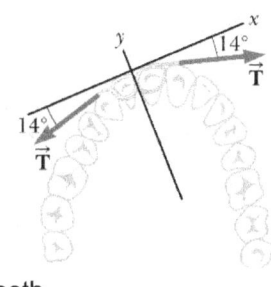

2. • Figure shows an acceleration-versus-force graph for three objects pulled by rubber bands. The mass of object 2 is 0.20 kg. What are the masses of objects 1 and 3? Explain your reasoning.

3. • FIGURE shows an acceleration-versus-force graph for a 200 g object. What force values go in the blanks on the horizontal scale?

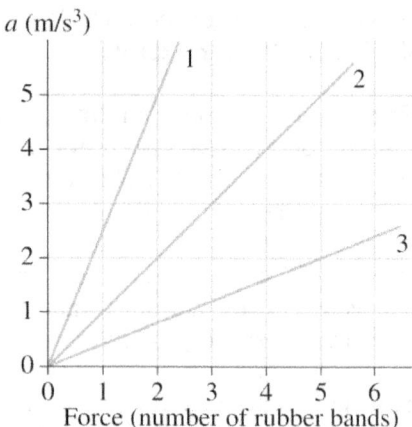

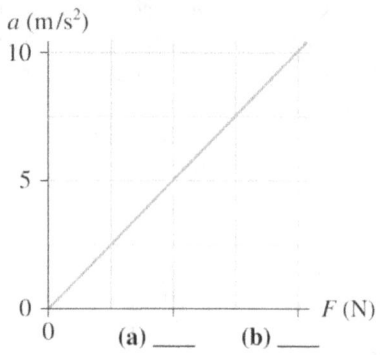

4. • FIGURE shows an acceleration-versus-force graph for a 500 g object. What acceleration values go in the blanks on the vertical scale?

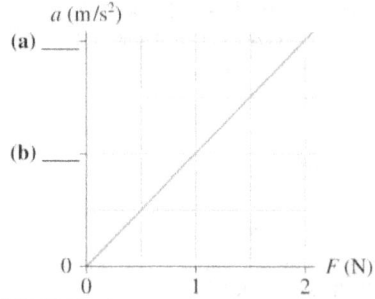

5. • FIGURE shows the acceleration of objects of different mass that experience the same force. What is the magnitude of the force?

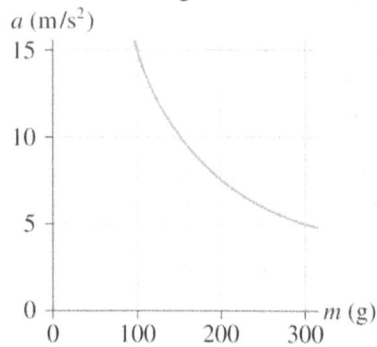

6. •• A 4.50-kg toy cart undergoes an acceleration in a straight line (the x-axis). The graph in adjoining FIGURE shows this acceleration as a function of time. (a) Find the maximum net force on this cart. When does this maximum force occur? (b) During what times is the net force on the cart a constant? (c) When is the net force equal to zero?

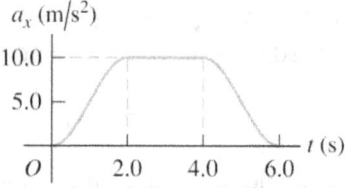

7. • Boxes A and B are in contact on a horizontal, frictionless surface, as shown in following figure. Box A has mass 20.0 kg and box B has mass 5.0 kg. A horizontal force of 100 N is exerted on box A. What is the magnitude of the force that box A exerts on box B?

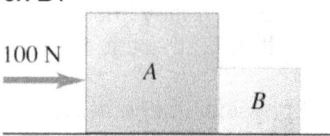

8. •• Two crates, one with mass 4.00 kg and the other with mass 6.00 kg, sit on the frictionless surface of a frozen pond, connected by a light rope (see figure). A woman wearing golf shoes (so she can get traction on the ice) pulls horizontally on the 6.00-kg crate with a force F that gives the crate an acceleration of 2.50 m/s²(a) What is the acceleration of the 4.00-kg crate? (b) Draw a free-body diagram for the 4.00-kg crate. Use that diagram and Newton's second law to find the tension T in the rope that connects the two crates. (c) Draw a free-body diagram for the 6.00-kg crate. What is the direction of the net force on the 6.00-kg crate? Which is larger in magnitude, force T or force F? (d) Use part (c) and Newton's second law to calculate the magnitude of the force F.

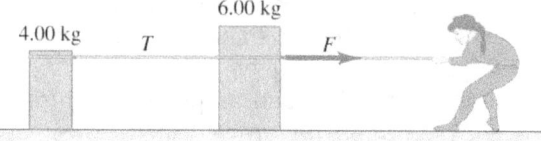

9. •• The two blocks are connected by a heavy uniform rope with a mass of 4.00 kg. An upward force of 200 N is applied as shown. (a) Draw three free-body diagrams: one for the 6.00-kg block, one for the 4.00-kg rope, and another one for the 5.00-kg block. For each force, indicate what body exerts that force. (b) What is the acceleration of the system? (c) What is the tension at the top of the heavy rope? (d) What is the tension at the midpoint of the rope?

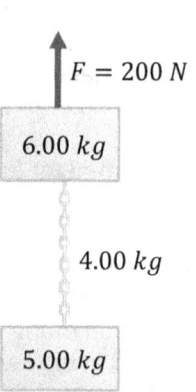

10. •• A hot-air balloon consists of a basket, one passenger, and some cargo. Let the total mass be M. Even though there is an upward lift force on the balloon, the balloon is initially accelerating

downward at a rate of $g/3$ (a) Draw a free-body diagram for the descending balloon. (b) Find the upward lift force in terms of the initial total weight Mg. (c) The passenger notices that he is heading straight for a waterfall and decides he needs to go up. What fraction of the total weight must he drop overboard so that the balloon accelerates *upward* at a rate of $g/2$. Assume that the upward lift force remains the same.

11. •• Two boxes, A and B, are connected to each end of a light vertical rope, as shown in adjoining figure. A constant upward force $F = 80 \, N$ is applied to box A. Starting from rest, box B descends 12.0 m in 4.00 s. The tension in the rope connecting the two boxes is 36.0 N. (a) What is the mass of box B? (b) What is the mass of box A?

12. •• The position of a $7.75 \times 10^5 N$ training helicopter under test is given by $\vec{r} = (2.2 \, m/s)t\hat{j} - (0.060 \, m/s^2)t^2\hat{k}$. Find the net force on the helicopter at $t = 5.0 \, s$.

13. •• An object of mass m is at rest in equilibrium at the origin. At a new force is applied that has components- $F_x(t) = k_1 + k_2 y,$ $F_y(t) = k_3 t$ where k_1, k_2 and k_3 are constants. Calculate the position $\vec{r}(t)$ and velocity $\vec{v}(t)$ vectors as functions of time.

14. •• Find the tension in each cord in following figure, if the weight of the suspended object is w.

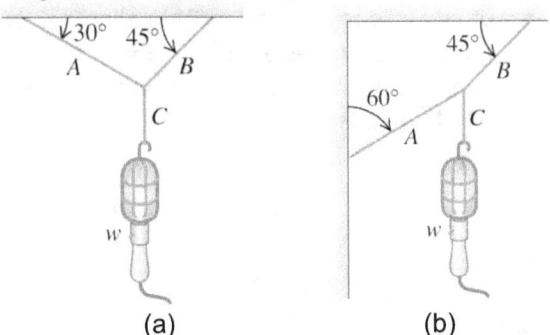

(a)   (b)

15. ••• In adjoining figure, the weight w is 60.0 N. (a) What is the tension in the diagonal string? (b) Find the magnitudes of the horizontal forces $\vec{F}_1$ and $\vec{F}_2$ that must be applied to hold the system in the position shown.

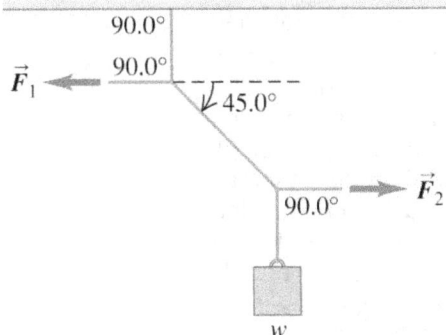

16. •• Three sleds are being pulled horizontally on frictionless horizontal ice using horizontal ropes (see following figure). The pull is of magnitude 125 N. Find (a) the acceleration of the system and (b) the tension in ropes A and B.

| 30.0 | B | 20.0 | A | 10.0 | Pull (P) |

17. •• A 8.00-kg block of ice, released from rest at the top of a 1.50-m-long frictionless ramp, slides downhill, reaching a speed of 2.50 m/s at the bottom. (a) What is the angle between the ramp and the horizontal? (b) What would be the speed of the ice at the bottom if the motion were opposed by a constant friction force of 10.0 N parallel to the surface of the ramp?

18. •• A box weighing 77.0 N rests on a table. A rope tied to the box runs vertically upward over a pulley and a weight is hung from the other end (Fig). Determine the force that the table exerts on the box if the weight hanging on the other side of the pulley weighs (a) 30.0 N, (b) 60.0 N, and (c) 90.0 N.

19. •• One 3.2-kg paint bucket is hanging by a massless cord from another 3.2-kg paint bucket, also hanging by a massless cord, as shown in adjoining Figure. (a) If the buckets are at rest, what is the tension in each cord? (b) If the two buckets are pulled upward with an acceleration of $1.25 \, m/s^2$ by the upper cord, calculate the tension in each cord.

20. •• Referring to adjoining figure, what are the tensions in all the cords?

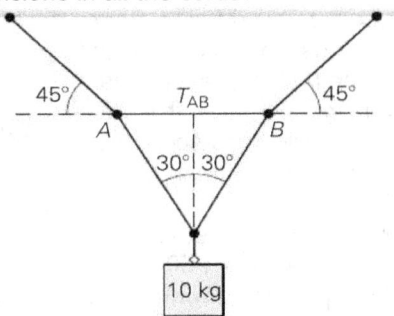

21. •• A 7180-kg helicopter accelerates upward at $0.80 \, m/s^2$ while lifting a 1080-kg frame at a construction site, Fig. 1. (a) What is the lift force exerted by the air on the helicopter rotors? (b) What is the tension in the cable (ignore its mass) which connects the frame to the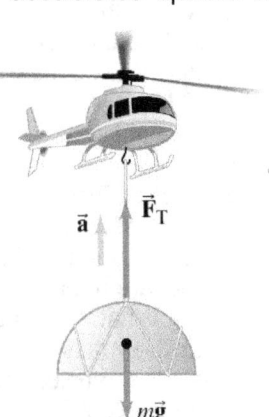

22. •• A 15.0-kg load of bricks hangs from one end of a rope that passes over a small, frictionless pulley. A 28.0 kg counterweight is suspended from the other end of the rope, as shown in adjoining figure. The system is released from rest. (a) What is the magnitude of the upward acceleration of the load of bricks? (b) What is the tension in the rope while the load is moving? How does the tension compare to the 15.0 kg weight of the load of bricks? To the weight of the counterweight?

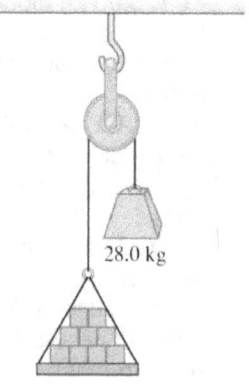

23. •• Three objects with masses $m_1 = 36.5$ kg, $m_2 = 19.2$ kg, and $m_3 = 12.5$ kg are hanging from ropes that run over pulleys. What is the acceleration of m_1?

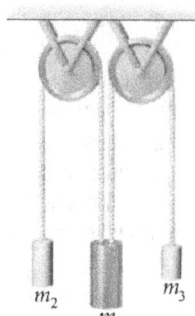

24. •• A window washer pulls herself upward using the bucket–pulley apparatus shown in adjoining figure. (a) How hard must she pull downward to raise herself slowly at constant speed? (b) If she increases this force by 15%, what will her acceleration be? The mass of the person plus the bucket is 72 kg.

25. •• An 8.00-kg block of ice, released from rest at the top of a 1.50-m-long frictionless ramp, slides downhill, reaching a speed of 2.50 m/s at the bottom. (a) What is the angle between the ramp and the horizontal? (b) What would be the speed of the ice at the bottom if the motion were opposed by a constant friction force of 10.0 N parallel to the surface of the ramp?

26. •• A light rope is attached to a block with mass 4.00 kg that rests on a frictionless, horizontal surface. The horizontal rope passes over a frictionless, massless pulley, and a block with mass m is suspended from the other end. When the blocks are released, the tension in the rope is 10.0 N. (a) What is the acceleration of either block? (b) Find the mass m of the hanging block. (c) How does the tension compare to the weight of the hanging block?

27. •• A transport plane takes off from a level landing field with two gliders in tow, one behind the other. The mass of each glider is 700 kg, and the total resistance (air drag plus friction with the runway) on each may be assumed constant and equal to 2500 N. The tension in the towrope between the transport plane and the first glider is not to exceed 12,000 N. (a) If a speed of 40 m/s is required for takeoff, what minimum length of runway is needed? (b) What is the tension in the towrope between the two gliders while they are accelerating for the takeoff?

28. •• A 750.0-kg boulder is raised from a quarry 125 m deep by a long uniform chain having a mass of 575 kg. This chain is of uniform strength, but at any point it can support a maximum tension no greater than 2.50 times its weight without breaking. (a) What is the maximum acceleration the boulder can have and still get out of the quarry, and (b) how long does it take to be lifted out at maximum acceleration if it started from rest?

29. •••A 1.00-kg glider on a horizontal air track is pulled by a string at an angle θ. The taut string runs over a pulley and is attached to a hanging object of mass 0.500 kg as shown in adjoining figure. (a) Show that the speed v_x of the glider and the speed v_y of the hanging object are related by $v_x = uv_y$, where $u = z(z^2 - h_0^2)^{-1/2}$. (b) The glider is released from rest. Show that at that instant the acceleration a_x of the glider and the acceleration a_y of the hanging object are related by $a_x = ua_y$. (c) Find the tension in the string at the instant the glider is released for $h_0 = 80.0$ cm and $\theta = 30.0°$.

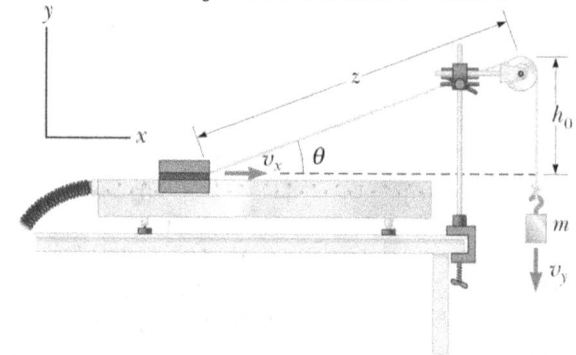

30. ••An object of mass M is held in place by an applied force $\vec{F}$ and a pulley system as shown in Figure. The pulleys are massless and frictionless. (a) Draw diagrams showing the forces on each pulley. Find (b) the tension in each section of rope, T_1, T_2, T_3, T_4, and T_5 and (c) the magnitude of $\vec{F}$.

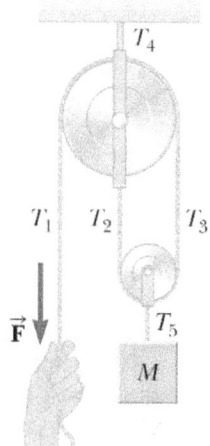

31. ••In terms of m_1, m_2, and g, find the acceleration of each block in the following figure.

There is no friction anywhere in the system. What will be your answer for and $m_2 = 2m_1$?

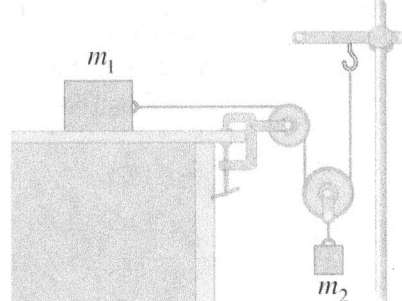

32. ••Consider the three connected objects shown in Figure. Assume first that the inclined plane is frictionless and that the system is in equilibrium. In terms of m, g, and θ, find (a) the mass M and (b) the tensions T_1 and T_2. Now assume that the value of M is double the value found in part (a). Find (c) the acceleration of each object and (d) the tensions T_1 and T_2. Next, assume that the coefficient of static friction between m and $2m$ and the inclined plane is m_s and that the system is in equilibrium. Find (e) the maximum value of M and (f) the minimum value of M. (g) Compare the values of T_2 when M has its minimum and maximum values.

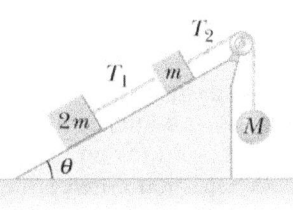

33. •••A wedge of mass m_2 sits at rest on a scale, as shown in Figure. A small block of mass m_1 slides down the frictionless incline of the wedge. Find the scale reading while the block slides. The wedge does not slide on the scale.

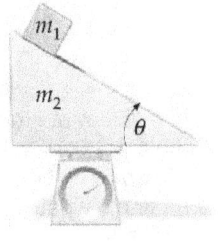

34. ••The bottom end of a massless, vertical spring of force constant k rests on a scale and the top end is attached to a massless cup, as in adjoining figure. Place a ball of mass m_b gently into the cup and ease it down into an equilibrium position where it sits at rest in the cup. (a) Draw the separate free-body diagrams for the ball and the spring. (b) Show that in this situation, the spring compression d is given by $\frac{m_b g}{k}$ (c) What is the scale reading under these conditions?

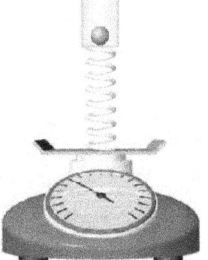

35. ••A 2540-kg test rocket is launched vertically from the launch pad. Its fuel (of negligible mass) provides a thrust force so that its vertical velocity as a function of time is given by $v(t) = At + Bt^2$, where A and B are constants and time is measured from the instant the fuel is ignited. At the instant of ignition, the rocket has an upward acceleration of 1.50 m/s² and 1.00 s later an upward velocity of 2.00 m/s. (a) Determine A and B, including their SI units. (b) At 4.00 s after fuel ignition, what is the acceleration of the rocket, and (c) what thrust force does the burning fuel exert on it, assuming no air resistance? Express the thrust in newtons and as a multiple of the rocket's weight. (d) What was the initial thrust due to the fuel?

36. ••A 2.00-kg box is moving to the right with speed 9.00 m/s on a horizontal, frictionless surface. At $t = 0$ a horizontal force is applied to the box. The force is directed to the left and has magnitude $F(t) = (6.00\ N/s^2)t^2$. (a) What distance does the box move from its position at $t = 0$ before its speed is reduced to zero? (b) If the force continues to be applied, what is the speed of the box at $t = 3.00$ s?

37. ••A 5.00-kg crate is suspended from the end of a short vertical rope of negligible mass. An upward force $F(t)$ is applied to the end of the rope, and the height of the crate above its initial position is given by $y(t) = (2.80\ m/s)t + (0.610\ m/s^3)t^3$. What is the magnitude of the force F when $t = 4.00$ s?

38. ••A U-shaped container has uniform cross sectional area S. It is suspended vertically with the help of a spring and two strings A and B as shown in the figure. The spring and strings are light. When water (density $= d$) is poured slowly into the container it was observed that the level of water remained unchanged with respect to the ground. Find the force constant of the spring.

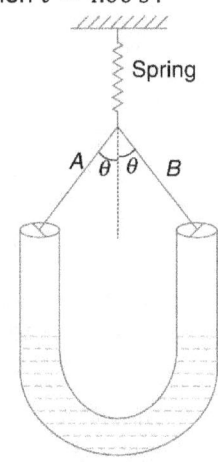

39. •• In emergencies with major blood loss, the doctor will order the patient placed in the Trendelenburg position, in which the foot of the bed is raised to get maximum blood flow to the brain. If the coefficient of static friction between the typical patient and the bedsheets is 1.20, what is the maximum angle at which the bed can be tilted with respect to the floor before the patient begins to slide?

40. •• In a laboratory experiment on friction, a 135-N block resting on a rough horizontal table is pulled by a horizontal wire. The pull gradually increases until the block begins to move and continues to increase thereafter. Figure shows a graph of the friction force on this block as a function of the pull. (a) Identify the regions of the graph where static and kinetic friction occur. (b) Find the coefficients of static and kinetic friction between the block and the table. (c) Why does the graph slant upward in the first part but then level out? (d) What would the

graph look like if a 135-N brick were placed on the box, and what would the coefficients of friction be in that case?

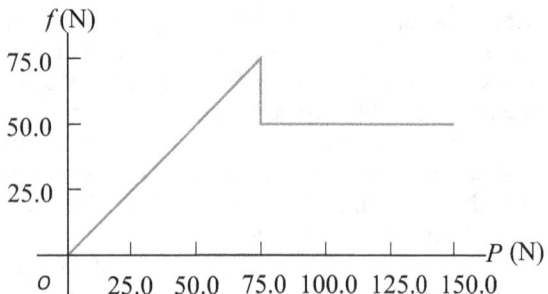

41. •• A box of bananas weighing 40.0 N rests on a horizontal surface. The coefficient of static friction between the box and the surface is 0.40, and the coefficient of kinetic friction is 0.20. (a) If no horizontal force is applied to the box and the box is at rest, how large is the friction force exerted on the box? (b) What is the magnitude of the friction force if a monkey applies a horizontal force of 6.0 N to the box and the box is initially at rest? (c) What minimum horizontal force must the monkey apply to start the box in motion? (d) What minimum horizontal force must the monkey apply to keep the box moving at constant velocity once it has been started? (e) If the monkey applies a horizontal force of 18.0 N, what is the magnitude of the friction force and what is the box's acceleration?

42. •• A crate of weight w is pushed by a force $\vec{P}$ on a horizontal floor as shown in Figure. The coefficient of static friction is μ_s, and $\vec{P}$ is directed at angle θ below the horizontal. (a) Show that the minimum value of P that will move the crate is given by
$$P = \frac{\mu_s w \sec\theta}{1 - \mu_s \tan\theta}$$
(b) Find the condition on θ in terms of μ_s for which motion of the crate is impossible for any value of P.

43. •• A 45.0-kg crate of tools rests on a horizontal floor. You exert a gradually increasing horizontal push on it and observe that the crate just begins to move when your force exceeds 313 N. After that you must reduce your push to 208 N to keep it moving at a steady 25.0 cm/s. (a) What are the coefficients of static and kinetic friction between the crate and the floor? (b) What push must you exert to give it an acceleration of 1.10 m/s²? (c) Suppose you were performing the same experiment on this crate but were doing it on the moon instead, where the acceleration due to gravity is 1.62 m/s². (i) What magnitude push would cause it to move? (ii) What would its acceleration be if you maintained the push in part (b)?

44. ••• You are lowering two boxes, one on top of the other, down the ramp shown in adjoining figure by pulling on a rope parallel to the surface of the ramp. Both boxes move together at a constant speed of 15.0 cm/s. The coefficient of kinetic friction between the ramp and the lower box is 0.444, and the coefficient of static friction between the two boxes is 0.800. (a) What force do you need to exert to accomplish this? (b) What are the magnitude and direction of the friction force on the upper box?

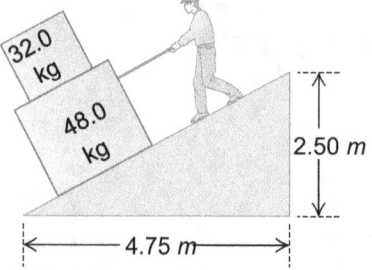

45. •• A pickup truck is carrying a toolbox, but the rear gate of the truck is missing, so the box will slide out if it is set moving. The coefficients of kinetic and static friction between the box and the bed of the truck are 0.355 and 0.650, respectively. Starting from rest, what is the shortest time this truck could accelerate uniformly to 30.0 m/s without causing the box to slide?

46. •• (a) If the coefficient of kinetic friction between tires and dry pavement is 0.80, what is the shortest distance in which you can stop an automobile by locking the brakes when traveling at 28.7 m/s? (b) On wet pavement the coefficient of kinetic friction may be only 0.25. How fast should you drive on wet pavement in order to be able to stop in the same distance as in part (a)? (Note: Locking the brakes is not the safest way to stop.)

47. • Consider the system shown in adjoining figure. Block A weighs 45.0 N and block B weighs 25.0 N. Once block B is set into downward motion, it descends at a constant speed. (a) Calculate the coefficient of kinetic friction between block A and the tabletop. (b) A cat, also of weight 45.0 N, falls asleep on top of block A. If block B is now set into downward motion, what is its acceleration (magnitude and direction)?

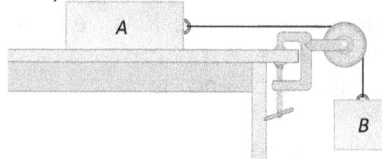

48. •• Two crates connected by a rope lie on a horizontal surface. Crate A has mass m_A and crate B has mass m_B. The coefficient of kinetic friction between each crate and the surface is μ_k. The crates are pulled to the right at constant velocity by a horizontal force $\vec{F}$. In terms of m_A, m_B, and μ_k, calculate (a) the magnitude of the force $\vec{F}$ and (b) the tension in the rope connecting the blocks. Include the free-body diagram or diagrams you used to determine each answer.

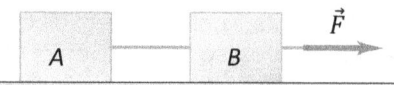

49. •• A 25.0-kg box of textbooks rests on a loading ramp that makes an angle α with the horizontal. The coefficient of kinetic friction is 0.25, and the coefficient of static friction is 0.35. (a) As the angle α is increased, find the minimum angle at which the box starts to slip. (b) At this angle, find the acceleration once the box has begun to move. (c) At this angle, how fast will the box be moving after it has slid 5.0 m along the loading ramp?

50. •• As shown in adjoining figure, block A (mass 2.25 kg) rests on a tabletop. It is connected by a horizontal cord passing over a light, frictionless pulley to a hanging block B (mass 1.30 kg). The coefficient of kinetic friction between block A and the tabletop is 0.450. After the blocks are released from rest, find (a) the speed of each block after moving 3.00 cm and (b) the tension in the cord. Include the free-body diagram or diagrams you used to determine the answers.

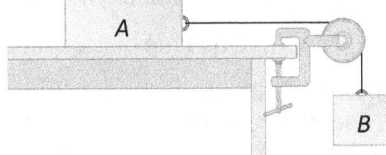

51. •• A large crate with mass m rests on a horizontal floor. The coefficients of friction between the crate and the floor are μ_s and μ_k. A woman pushes downward at an angle θ below the horizontal on the crate with a force $\vec{F}$. (a) What magnitude of force $\vec{F}$ is required to keep the crate moving at constant velocity? (b) If μ_s is greater than some critical value, the woman cannot start the crate moving no matter how hard she pushes. Calculate this critical value of μ_s.

52. •• You throw a baseball straight up. The drag force is proportional to v^2. In terms of g, what is the y-component of the ball's acceleration when its speed is half its terminal speed and (a) it is moving up? (b) It is moving back down?

53. •• An adventurous archaeologist crosses between two rock cliffs by slowly going hand over hand along a rope stretched between the cliffs. He stops to rest at the middle of the rope (Fig. P5.56). The rope will break if the tension in it exceeds 2.50×10^4 N, and our hero's mass is 90.0 kg. (a) If the angle θ is 30.0°, find the tension in the rope. (b) What is the smallest value the angle θ can have if the rope is not to break?

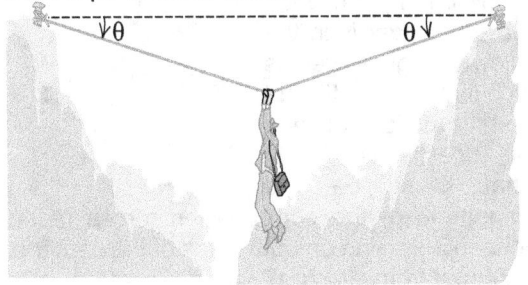

54. •• In adjoining figure a worker lifts a weight w by pulling down on a rope with a force $\vec{F}$. The upper pulley is attached to the ceiling by a chain, and the lower pulley is attached to the weight by another chain. In terms of w, find the tension in each chain and the magnitude of the force $\vec{F}$ if the weight is lifted at constant speed. Include the free-body diagram or diagrams you used to determine your answers. Assume that the rope, pulleys, have negligible weights and chains all.

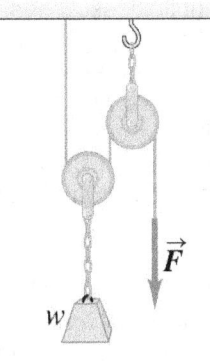

55. •• A solid uniform 45.0-kg ball of diameter 32.0 cm is supported against a vertical, frictionless wall using a thin 30.0-cm wire of negligible mass, as shown in adjoining figure. (a) Draw a free-body diagram for the ball and use it to find the tension in the wire. (b) How hard does the ball push against the wall?

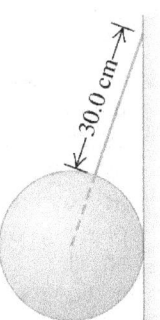

56. •• A horizontal wire holds a solid uniform ball of mass m in place on a tilted ramp that rises 35.0° above the horizontal. The surface of this ramp is perfectly smooth, and the wire is directed away from the center of the ball (adjoining figure). (a) How hard does the surface of the ramp push on the ball? (b) What is the tension in the wire?

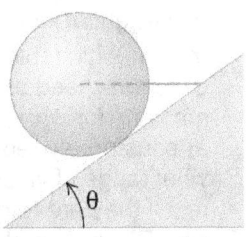

57. •• A box of mass m sits on a ramp that is inclined at θ above the horizontal. The coefficient of kinetic friction between the box and the ramp is μ_k. What *horizontal* force is required to move the box up the incline with a constant acceleration of a?

58. ••• Two identical balls, each having mass m, are suspended by three wires as shown in Fig. P5.87. The surfaces of the balls are perfectly smooth. (a) Find the tension in each of the three wires. (b) How hard does each ball push on the other one?

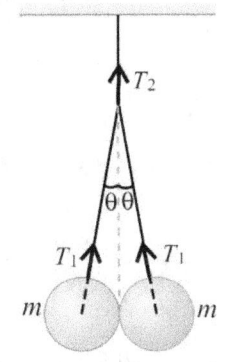

59. •• In Fig. of problem 50, block A has mass m and block B has mass 6.00 kg. The coefficient of kinetic friction between block A and the tabletop is $\mu_k = 0.40$. The mass of the rope connecting the

blocks can be neglected. The pulley is light and frictionless. When the system is released from rest, the hanging block descends 5.00 m in 3.00 s. What is the mass m of block A?

60. •• Two bicycle tires are set rolling with the same initial speed of 3.50 m/s on a long, straight road, and the distance each travels before its speed is reduced by half is measured. One tire is inflated to a pressure of 40 psi and goes 18.1 m; the other is at 105 psi and goes 92.9 m. What is the coefficient of rolling friction μ_r for each? Assume that the net horizontal force is due to rolling friction only.

61. •• A block with mass M is attached to the lower end of a vertical, uniform rope with mass m and length L. A constant upward force $\vec{F}$ is applied to the top of the rope, causing the rope and block to accelerate upward. Find the tension in the rope at a distance x from the top end of the rope, where x can have any value from 0 to L.

62. •• A block with mass m_1 is placed on an inclined plane with slope angle α and is connected to a second hanging block with mass m_2 by a cord passing over a small, frictionless pulley (see adjoining figure). The coefficient of static friction is μ_s and the coefficient of kinetic friction is μ_k. (a) Find the mass m_2 for which block m_1 moves up the plane at constant speed once it is set in motion. (b) Find the mass m_2 for which block m_1 moves down the plane at constant speed once it is set in motion. (c) For what range of values of m_2 will the blocks remain at rest if they are released from rest?

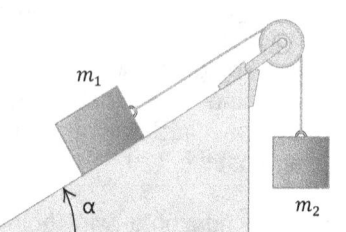

63. ••• Block A, in the following Figure, weighs 60.0 N. The coefficient of static friction between the block and the surface on which it rests is 0.25. The weight w is 12.0 N and the system is in equilibrium. (a) Find the friction force exerted on block A. (b) Find the maximum weight w for which the system will remain in equilibrium.

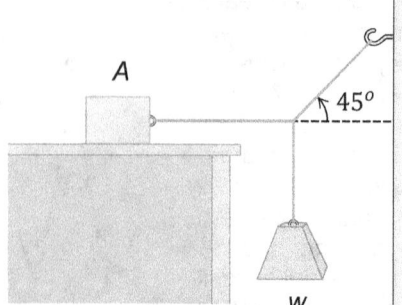

64. •• Block A in adjoining Fig. weighs 2.40 N and block B weighs 3.60 N. The coefficient of kinetic friction between all surfaces is 0.300. Find the magnitude of the horizontal force F necessary to drag block B to the left at constant speed (a) if A rests on B and moves with it (Fig.(a)). (b) If A is held at rest (Fig.(b)).

(a) 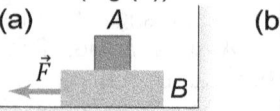   (b)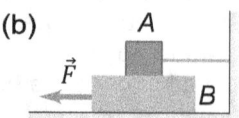

65. •• Consider the block in adjoining figure. Show that, in general, the following results hold for a block of mass m resting on a horizontal surface whose coefficient of static friction is μ_s (a) If you want to apply the minimum possible force to move the block, you should apply it with the force pulling upward at an angle $\theta = \tan^{-1} \mu_s$ (b) The minimum force necessary to start the block moving is $F_{min} = \left(\mu_s / \sqrt{1 + \mu_s^2}\right) mg$ (c) Once the block starts moving, if you want to apply the least possible force to keep it moving, should you keep the angle at which you are pulling the same, increase it, or decrease it?

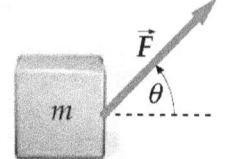

66. •• Answer the questions in Problem 65, but for a force that pushes down on the block at an angle θ below the horizontal.

67. •• A 100-kg mass is pulled along a frictionless surface by a horizontal force $\vec{F}$ such that its acceleration is $a_1 = 6.00 \, m/s^2$ (See adjoining figure). A 20.0-kg mass slides along the top of the 100-kg mass and has an acceleration of $a_2 = 4.00 \, m/s^2$ (It thus slides backward relative to the 100-kg mass.) (a) What is the frictional force exerted by the 100-kg mass on the 20.0-kg mass? (b) What is the net force acting on the 100-kg mass? What is the force F? (c) After the 20.0-kg mass falls off the 100-kg mass, what is the acceleration of the 100-kg mass? (Assume that the force F does not change.)

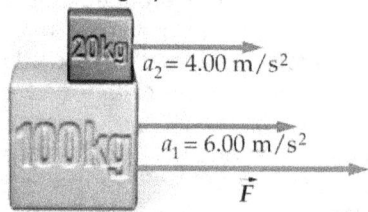

68. •• The board sandwiched between two other boards in the following figure weighs 95.5 N. If the coefficient of static friction between the boards is 0.663, what must be the magnitude of the compression forces (assumed horizontal) acting on both sides of the center board to keep it from slipping?

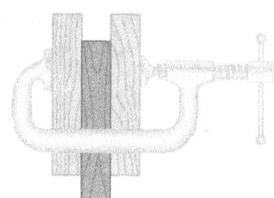

69. •• The 70.0-kg climber in adjoining figure is supported in the "chimney" by the friction forces exerted on his shoes and back. The static coefficients of friction between his shoes and the wall, and between his back and the wall, are 0.80 and 0.60, respectively. What is the minimum normal force he must exert? Assume the walls are vertical and that the static friction forces are both at their maximum. Ignore his grip on the rope.

70. ••• A block of mass 2.20 kg is accelerated across a rough surface by a light cord passing over a small pulley as shown in adjoining figure. The tension T in the cord is maintained at 10.0 N, and the pulley is 0.100 m above the top of the block. The coefficient of kinetic friction is 0.400.
(a) Determine the acceleration of the block when $x = 0.400$ m. (b) Describe the general behavior of the acceleration as the block slides from a location where x is large to $x = 0$. (c) Find the maximum value of the acceleration and the position x for which it occurs. (d) Find the value of x for which the acceleration is zero.

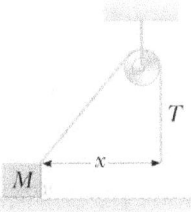

71. •• A block of mass $m = 2.00$ kg rests on the left edge of a block of mass $M = 8.00$ kg. The coefficient of kinetic friction between the two blocks is 0.300, and the surface on which the 8.00-kg block rests is frictionless. A constant horizontal force of magnitude $F = 10.0$ N is applied to the 2.00-kg block, setting it in motion as shown in Figure a. If the distance L that the leading edge of the smaller block travels on the larger block is 3.00 m, (a) in what time interval will the smaller block make it to the right side of the 8.00-kg block as shown in Figure b? (Note: Both blocks are set into motion when $\vec{F}$ is applied.) (b) How far does the 8.00-kg block move in the process?

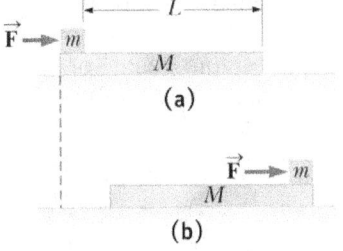

72. •• The two blocks, $m = 16$ kg and $M = 88$ kg, shown in adjoining figure, are free to move. The coefficient of static friction between the blocks is $\mu_s = 0.38$ but the surface beneath M is frictionless. What is the minimum horizontal force F required to hold m against M?

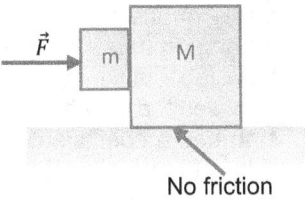

73. ••• A 25,000-kg rocket blasts off vertically from the earth's surface with a constant acceleration. During the motion considered in the problem, assume that g remains constant. Inside the rocket, a 15.0-N instrument hangs from a wire that can support a maximum tension of 45.0 N. (a) Find the minimum time for this rocket to reach the sound barrier (330 m/s) without breaking the inside wire and the maximum vertical thrust of the rocket engines under these conditions. (b) How far is the rocket above the earth's surface when it breaks the sound barrier?

74. ••• You are standing on a bathroom scale in an elevator in a tall building. Your mass is 64 kg. The elevator starts from rest and travels upward with a speed that varies with time according to $v(t) = (3.0 \, m/s^2)t + (0.20 \, m/s^2)t^2$. When $t = 4.0 \, s$, what is the reading of the bathroom scale?

75. ••• You are designing an elevator for a hospital. The force exerted on a passenger by the floor of the elevator is not to exceed 1.60 times the passenger's weight. The elevator accelerates upward with constant acceleration for a distance of 3.0 m and then starts to slow down. What is the maximum speed of the elevator?

76. ••• You are working for a shipping company. Your job is to stand at the bottom of a 8.0-m-long ramp that is inclined at 37° above the horizontal. You grab packages off a conveyor belt and propel them up the ramp. The coefficient of kinetic friction between the packages and the ramp is $\mu_k = 0.30$.
(a) What speed do you need to give a package at the bottom of the ramp so that it has zero speed at the top of the ramp? (b) Your coworker is supposed to grab the packages as they arrive at the top of the ramp, but she misses one and it slides back down. What is its speed when it returns to you?

77. •• If the coefficient of static friction between a table and a uniform massive rope is μ_s, what fraction of the rope can hang over the edge of the table without the rope sliding?

78. •• A 40.0-kg packing case is initially at rest on the floor of a 1500-kg pickup truck. The coefficient of static friction between the case and the truck floor is 0.30, and the coefficient of kinetic friction is 0.20. Before each acceleration given below, the truck is traveling due north at constant speed. Find the magnitude and direction of the friction force acting on the case (a) when the truck accelerates at 2.20 m/s² northward and (b) when it accelerates at 3.40 m/s² southward.

79. •• A 12.0-kg box rests on the flat floor of a truck. The coefficients of friction between the box and floor are $\mu_s = 0.19$ and $\mu_k = 0.15$. The truck stops at a stop sign and then starts to move with an acceleration of 2.20 m/s². If the box is 1.80 m from

80. •• Block A in adjoining figure weighs 1.90 N, and block B weighs 4.20 N. The coefficient of kinetic friction between all surfaces is 0.30. 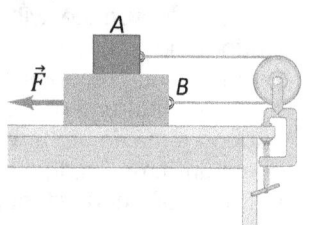 Find the magnitude of the horizontal force $\vec{F}$ necessary to drag block B to the left at constant speed if A and B are connected by a light, flexible cord passing around a fixed, frictionless pulley.

81. •• Block A in following figure has a mass of 4.00 kg, and block B has mass 12.0 kg. The coefficient of kinetic friction between block B and the horizontal surface is 0.25. (a) What is the mass of block C if block B is moving to the right and speeding up with an acceleration of 2.00 m/s²? (b) What is the tension in each cord when block B has this acceleration?

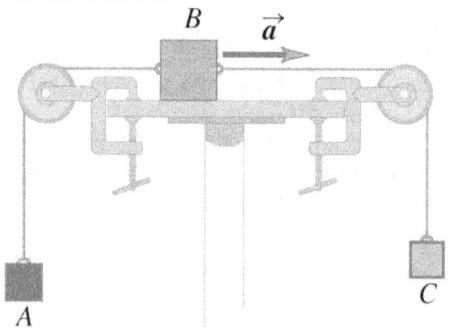

82. •• Two blocks connected by a cord passing over a small, frictionless pulley rest on frictionless planes (see adjoining figure). (a) Which way will the system move when the blocks are released from rest? (b) What is the acceleration of the blocks? (c) What is the tension in the cord?

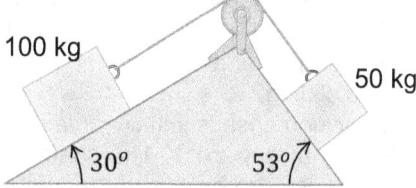

83. •• Block B, with mass 5.00 kg, rests on block A, with mass 8.00 kg, which in turn is on a horizontal tabletop (see adjoining figure). There is no friction between block A and the tabletop, but the coefficient of static friction between block A and block B is 0.750. A light string attached to block A passes over a frictionless, massless pulley, and block C is suspended from the other end of the string. What is the largest mass that block C can have so that blocks A and B still slide together when the system is released from rest?

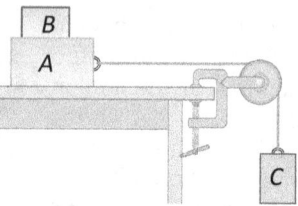

84. •• As shown in the figure, two masses, $m_1 = 3.50$ kg and $m_2 = 5.00$ kg, are on a frictionless tabletop and mass $m_3 = 7.60$ kg is hanging from m_1. The coefficients of static and kinetic friction between m_1 and m_2 are 0.600 and 0.500, respectively.
a) What are the accelerations of m_1 and m_2?
b) What is the tension in the string between m_1 and m_3?

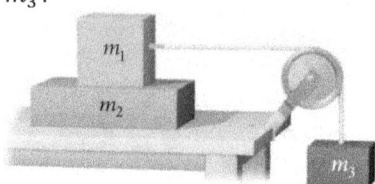

85. •• As shown in the figure, a block of mass $M_1 = 0.450$ kg is initially at rest on a slab of mass $M_2 = 0.820$ kg, and the slab is initially at rest on a level table. A string of negligible mass is connected to the slab, runs over a frictionless pulley on the edge of the table, and is attached to a hanging mass M_3. The block rests on the slab but is not tied to the string, so friction provides the only horizontal force on the block. The slab has a coefficient of kinetic friction $\mu_k = 0.340$ and a coefficient of static friction $\mu_s = 0.560$ with both the table and the block. When released, M3 pulls on the string and accelerates the slab, which accelerates the block. Find the maximum mass of M_3 that allows the block to accelerate with the slab, without sliding on top of the slab.

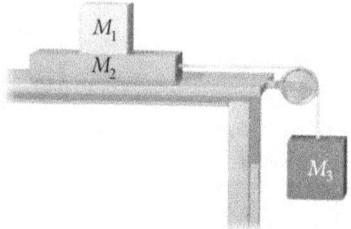

86. •• As shown in the figure with Problem 4.94, a block of mass $M_1 = 0.250$ kg is initially at rest on a slab of mass $M_2 = 0.420$ kg, and the slab is initially at rest on a level table. A string of negligible mass is connected to the slab, runs over a frictionless pulley on the edge of the table, and is attached to a hanging mass $M_3 = 1.80$ kg. The block rests on the slab but is not tied to the string, so friction provides the only horizontal force on the block. The slab has a coefficient of kinetic friction $\mu_k = 0.340$ with both the table and the block. When released, M_3 pulls on the string, which accelerates the slab so quickly that the block starts to slide on the slab. Before the block slides off the top of the slab:

a) Find the magnitude of the acceleration of the block.
b) Find the magnitude of the acceleration of the slab.

87. •• Two objects with masses 5.00 kg and 2.00 kg hang 0.600 m above the floor from the ends of a cord 6.00 m long passing over a frictionless pulley. Both objects start from rest. Find the maximum height reached by the 2.00-kg object.

88. •• You are riding in an elevator on the way to the 18th floor of your dormitory. The elevator is accelerating upward with $a = 1.90$ m/s². Beside you is the box containing your new computer; the box and its contents have a total mass of 28.0 kg. While the elevator is accelerating upward, you push horizontally on the box to slide it at constant speed toward the elevator door. If the coefficient of kinetic friction between the box and the elevator floor is $\mu_k = 0.32$, what magnitude of force must you apply?

89. •• A block is placed against the vertical front of a cart as shown in following figure. What acceleration must the cart have so that block A does not fall? The coefficient of static friction between the block and the cart is μ_s. How would an observer on the cart describe the behaviour of the block?

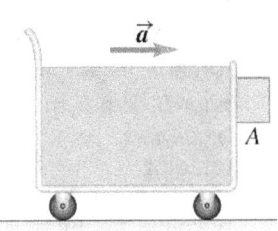

90. •• A 28.0-kg block is connected to an empty 2.00-kg bucket by a cord running over a frictionless pulley (see adjoining figure). The coefficient of static friction between the table and the block is 0.45 and the coefficient of kinetic friction between the table and the block is 0.32. Sand is gradually added to the bucket until the system just begins to move. (a) Calculate the mass of sand added to the bucket. (b) Calculate the acceleration of the system. Ignore mass of cord

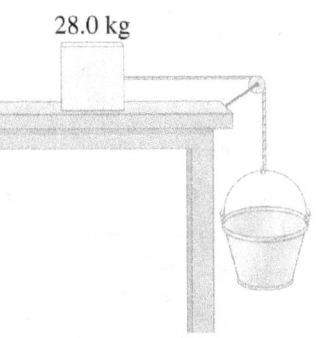

91. •• Two blocks with masses 4.00 kg and 8.00 kg are connected by a string and slide down a 30.0° inclined plane. The coefficient of kinetic friction between the 4.00-kg block and the plane is 0.25; that between the 8.00-kg block and the plane is 0.35. (a) Calculate the acceleration of each block. (b) Calculate the tension in the string. (c) What happens if the positions of the blocks

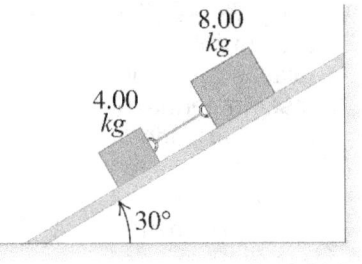

are reversed, so the 4.00-kg block is above the 8.00-kg block?

92. ••• Block A, with weight $3w$, slides down an inclined plane S of slope angle 37° at a constant speed while plank B, with weight w, rests on top of A. The plank is attached by a cord to the wall (see figure). If the coefficient of kinetic friction is the same between A and B and between S and A, determine its value.

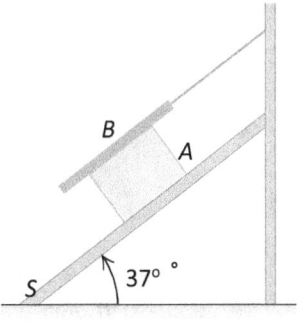

93. ••• The system, shown in following figure, can be used to measure the acceleration of the system. An observer riding on the platform measures the angle θ that the thread supporting the light ball makes with the vertical. There is no friction anywhere. (a) How is θ related to the acceleration of the system? (b) If $m_1 = 250$ kg and $m_2 = 1250$ kg, what is θ? (c) If you can vary m_1 and m_2, what is the largest angle θ you could achieve? Explain how you need to adjust m_1 and m_2 to do this.

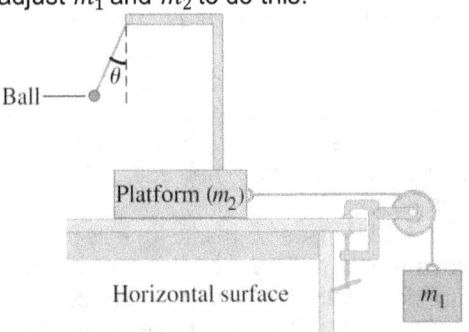

94. •• A platform is placed on an incline having inclination θ. There is a rigid L shaped frame fixed to the platform. A plumb line (a ball connected to a thread) is attached to the end A of the frame. The system is released on the inline. Find the angle that the plumb line will make with vertical in its equilibrium position relative to the platform when (a) the incline is smooth
(b) there is friction and the acceleration of the block is half its value when the incline is smooth

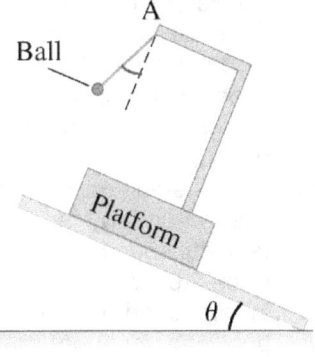

95. ••• Blocks A, B, and C are placed as in adjoining figure and connected by ropes of negligible mass. Both A and B weigh 25.0 N each, and the coefficient of kinetic friction between each block and the surface is 0.35. Block C descends with

constant velocity. (a) Find the tension in the rope connecting blocks A and B. (b) What is the weight of block C? (c) If the rope connecting A and B were cut, what would be the acceleration of C?

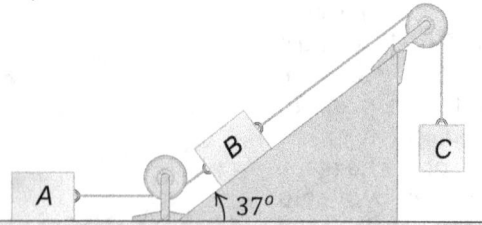

96. •• A 20-kg monkey has a firm hold on a light rope that passes over a frictionless pulley and is attached to a 20-kg bunch of bananas (see adjoining figure). The monkey looks up, sees the bananas, and starts to climb the rope to get them. (a) As the monkey climbs, do the bananas move up, down, or remain at rest? (b) As the monkey climbs, does the distance between the monkey and the bananas decrease, increase, or remain constant? (c) The monkey releases her hold on the rope. What happens to the distance between the monkey and the bananas while she is falling? (d) Before reaching the ground, the monkey grabs the rope to stop her fall. What do the bananas do?

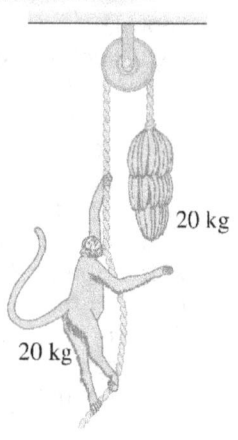

97. ••• A rock with mass $m = 3.00$ kg falls from rest in a viscous medium. The rock is acted on by a net constant downward force of $18.0\ N$ (a combination of gravity and the buoyant force exerted by the medium) and by a fluid resistance force $f = kv$, where v is the speed in m/s and $k = 2.20$ N s/m. (a) Find the initial acceleration a_0. (b) Find the acceleration when the speed is 3.00 m/s. (c) Find the speed when the acceleration equals $0.1 a_0$. (d) Find the terminal speed v_t. (e) Find the coordinate, speed, and acceleration 2.00 s after the start of the motion. (f) Find the time required to reach a speed of $0.9 v_t$.

98. ••• A small rock moves in water, and the force exerted on it by the water is given by $f = kv$. The terminal speed of the rock is measured and found to be 2.0 m/s. The rock is projected *upward* at an initial speed of 6.0 m/s. You can ignore the buoyancy force on the rock. (a) In the absence of fluid resistance, how high will the rock rise and how long will it take to reach this maximum height? (b) When the effects of fluid resistance are included, what are the answers to the questions in part (a)?

99. •• A wedge with mass M rests on a frictionless, horizontal tabletop. A block with mass m is placed on the wedge (see following figure). There is no friction between the block and the wedge. The system is released from rest. (a) Calculate the acceleration of the wedge and the horizontal and vertical components of the acceleration of the block. (b) Do your answers to part (a) reduce to the correct results when M is very large? (c) As seen by a stationary observer, what is the shape of the trajectory of the block?

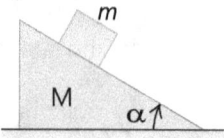

100. •• A wedge with mass M rests on a frictionless horizontal tabletop. A block with mass m is placed on the wedge and a horizontal force $\vec{F}$ is applied to the wedge (see adjoining figure). What must the magnitude of $\vec{F}$ be if the block is to remain at a constant height above the tabletop?

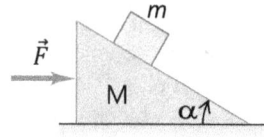

101. ••• You drop a baseball from the roof of a tall building. As the ball falls, the air exerts a drag force proportional to the square of the ball's speed ($f = Dv^2$). (a) In a diagram, show the direction of motion and indicate, with the aid of vectors, all the forces acting on the ball. (b) Apply Newton's second law and infer from the resulting equation the general properties of the motion. (c) Show that the ball acquires a terminal speed given $v_t = \sqrt{\frac{mg}{D}}$ (d) Derive the equation for the speed at any time. (Note: $\int \frac{dx}{a^2-x^2} = \frac{1}{a} \tanh\left(\frac{x}{a}\right)$ where

$$\tanh x = \frac{e^x - e^{-x}}{e^x + e^{-x}} = \frac{e^{2x}-1}{e^{2x}+1}$$

defines the hyperbolic tangent.)

102. ••• The double Atwood machine shown in adjoining figure has frictionless, massless pulleys and cords. The system is released from rest. In terms of m_1, m_2, m_3, and g, (a) what is the acceleration of block m_3; (b) the acceleration of pulley B; (c) the acceleration of block m_1; (d) the acceleration of block m_2; (e) the tension in string A; (f) the tension in string C? (g) What do your

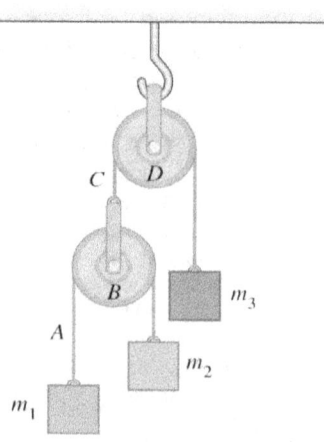

expressions give for the special case of $m_1 = m_2$ and $m_3 = m_1 + m_2$? Is this sensible?

103. ••• The masses of blocks A and B in Fig. are 20.0 kg and 10.0 kg, respectively. The blocks are initially at rest on the floor and are connected by a massless string passing over a massless and frictionless pulley. An upward force $\vec{F}$ is applied to the pulley. Find the accelerations $\vec{a}_A$ of block A and $\vec{a}_B$ of block B when F is (a) 124 N; 294 N; (c) 424 N.

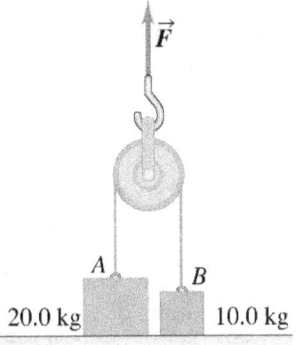

104. ••• A ball is held at rest at position A (see adjoining figure), by two light strings. The horizontal string is cut and the ball starts swinging as a pendulum. Point B is the farthest to the right the ball goes as it swings back and forth. What is the ratio of the tension in the supporting string at position B to its value at A before the horizontal string was cut?

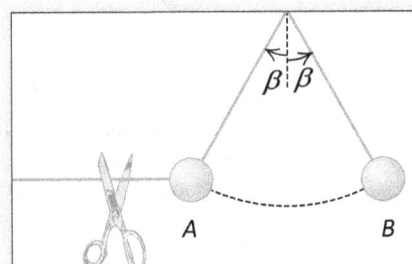

105. ••• A worker wishes to pile sand onto a circular area in his yard. The radius of the circle is R. No sand is to spill onto the surrounding area; see adjoining figure. Show that the greatest volume of sand that can be stored in this manner is $\pi \mu_s R^3/3$, where μ_s is the coefficient of static friction of sand on sand. (The volume of a cone is $Ah/3$, where A is the base area and h is the height.)

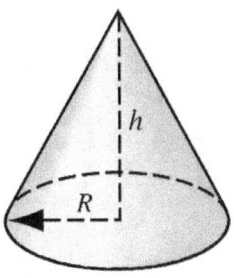

106. ••• A massless rope is tossed over a wooden dowel of radius r in order to lift a heavy object of weight W off of the floor, as shown in following figure. The coefficient of sliding friction between the rope and the dowel is μ. Show that the minimum downward pull on the rope necessary to lift the object is

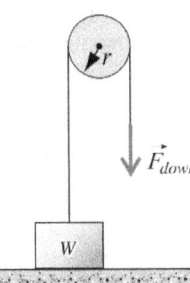

$F_{down} = W e^{\pi \mu}$

107. ••• In the situation shown in the following figure, if mass M is going down along the incline at an acceleration of 5 m/s² and m is moving toward right relative to M horizontally with 3m/s². Find the net acceleration of m.

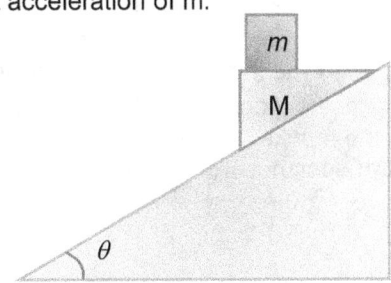

108. •• Figure shows a 65 kg man standing stationary with respect to a horizontal conveyor belt that is accelerating with $1 m\, s^{-2}$. What is the net force on the man? If the coefficient of static friction between the man's shoes and the belt is 0.2, up to what acceleration of the belt can the man continue to be stationary relative to the belt?

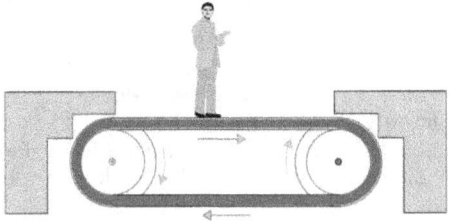

109. •• Find the acceleration of the block of mass M in the situation of adjoining figure μ_1. The coefficient of friction between the two blocks is μ_1 and that between the bigger block and the ground is μ_2.

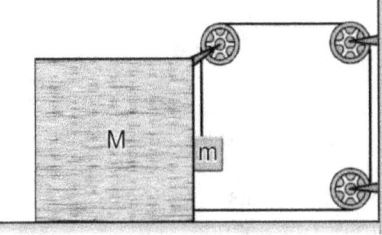

110. •• If in the above problem $M = 2m$. The coefficient of friction at all surfaces is μ, string is light and inextensible and all the pulleys are smooth, then
(A) The free body diagram for block M will be as follows (True/False).

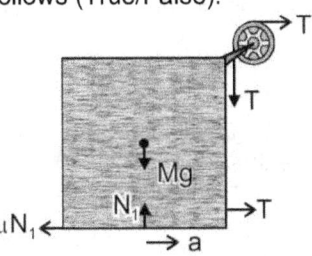

(B) The free body diagram for block m will be as follows (True/False).

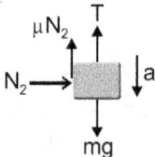

111. •• In the adjoining Figure, the combined mass of the bananas and platform is m, the mass of the monkey is M. If the friction coefficient between the platform and the floor, is μ. Find the maximum force that the monkey can exert on the rope so that the platform does not slip on the floor

112. •• Figure shows a small block A of mass m kept at the left end of a plank B of mass $M = 2m$ and length l. The system can slide on a horizontal road. The system is started towards right with the initial velocity v. The friction coefficients between the road and the plank is 1/2 and that between the plank and the block is 1/4. Find
(a) the time elapsed before the block separate from the plank.
(b) displacement of block and plank relative to ground till that moment.

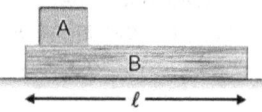

113. ••• A 1 kg block 'B' rests as shown on a bracket 'A' of same mass. Constant forces $F_1 = 20\ N$ and $F_2 = 8\ N$ start to act at time $t = 0$ when the distance of block B from pulley is 50 cm. Time when block B reaches the pulley is

114. •• Three equal balls 1, 2, 3 are suspended on springs on below the other as shown in the FIGURE. OA is a weightless thread.
(a) If the string is cut, the system starts falling. Find the acceleration of all the balls at the initial instant.
(b) Find the initial accelerations of all the balls if we cut the spring BC which is supporting ball 3 instead of cutting the thread.

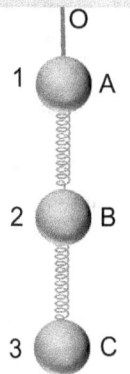

115. ••• A uniform flexible chain of length 1.50 m rest on a fixed smooth sphere of radius $R = \dfrac{2}{\pi m}$ such that one end A of chain is at top of the sphere while the other end B is hanging freely. Chain is held stationary by a horizontal thread PA as shown in FIGURE. Calculate the acceleration of the chain when the thread is burnt

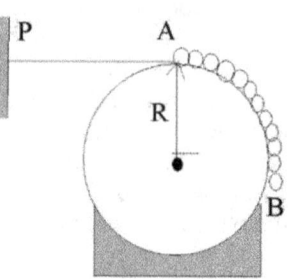

116. ••• A wedge of mass 'M' and angle of inclination 'θ' and of mass 'm' is arranged in a manner shown in the FIGURE. The spring of force constant 'k' attached to the wedge. Assuming the pulleys to be massless and all surfaces to be frictionless. Find the compression of the spring under equilibrium condition.

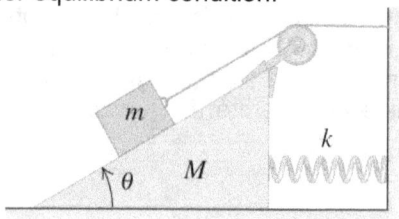

117. •• The pulley block system shown in the adjoining figure is released from rest. Assuming the pulleys to be light and frictionless and the string to be light and inextensible, find (a) the acceleration of the blocks A, B and C. (b) the tension in the string connecting the blocks. Given that $m_1 = m_2 = m$, $m_3 = 2m$

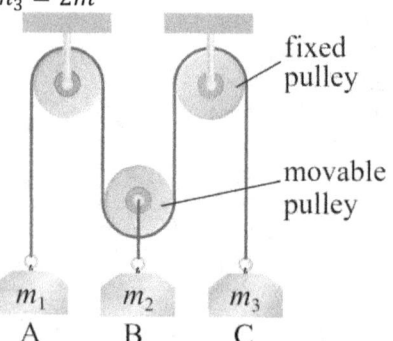

118. •• In adjoining figure, $m_2 = nm_1$, find the acceleration of m_2. The masses of the pulleys and the threads, as well as the friction are assumed to be negligible.

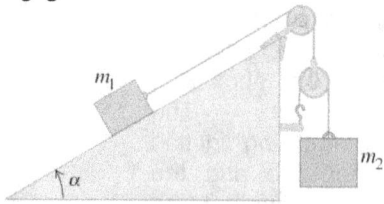

119. ••• In the arrangement shown in FIGURE pulleys are small, light and frictionless, threads are

inextensible and mass of block A, B and C is $m_1 = 5$ kg, $m_2 = 4$ kg and $m_3 = 2.5$ kg respectively. Co-efficient of friction for both the planes is . Calculate acceleration of each block when system is released from rest.

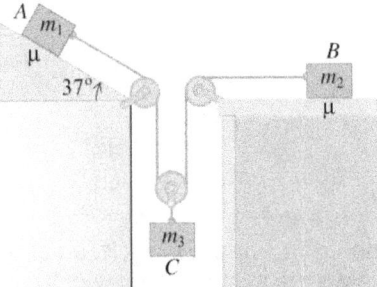

120. ••• In the arrangement shown in the adjoining figure, the masses m of the bar and M of the wedge, as well as the wedge angle, are known. The masses of the pulley and thread are negligible. Friction is absent. Find the acceleration of the wedge M.

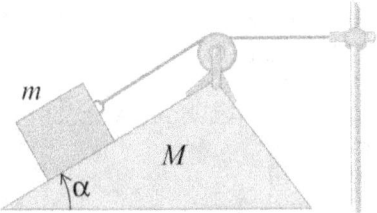

121. •• Find the acceleration of B w.r.t ground

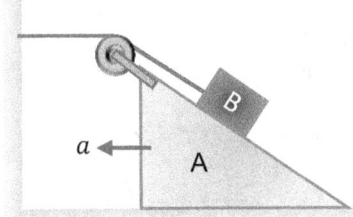

122. •• The adjoining figure shows one end of a string being pulled down at constant velocity v. Find the velocity of mass "m" as a function of "x".

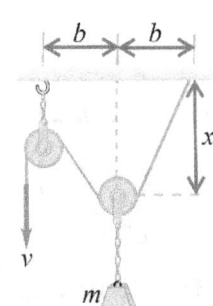

123. •• Two blocks 'A' and 'B' of same mass m attached with a light string are suspended by a spring as shown in adjoining figure Find the acceleration of block 'A' and 'B' just after the string is cut.

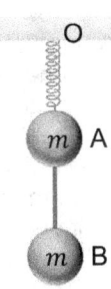

124. •• In the system shown in the adjoining figure ($M = m$), the block A is released from rest. Find:

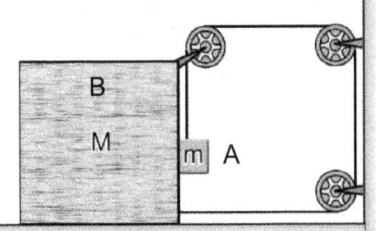

(i) the acceleration of both blocks 'A' and 'B'.
(ii) Tension in the string.
(iii) Contact force between 'A' and 'B'.

125. ••• In the following figure, find the acceleration of B.

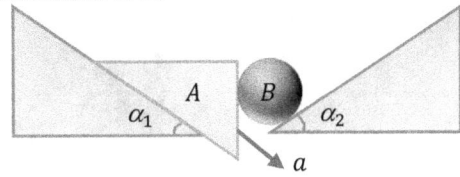

126. ••• Find the acceleration of A w.r.t. ground.

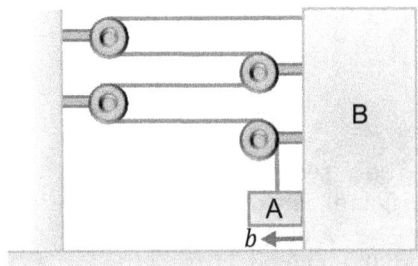

127. ••• Find the acceleration of C w.r.t. ground

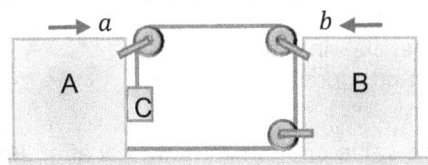

128. ••• Find the velocity of B w.r.t. ground.

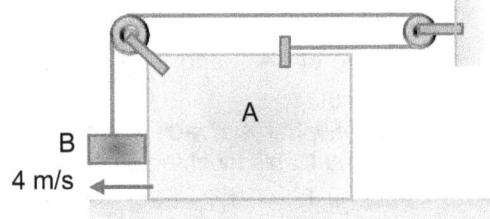

129. ••• Find the acceleration of B

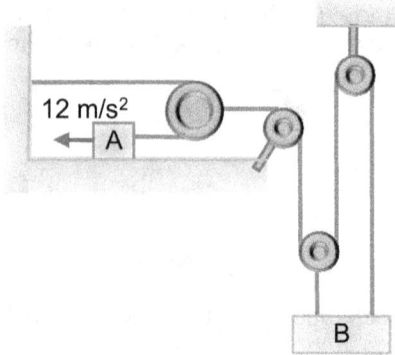

130. •• If block A is released from rest, then find the extension in spring and accelerations of the blocks of mass m just after release.

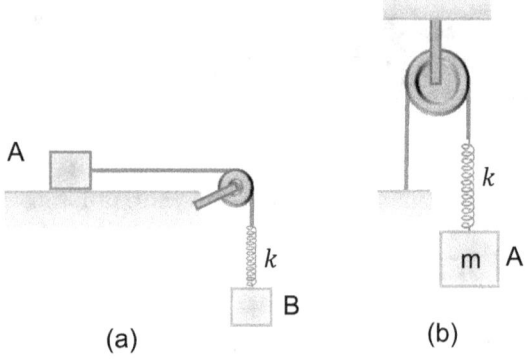

(a)   (b)

131. ••• The system shown in figure is in equilibrium. Pulley, springs and the strings are massless. The three blocks A, B and C have equal masses. x_1 and x_2 are extensions in the spring 1 and spring 2 respectively.

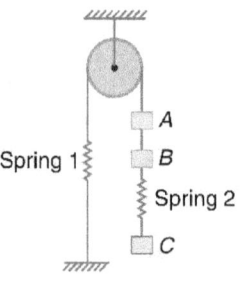

(a) Find the value of $\left|\frac{d^2x_2}{dt^2}\right|$ immediately after spring 1 is cut.

(b) Find the value of $\left|\frac{d^2x_1}{dt^2}\right|$ and $\left|\frac{d^2x_2}{dt^2}\right|$ immediately after string AB is cut.

(c) Find the value of $\left|\frac{d^2x_1}{dt^2}\right|$ and $\left|\frac{d^2x_2}{dt^2}\right|$ immediately after spring 2 is cut.

132. ••• Two hemispheres of radii R and r (< R) are fixed on a horizontal table touching each other (see figure). A uniform rod rests on two spheres as shown. The coefficient of friction between the rod and two spheres is μ. Find the minimum value of the ratio r/R for which the rod will not slide.

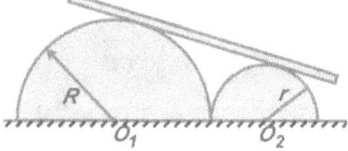

133. ••• Determine the force P required to impend the motion of the block B shown in figure. Take coefficient of friction = 0.3 for all surfaces in contact.

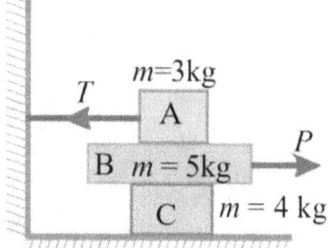

134. ••• In order to lift a heavy block A, an engineer has designed a wedge system as shown. Wedge C is fixed. A horizontal force F is applied to B to lift block A. Wedge B itself has negligible mass and mass of A is M. The coefficient of friction at all surfaces is μ. Find the value of applied force F at which the block A just begins to rise.

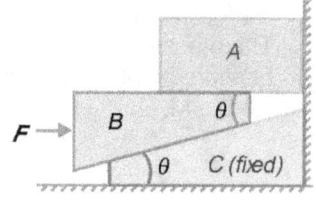

135. A 10 kg block A resting against 50 kg block B is shown in figure. The coefficient of static friction between block A and the wall is negligible. If P = 100N, determine the value of μ_s (as shown) for which motion is impending.

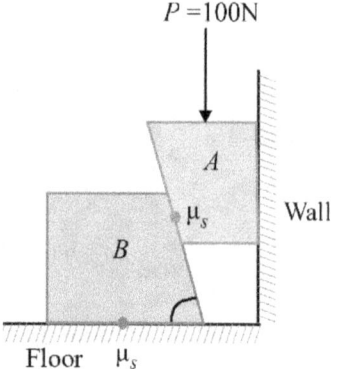

49.3. MULTIPLE CHOICE ASSIGNMENTS

49.3.1. NEWTON'S LAWS

49.3.1.1. LEVEL 1

QUESTIONS BASED ON GENERAL CONCEPT

1. The Newton's laws of motion are valid in-
   (A) inertial frames
   (B) non-inertial frames
   (C) rotating frames
   (D) accelerated frames

2. The incorrect statement about Newton's second law of motion is-
   (A) it provides a measure of inertia
   (B) it provides a measure of force
   (C) it relates force and acceleration
   (D) it relates momentum and force

3. Newton's third law is equivalent to the-
   (A) law of conservation of linear momentum
   (B) law of conservation of angular momentum
   (C) law of conservation of energy
   (D) law of conservation of energy and mass
4. We can derive Newton's-
   (A) second and third laws from the first law
   (B) first and second laws from the third law
   (C) third and first laws from the second law
   (D) All the three laws are independent of each other
5. Ratio of inertial mass to gravitational mass is-
   (A) 1 : 2   (B) 1 : 1
   (C) 2 : 1   (D) No fixed number
6. A rider on horse back falls when horse starts running all of a sudden because-
   (A) rider is taken back
   (B) rider is suddenly afraid of falling
   (C) inertia of rest keeps the upper part of body at rest where as lower part of the body moves forward with the horse
   (D) None of the above
7. A man getting down a running bus, falls forward because-
   (A) due to inertia of rest, road is left behind and man reaches forward
   (B) due to inertia of motion upper part of body continues to be in motion in forward direction while feet come to rest as soon as they touch the road
   (C) he leans forward as a matter of habit
   (D) of the combined effect of all the three factors stated in (A), (B) and (C)
8. When we jump out a boat standing in water it moves-
   (A) forward   (B) backward
   (C) side ways   (D) none of these
9. A man is at rest in the middle of a pond on perfectly smooth ice. He can get himself to the shore by making use of Newton's-
   (A) first law   (B) second law
   (C) third law   (D) all the laws
10. You are on a friction less horizontal plane. How can you get off if no horizontal force is exerted by pushing against the surface?
    (A) by jumping
    (B) by spitting or sneezing
    (C) by rolling your body on the surface
    (D) by running on the plane
11. Swimming is possible on account of -
    (A) first law of motion
    (B) second law of motion
    (C) third law of motion
    (D) Newton's law of gravitation
12. A boy sitting on the top most berth in the compartment of a train which is just going to stop on a railway station, drops an apple aiming at the open hand of his brother sitting vertically below his hands at a distance of about 2 meter. The apple will fall-
    (A) precisely on the hand of his brother
    (B) slightly away from the hand of his brother in the direction of motion of the train
    (C) slightly away from the hand of his brother in the direction opposite to the direction of motion of the train
    (D) none of the above
13. The incorrect relation is-
    (A) $F = ma$ (B) $F = m\frac{dv}{dt}$
    (C) $F = \frac{dp}{dt}$ (D) $F = mv$
14. A heavy block of mass m is supported by a cord C from the ceiling, and another cord D is attached to the bottom of the block. If a sudden jerk is given to D, then-
    (A) cord C breaks
    (B) cord D breaks
    (C) cord C and D both break
    (D) none of the cords breaks
15. ABCD is a rectangle forces of 9N, 8N, 3N act along the lines DC, CB and BA, respectively, in the directions indicated by the order of the letters. Then the resultant force is
    (A) 8 N
    (B) 5 N
    (C) 20 N
    (D) 10 N

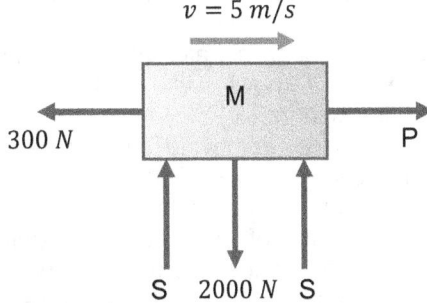

16. The forces acting on an object are shown in the fig. If the body moves horizontally at a constant speed of 5 m/s, then the values of the forces P and S are, respectively -
    (A) 0 N, 0 N   (B) 300 N, 200 N
    (C) 300 N, 1000 N   (D) 2000 N, 300 N

17. A person says that he measured the acceleration of a particle to be non-zero while no force was acting on the article-
    (A) He is a liar
    (B) His clock might have run slow
    (C) His meter scale might have been longer than the standard
    (D) He might have non-inertial frame

QUESTIONS BASED ON 2nd AND 3rd LAW

18. Three books are at rest on a horizontal table, as shown in adjoining figure. The *net* force on the middle book is
    (A) 5 N downward.
    (B). 15 N upward.
    (C) 15 N downward.
    (D). 0 N.

19. In outer space, where there is no gravity or air, an astronaut pushes with an equal force of 12 N on a 2 N moon rock and on a 4 N moon rock.
    (A) Since both rocks are weightless, they will have the same acceleration.
    (B) The 4 N rock pushes back on the astronaut twice as hard as the 2 N rock.
    (C) Both rocks push back on the astronaut with 12 N.
    (D) Since both rocks are weightless, they do not push back on the astronaut.

20. A rocket firing its engine and accelerating in outer space (no gravity, no air resistance) suddenly runs out of fuel. Which of the following best describes its motion after burnout?
    (A) It continues to accelerate at the same rate.
    (B) It continues to accelerate but at a gradually decreasing rate, until it reaches a constant velocity.
    (C) It immediately stops accelerating, and continues moving at the velocity it had when burnout occurred.
    (D) It immediately begins slowing down and gradually approaches zero velocity.

21. A person pushes horizontally with constant force P on a 250 N box resting on a frictionless horizontal floor. Which of the following statements about this box is correct?
    (A) The box will accelerate no matter how small P is.
    (B) The box will not accelerate unless $P > 250 N$
    (C) The box will move with constant velocity because P is constant.
    (D) Once the box is set moving, it will come to rest after P is removed.

22. Suppose the sun, including its gravity, suddenly disappeared. Which of the following statements best describes the subsequent motion of the earth?
    (A) The earth would continue moving in a straight line tangent to its original direction, but would gradually slow down.
    (B) The earth would speed up because the sun's gravity would not be able to slow it down.
    (C) The earth would continue moving in a straight line tangent to its original direction and would not change its speed.
    (D) The earth would move directly outward with constant speed away from the sun's original position.

23. Three weights hang by very light wires as shown in adjoining figure. What must be true about the tensions in these wires? (There may be more than one correct choice.)
    (A) The tension in *A* is the greatest.
    (B) The tension in *C* is the greatest.
    (C) All three wires have the same tension because the system is in equilibrium.
    (D) The tension in *C* is the least.

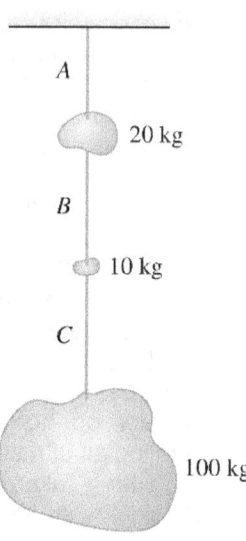

24. A person pushes two boxes with a horizontal 100 N force on a frictionless floor, as shown in adjoining figure. Box *A* is heavier than box *B*. Which of the following statements about these boxes is correct?

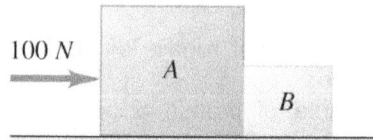

    (A) Box *A* pushes on box *B* with a force of 100 N, and box *B* pushes on box *A* with a force of 100 N.
    (B) Box *A* pushes on box *B* harder than box *B* pushes on box *A*.
    (C) Boxes *A* and *B* push on each other with equal forces of less than 100 N.
    (D) The boxes will not begin to move unless the total weight of the two boxes is less than 100 N.

25. A horizontal pull P pulls two wagons over a horizontal frictionless floor, as shown in adjoining figure. The tension in the light horizontal rope connecting the wagons is

    (A) equal to P, by Newton's third law.
    (B) equal to 2000 N.
    (C) greater than P.
    (D) less than P.

26. An object is moving north. From only this information one can conclude
    (A) that there is a single force on the object directed north.
    (B) that there is a net force on the object directed north.
    (C) that there may be several forces on the object, but the largest must be directed north.
    (D) nothing about the forces on the object.

27. When a 1 newton force acts on a $1\,kg$ body that is able to move freely, the body receives-
    (A) A speed of $1\,m/sec$
    (B) An acceleration of $1\,m/sec^2$

(C) An acceleration of $980\ cm/s^2$
(D) An acceleration of $1\ cm/s^2$

28. A force of 10 newton acts on a body of mass 20 kg for 10 seconds. The change produced in momentum is given by-
    (A) 5 kg m/sec
    (B) 100 kg m/sec
    (C) 200 kg m/sec
    (D) 2000 kg m/sec

29. A car travelling at a speed of 30 kilometer per hour is brought to a halt in 8 meters by applying brakes. If the same car is travelling at 60 km per hour, it can be brought to a halt with same braking power in-
    (A) 8 meters
    (B) 16 meters
    (C) 24 meters
    (D) 32 meters

30. A bullet of 5 g, travelling at a speed of 100 m/s penetrates a wooden block up to 6.0 cm. Then the average force applied by the bullet on the block is-
    (A) 417 N
    (B) 8333 N
    (C) 83.3 N
    (D) zero

31. A force-time graph for a linear motion is shown in figure where the segments are circular. The linear momentum gained between zero and 8 seconds in -

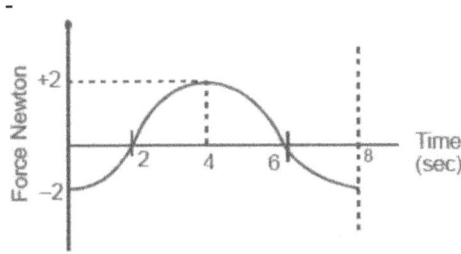

    (A) $-2\pi\ N.s$
    (B) $0\ N.s$
    (C) $4\pi\ N.s$
    (D) $-6\pi\ N.s$

32. A particle moves in the xy plane under the action of a force $\vec{F}$ such that the value of its linear momentum $(\vec{P})$ at any time t is, $P_x = 2\cos t$, $P_y = 2\sin t$. The angle θ between $\vec{P}$ and $\vec{F}$ at that time t will be -
    (A) 0°   (B) 30°   (C) 90°   (D) 180°

33. The linear momentum P of a body moving in one dimension varies with time according to the equation $P = at^3 + bt$ where a and b are positive constants. The net force acting on the body is
    (A) proportional to t^2
    (B) a constant
    (C) proportional to t
    (D) inversely proportional to t

34. A player catches a ball of $200\ g$ moving with a speed of $20\ m/s$. If the time taken to complete the catch is $0.5\ s$, the force exerted on the players hand is -
    (A) 8 N
    (B) 4 N
    (C) 2 N
    (D) 0 N

QUESTIONS BASED ON MOTION OF THE BLOCKS

35. Blocks are in contact on a frictionless table. A horizontal force F = 3N is applied to one block as shown. The force exerted by the smaller block m₂ on block m₁ is-

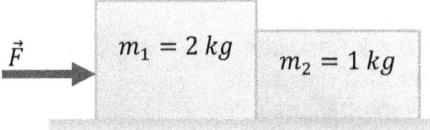

    (A) 1 N
    (B) 2 N
    (C) 3 N
    (D) 6 N

36. Three block are connected as shown, on a horizontal frictionless table and pulled to the right with a force $T_3 = 60\ N$. If $m_1 = 10\ kg$, $m_2 = 20\ kg$ and $m_3 = 30\ kg$, the tension T_2 is-

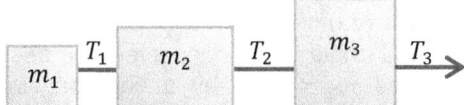

    (A) 10 N   (B) 20 N   (C) 30 N   (D) 60 N

37. A block of mass M is pulled along a horizontal frictionless surface by a rope of mass m. A force P is being applied to one end of the rope, the force that the rope exerts on the block M is-
    (A) $\left(\frac{M}{M+m}\right)P$
    (B) $\left(\frac{m}{M+m}\right)P$
    (C) $\left(\frac{M+m}{m}\right)P$
    (D) $\left(\frac{M+m}{M}\right)P$

38. Two masses are hanging vertically over frictionless pulley. The acceleration of the two masses is-
    (A) $\frac{m_1}{m_2}g$
    (B) $\frac{m_2}{m_1}g$
    (C) $\left(\frac{m_2-m_1}{m_1+m_2}\right)g$
    (D) $\left(\frac{m_1+m_2}{m_2-m_1}\right)g$

39. Two bodies of 5 kg and 4 kg are tied to a string as shown in the figure. If the table and pulley both are smooth, acceleration of 5 kg body will be equal to-

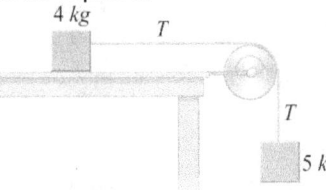

    (A) g
    (B) $g/9$
    (C) $4g/9$
    (D) $5g/9$

40. Three equal weights A, B, C of mass 2 kg each are hanging on a string passing over a fixed frictionless pulley as shown in the fig. The tension in the string connecting weights B and C is-
    (A) zero
    (B) 13 Newton
    (C) 3.3 Newton

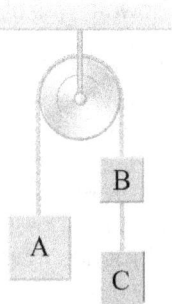

41. Two bodies of mass 0.3 kg and 0.4 kg are tied to the ends of a weightless string which passes over a smooth pulley as shown in the figure. The tension in the string is-
    (A) 3.06 Newton
    (B) 3.36 Newton
    (C) 4.05 Newton
    (D) 3.0 Newton

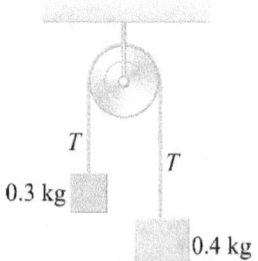

42. A block of mass $m_1 = 2\ kg$ on a smooth inclined plane at angle 30° is connected to a second block of mass $m_2 = 3$ kg by a cord passing over a frictionless pulley as shown in figure. The acceleration of each block is-(Assume $g = 10$ m/sec²)

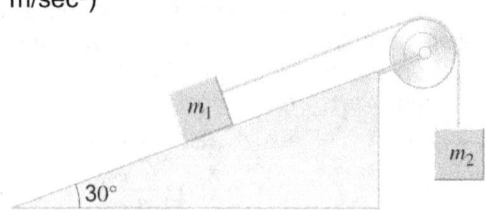

    (A) 2 m/sec²   (B) 4 m/sec²
    (C) 6 m/sec²   (D) 8 m/sec²

43. A body floats in liquid contained in a beaker. If the whole system as shown in figure falls under gravity then the upthrust on the body is
    (A) 2 mg   (B) zero
    (C) mg   (D) less than mg

44. Two blocks are connected by a cord passing over a small frictionless pulley and resting on frictionless planes as shown in the figure. The acceleration of the blocks is-

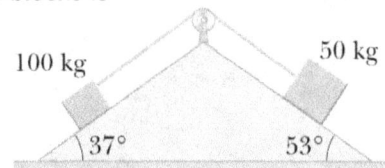

    (A) 0.33 m/s²   (B) 0.66 m/s²
    (C) 1 m/s²   (D) 1.32 m/s²

45. A thief stole a box full of valuable articles of weight W and while carrying it on his back, he jumped down a wall of height h from the ground. Before he reached the ground, he experienced a load of
    (A) 2W   (B) W
    (C) W/2   (D) zero

46. A block of mass m is placed on a smooth wedge of inclination θ. The whole system is accelerated horizontally so that the block does not slip on the wedge. The force exerted by the wedge on the block has magnitude -
    (A) mg   (B) mg/cos θ
    (C) mg cos θ   (D) mg tan θ

47. Two objects of masses m_1 and m_2 are attached by strings as shown in adjoining figure. If they are given an upward acceleration 'a', then the ratio of tension $T_1 : T_2$ is –
    (A) $(m_1 + m_2)/m_2$
    (B) $(m_1 + m_2)/m_1$
    (C) $\dfrac{m_1+m_2}{m_1-m_2}$
    (D) $\dfrac{m_1-m_2}{m_1+m_2}$

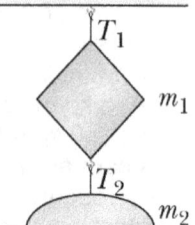

48. If the arrangement in above figure is given a downward acceleration (a) then the ratio of tensions $T_1 : T_2$ in strings, is –
    (A) $(m_1 + m_2)/m_2$ (B) $(m_1 + m_2)/m_1$
    (C) $\dfrac{m_1-m_2}{m_1+m_2}$ (D) $\dfrac{m_1+m_2}{m_1-m_2}$

49. In given figure find out the acceleration of any of the particle-

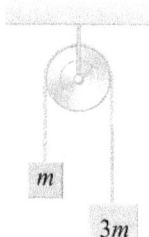

    (A) (1/2)g
    (B) g
    (C) (1/3) g
    (D) (1/4) g

50. In the figure a smooth pulley of negligible weight is suspended by a spring balance. Weights of 1kg and 5 kg are attached to the opposite ends of a string passing over the pulley and move with acceleration because of gravity. During the motion, the spring balance reads a weight of -
    (A) 6 kg
    (B) less than 6 kg
    (C) more than 6 kg
    (D) may be more or less than 6 kg

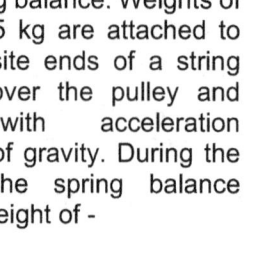

QUESTIONS BASED ON MOTION OF THE LIFT

51. A lift moves downwards with an acceleration a A passenger in the lift drops a book. The acceleration of the book with respect to the floor of lift is- (assume acceleration due to gravity = g)
    (A) g (B) a (C) $g + a$ (D) $g + a$

52. The ratio of the weight of a man in a stationary lift and in a lift accelerating downwards with a uniform acceleration 'a' is 3 : 2. The acceleration of the lift is-
    (A) $g/3$ (B) $g/2$ (C) g (D) $2g$

53. A lift is moving up with an acceleration of 3.675 m/sec2. The weight of a man-
    (A) increases by 37.5%
    (B) decreases by 37.5%
    (C) increases by 137.5%

(D) remains the same
54. If the tension in the cable supporting an elevator is equal to the weight of the elevator, the elevator may be -
(a) going up with increasing speed
(b) going down with increasing speed
(c) going up with uniform speed
(d) going down with uniform speed
(A) a, d (B) a, b, c
(C) c, d (D) a, b

55. The mass of a lift is 600 kg and it is moving upwards with a uniform acceleration of 2 m/s². Then the tension in the cable of the lift is-
(A) 7080 N (B) 5880 N
(C) 4680 N (D) zero N

QUESTIONS BASED ON EQUILIBRIUM OF THE SYSTEM

56. A metal sphere is hung by a string fixed to a wall. The forces acting on the sphere are shown in adjoining figure. Which of the following statements is/are correct?
(a) $N + T + W = 0$
(b) $T^2 = N^2 + W^2$
(c) $T = N + W$
(d) $N = W \tan\theta$

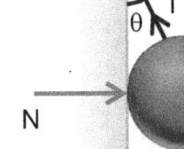

(A) a, b, c (B) b, c, d
(C) a, b, d (D) a, b, c, d

57. A body of mass 5 kg is suspended by the strings making angles 30º and 60º with the horizontal

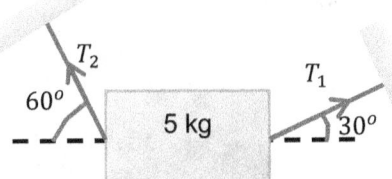

(a) $T_1 = 25$ N (b) $T_2 = 25$ N
(c) $T_1 = 25\sqrt{3}$ N (d) $T_2 = 25\sqrt{3}$ N
(A) a, b (B) a, d
(C) c, d (D) b, c

58. A ball D weighing 300 kg is suspended by means of two cords A and B as shown in the figure. W is a vertical wall and R a horizontal rigid beam. The tension in the string A in kg is
(A) zero
(B) 150
(C) 300
(D) 400

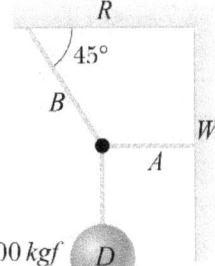

59. Two weights of 15 kg each are attached by means of two strings to the two ends of a spring balance, as shown in the diagram. The pulleys are frictionless. The reading of the balance would be

(A) zero (B) 15 kg
(C) 30 kg (D) 75 kg

60. As an inclined plane is made slowly horizontal by reducing the value of angle θ with horizontal. The component of weight parallel to the plane of a block resting on the inclined plane-
(A) decreases
(B) remains same
(C) increases
(D) increases if the plane is smooth

49.3.1.2. LEVEL-2

1. The linear momentum P of a body varies with time and is given by the equation $P = x + yt^2$, where x and y are constants. The net force acting on the body for a one dimensional motion is proportional to-
(A) t^2 (B) a constant
(C) $1/t$ (D) t

2. Two blocks of masses m_1 and m_2 are connected by a light spring and put on a horizontal frictionless table. The ratio of their acceleration after they are pulled apart and then released, is-
(A) m_1/m_2 (B) m_2/m_1
(C) $\frac{m_1-m_2}{m_1+m_2}$ (D) $\frac{4m_1m_2}{(m_1+m_2)^2}$

3. A rope of length L is pulled by a constant force F. What is the tension in the rope at a distance x from the end where the force is applied?
(A) $\frac{Fx}{L-x}$ (B) $F\frac{L}{L-x}$
(C) FL/x (D) $F(L-x)/L$

4. A spring toy weighing 1 kg on a spring balance suddenly jumps upward. A boy standing near the toy notices that the scale of the balance reads 1.05 kg. In this process the maximum acceleration of the toy is (g = 10 m sec⁻²) -
(A) 0.05 m sec⁻² (B) 0.5 m sec⁻²
(C) 1.05 m sec⁻² (D) 1 m sec⁻²
[Hint: Extra force on toy = $(1.05 - 1)g = 1a \Rightarrow a = 0.5\ m/s^2$]

5. The acceleration with which an object of mass 100 kg be lowered from a roof using a cord with a breaking strength of 60 kg weight without breaking the rope is-
(assume g = 10 m/sec²)
(A) 2 m/sec² (B) 4 m/sec²

(C) 6 m/sec²   (D) 10 m/sec²
6. A girl, of weight W, is sitting on an electric swing rotating in a vertical plane. She feels her weight to have increased by 25% as the swing goes up. What weight she would experience when the swing comes down?
(A) 3/2 W   (B) 5/4 W
(C) 3/4 W   (D) W/2
7. A weight W is tied to two strings passing over the frictionless pulleys A and B as shown in the figure. If weights P and Q move downwards with speed v, the weight W at any instant rises with the speed-

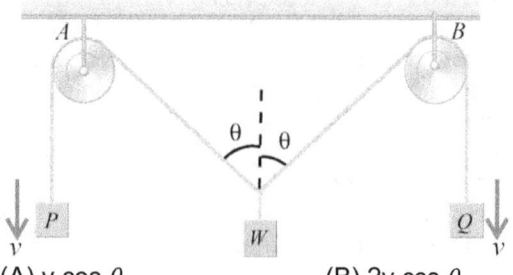

(A) v cos θ   (B) 2v cos θ
(C) v/cos θ   (D) 2v/cos θ
8. A mass is suspended from the roof of a car by a string. While the car has a constant acceleration a, the string makes an angle of 60° with the vertical. If g = 10 m/s², the value of a is-
(A) 10(3)1/2 m/s²   (B) 10/(3)1/2 m/s²
(C) 5 m/s²   (D) 5(3)1/2 m/s²
9. Two blocks are in contact on a frictionless table one has a mass m and the other $2m$. A force $\vec{F}$ is applied on $2m$ as shown is Figure. Now the same force $\vec{F}$ is applied on m. In the two cases respectively the ratio of force of contact between the two blocks will be-

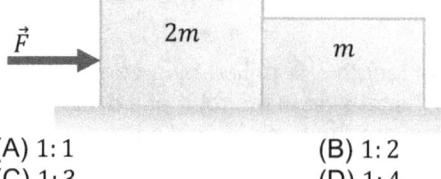

(A) 1 : 1   (B) 1 : 2
(C) 1 : 3   (D) 1 : 4

10. **Step Pulley Problems:** In the figure at the free end a force F is applied to keep the suspended mass of 18 kg at rest. The value of F is
(A) 180 N
(B) 90 N
(C) 60 N
(D) 30 N

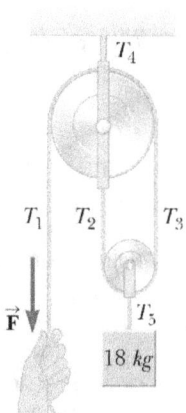

11. Figure shows a uniform rod of mass 3 kg and of length 30 cm. The strings shown in figure are pulled by constant forces of 20 N and 32 N. The acceleration of the rod is-

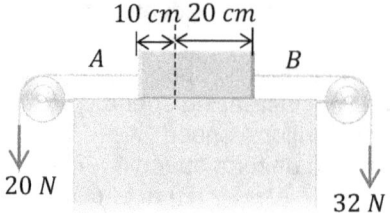

(A) 2 m/s²   (B) 3 m/s²
(C) 4 m/s²   (D) 6 m/s²
12. In the above question tension in rod at a distance 10 cm from end A is-
(A) 18 N   (B) 20 N
(C) 24 N   (D) 36 N
13. A balloon of mass M and a fixed size starts coming down with an acceleration f (f < g). The ballast mass m to be dropped from the ballon to have it go up with an acceleration f. Assuming negligible air resistance is find the value of m
(A) $\left(\dfrac{M}{g+f}\right)f$ (B) $\dfrac{Mf}{2(g+f)}$
(C) $\left(\dfrac{2Mf}{g+f}\right)$ (D) $\dfrac{M(g+a)}{g}$
14. A conveyor belt is moving horizontally with a uniform velocity of 2 m/sec. Material is dropped at one end at the rate of 5 kg/sec and discharged at the other end. Neglecting the friction, the power required to move the belt is-
(A) 10 watts   (B) 15 watts
(C) 20 watts   (D) 40 watts
15. Calculate the acceleration of the masses 12 kg shown in the setup of fig. Also calculate the tension in the string connecting the 12 kg mass. The string are weightless and inextensible, the pulleys are weightless and frictionless-

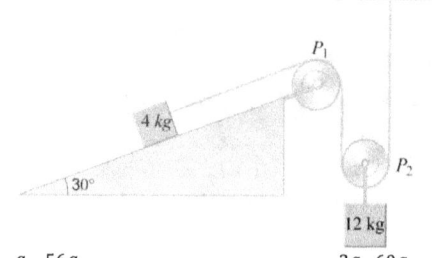

(A) $\dfrac{g}{10}, \dfrac{56g}{5}N$ (B) $\dfrac{2g}{7}, \dfrac{60g}{17}N$
(C) $\dfrac{10}{g}, \dfrac{5}{56g}N$ (D) $\dfrac{g}{14}, \dfrac{5}{56g}N$

16. In fig, a mass 5 kg slides without friction on an inclined plane making an angle 30° with the horizontal. Then the acceleration of this mass when it is moving upwards, the other mass is 10 kg. The pulleys are massless and frictionless. Take g = 10 m/sec².-

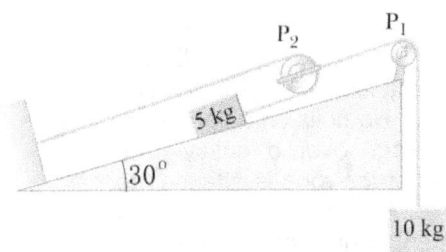

(A) .33 m/sec² (B) 3.3 m/sec²
(C) 33 m/sec² (D) None of these

17. Two masses m₁ and m₂ are connected by light string, which passes over the top of a smooth plane inclined at 30° to the horizontal, so that one mass rests on the plane and the other hangs vertically as shown in fig. It is found that m₁, hanging vertically can draw m₂ up the full length of the plane in half the time in which m₂ hanging vertically draws m₁ up. Find m₁/m₂. Assume pulley to be smooth-

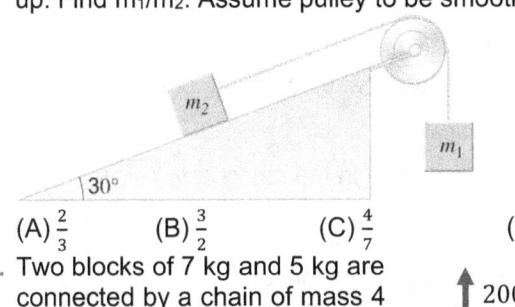

(A) $\frac{2}{3}$ (B) $\frac{3}{2}$ (C) $\frac{4}{7}$ (D) $\frac{7}{4}$

18. Two blocks of 7 kg and 5 kg are connected by a chain of mass 4 kg. An upward force of 200N is applied as shown in the diagram. The tension at the top of the chain at point P is: ($g = 10$ m/s²)
(A) 2.27 N
(B) 112.5 N
(C) 87.5 N
(D) 360 N

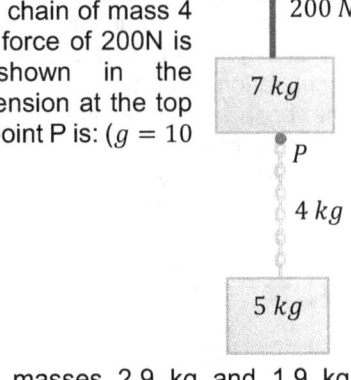

19. Two blocks of masses 2.9 kg and 1.9 kg are suspended from a rigid support S by two inextensible wires each of length 1 m. The upper wire has negligible mass and the lower wire has a uniform mass of 0.2 kg/m. The whole system of block, wire and support have an upward acceleration of 0.2 m/s². $g = 9.8$ m/s². The tension at the mid-point P of lower wire is-
(A) 10 N (B) 20 N
(C) 30 N (D) 50 N

20. Body A is placed on frictionless wedge making an angle θ with the horizon. The horizontal acceleration towards left to be imparted to the wedge for the body A to freely fall vertically, is-

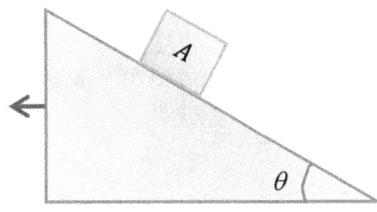

(A) g sin θ (B) g cos θ
(C) g tan θ (D) g cot θ

21. A triangular block of mass M with angle 30°, 60°, 90° rests with its 30°– 90° side on a horizontal smooth fixed table. A cubical block of mass m rests on the 60° – 30° side of the triangular block. What horizontal acceleration a must M have relative to the stationary table so that m remains stationary with respect to the triangular block [$M = 9$ kg, $m = 1$ kg]

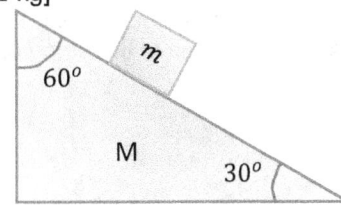

(A) 2.8 m/s² (B) 5.6 m/s²
(C) 8.4 m/s² (D) Zero

22. In the arrangement of figure assume negligible friction between the blocks and table. if F the pulling force and m₁ and m₂ the masses are known, then the tension in the string is–

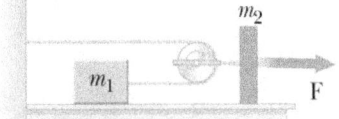

(A) m₁F/ (m₁ + m₂)
(B) 2m₁F / (m₁ + m₂)
(C) 2m₁F/ (4m₁ + m₂)
(D) None of the above

23. An empty plastic box of mass m is found to accelerate up at the rate of g/6, when placed deep inside water. How much sand should be put inside the box so that it may accelerate down at the rate of g/6 ?
(A) 2m/3 (B) 2m/5
(C) m/5 (D) 6m/7

24. A body of mass 8 kg is hanging from another body of mass 12 kg. The combination is being pulled up by a string with an acceleration of 2.2 m/sec². The tension T₁ will be -
(A) 260 N (B) 240 N
(C) 220 N (D) 200 N

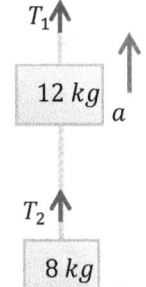

49.3.1.3. LEVEL-3

1. In the given figure, find the speed of pulley P –

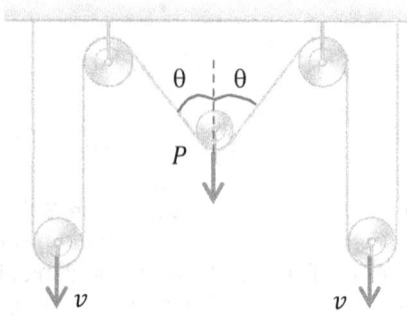

   (A) $\frac{v}{2}$
   (B) $2v \cos\theta$
   (C) $\frac{2v}{\cos\theta}$
   (D) $\frac{v}{2\sin\theta}$

2. In the given arrangement, n number of equal masses are connected by strings of negligible masses. The tension in the string connected to nth mass is –

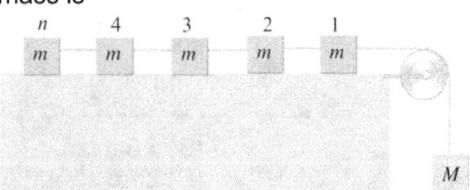

   (A) $\frac{mMg}{nm+M}$
   (B) $\frac{mMg}{nmM}$
   (C) mg
   (D) mng

3. In the given figure, pulleys and strings are massless. For equilibrium of the system, the value of θ is –

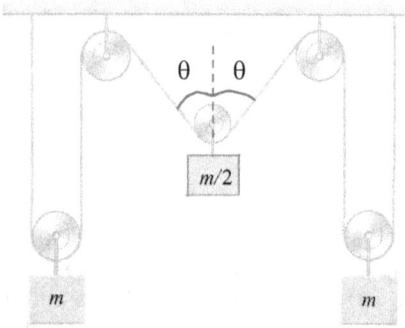

   (A) 60°
   (B) 30°
   (C) 90°
   (D) 120°

4. In the figure, the blocks A, B and C each of mass m have accelerations a_1, a_2 and a_3 respectively. F_1 and F_2 are external forces of magnitude $2mg$ and mg respectively, then –

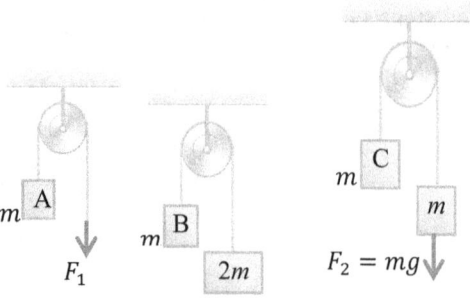

   (A) $a_1 = a_2 = a_3$
   (B) $a_1 > a_3 > a_2$
   (C) $a_1 = a_2, a_2 > a_3$
   (D) $a_1 > a_2, a_2 = a_3$

5. A man of mass m stands on a frame of mass M. He pulls on a light rope, which passes over a pulley. The other end of the rope is attached to the frame. For the system to be in equilibrium, what force must the man exert on the rope?

   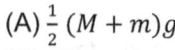

   (A) $\frac{1}{2}(M+m)g$
   (B) $(M+m)g$
   (C) $(M-m)g$
   (D) $(M+2m)g$

6. A simple pendulum with a bob of mass m is suspended from the roof of a car moving with a horizontal acceleration a

   (A) The string makes an angle of $\tan^{-1}\left(\frac{a}{g}\right)$ with the vertical
   (B) The string makes an angle of $\tan^{-1}\left(1-\frac{a}{g}\right)$ with the vertical
   (C) The tension in the string is $m\sqrt{a^2+g^2}$
   (D) The tension in the string is $m\sqrt{g^2-a^2}$

7. A particle slides down a smooth inclined plane of elevation θ, fixed in an elevator going up with an acceleration a_0 (figure). The base of the incline has a length L. Find the time taken by the particle to each to the bottom –

   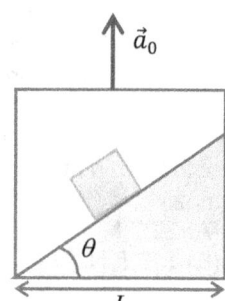

   (A) $\left[\frac{2L}{g\sin\theta\cos\theta}\right]^{1/2}$
   (B) $\left[\frac{2L}{(g-a_0)\sin\theta\cos\theta}\right]^{1/2}$
   (C) $\left[\frac{2L\sin\theta}{(g+a_0)\cos\theta}\right]^{1/2}$
   (D) $\left[\frac{2L}{(g+a_0)\sin\theta\cos\theta}\right]^{1/2}$

8. Find the mass M of the hanging block in figure which will prevent the smaller block from slipping over the triangular block. All the surfaces are frictionless and the strings and the pulleys are light

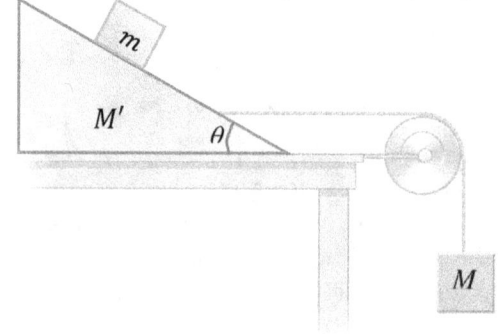

(A) $\frac{M'+m}{\tan\theta}$ (B) $(M'+m)\tan\theta$
(C) $\frac{M'+m}{1-\tan\theta}$ (D) $\frac{M'+m}{\cot\theta-1}$

9. The masses of 10 kg and 20 kg respectively are connected by a massless spring as shown in fig. A force of 200 newton acts on the 20 kg mass. At the instant shown the 10 kg mass has acceleration of 12 m/sec². What is the acceleration of 20 kg mass?

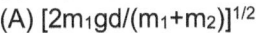

(A) 12 m/sec² (B) 4 m/sec²
(C) 10 m/sec² (D) zero

10. A chain has five rings. The mass of each ring is 0.1 kg. This chain is pulled upwards by a force F producing an acceleration of 2.50 m/sec² in the chain. Then the force of action (reaction) on the joint of second and third ring from the top is –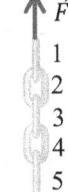
(A) 0.25 N (B) 1.23 N
(C) 3.69 N (D) 6.15 N

11. If the masses are released from the position shown in figure then the speed of mass m_1 just before it strikes the floor is –
(A) $[2m_1gd/(m_1+m_2)]^{1/2}$
(B) $[2(m_1-m_2)gd/(m_1+m_2)]^{1/2}$
(C) $[2(m_1-m_2)gd/m_1]^{1/2}$
(D) none of the above

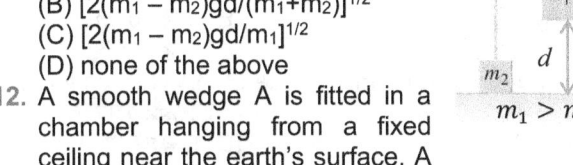

12. A smooth wedge A is fitted in a chamber hanging from a fixed ceiling near the earth's surface. A block B placed at the top of the wedge takes a time T to slide down the length of the wedge. if the block is placed at the top of the wedge and the cable supporting the chamber is broken at the same instant the block will -
(A) take a time longer than T to slide down the wedge
(B) take a time shorter than T to slide down the wedge
(C) remain at the top of the wedge
(D) jump off the wedge

13. A block can slide on a smooth inclined plane of inclination θ kept on the floor of a lift. When the lift is descending with a retardation a. the acceleration of the block relative to the incline is:
(A) $(g+a)\sin\theta$
(B) $(g-a)$
(C) $g\sin\theta$
(D) $(g-a)\sin\theta$

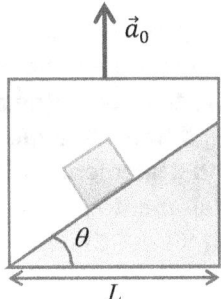

14. A house painter uses the chair-and-pulley arrangement of (see adjoining FIGURE) to lift himself up the side of a house. The painter's mass is 70 kg and the chair's mass is 10 kg. With what force must he pull down on the rope in order to accelerate upward at 0.20 m/s²?
(A) 3.0×10^2 N (B) 8.0×10^2 N
(C) 4.0×10^2 N (D) 2.0×10^2 N

15. In the system shown, the acceleration of the wedge of mass $5M$ is (there is no friction anywhere)

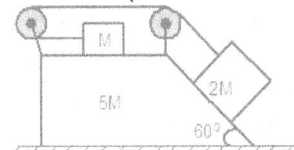

(A) zero (B) g/2
(C) g/3 (D) g/4

16. A flexible chain of weight W hangs between two fixed points A and B at the same level. The inclination of the chain with the horizontal at the two points of support is θ. What is the tension of the chain at the end point.
(A) $\frac{W}{2}\csc\theta$ (B) $\frac{W}{2}\sec\theta$
(C) $W\cos\theta$ (D) $\frac{W}{3}\sin\theta$

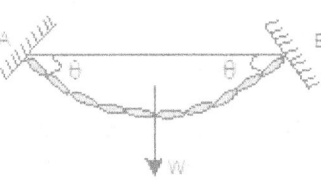

17. An inclined plane makes an angle 30° with the horizontal. A groove OA = 5m cut in the plane makes an angle 30° with OX. A short smooth cylinder is free to slide down the influence of gravity. The time taken by the cylinder to reach from A to O is (g = 10 m/s²)

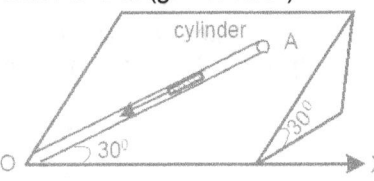

(A) 4 s (B) 2 s
(C) $2\sqrt{2}$ s (D) 1 s

18. System is shown in the FIGURE and man is pulling the rope from both sides with constant speed ' u'. Then the velocity of the block will be:

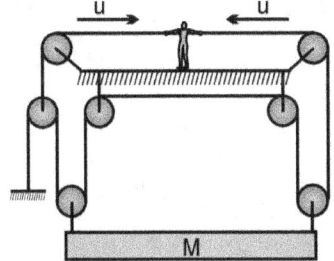

(A) $\frac{3u}{4}$ (B) $\frac{3u}{2}$
(C) $\frac{u}{4}$ (D) none of these

19. A system is shown in the following figure. Assume that cylinder remains in contact with the two wedges. The velocity of cylinder is

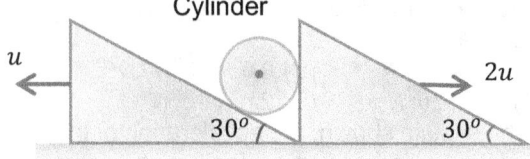

(A) $\sqrt{19-4\sqrt{3}}\frac{u}{2}$ m/s (B) $\frac{\sqrt{13}\,u}{2}$ m/s
(C) $\sqrt{3}\,u$ m/s (D) $\sqrt{7}u$ m/s

20. Two smooth spheres each of radius 5 cm and weight W rest one on the other inside a fixed smooth cylinder of radius 8 cm. The reactions between the spheres and the vertical side of the cylinder are: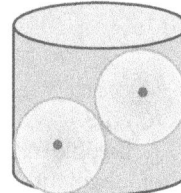
(A) W/4 & 3W/4 (B) W/4 & W/4
(C) 3W/4 & 3W/4 (D) W & W

21. Two blocks 'A' and 'B' each of mass 'm' are placed on a smooth horizontal surface. Two horizontal force F and $2F$ are applied on the 2blocks 'A' and 'B' respectively as shown in adjoining figure. The block A does not slide on block B. Then the normal reaction acting between the two blocks is : (A and B are smooth)

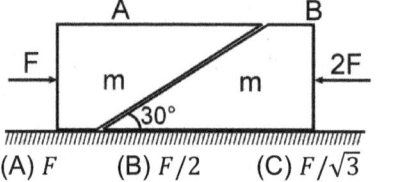

(A) F (B) $F/2$ (C) $F/\sqrt{3}$ (D) $3F$

PASSAGE BASED QUESTIONS

(I) In the systems shown in figure (A), (B), (C) and (D), the scales of the springs are calibrated in newton.

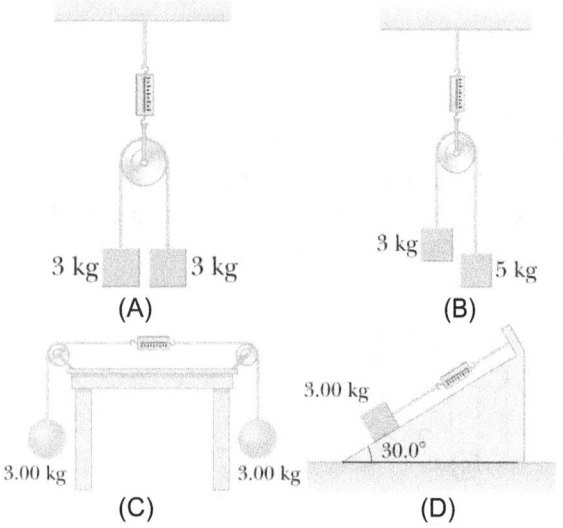

Assume that
- Pulleys are massless and frictionless
- Strings are massless
- The surface in figure (D) is frictionless

22. Reading of the spring scale in figure (A) is
(A) 30 N (B) 45 N
(C) 60 N (D) 22.5 N

23. Reading of the spring scale in figure (B) is
(A) 90 N (B) 62.5 N
(C) 55 N (D) 75 N

24. Reading of the spring scale in figure (C) is
(A) 30 N (B) 45 N
(C) 15 N (D) 22.5 N

25. Reading of the spring scale in figure (D) is
(A) 30 N (B) 45 N
(C) 15 N (D) 22.5 N

(II) A bacterium using its flagellum as propulsion can move through liquids at a rate of 0.003 m/s. For a 50μm long bacterium, that is the equivalent of 60 cell lengths per second. Bacteria of that size have a mass of approximately $1 \times 10^{-12}g$.
The viscous drag on a swimming bacterium is so great that if it stops beating its flagellum it will stop within a distance of 0.01 nm.

26. What is the acceleration that stops the bacterium?
(A) $1.2 \times 10^4 m/s^2$ (B) $5 \times 10^5 m/s^2$
(C) $6 \times 10^5 m/s^2$ (D) $9 \times 10^5 m/s^2$

27. What is average magnitude of this viscous force?
(A) $4 \times 10^{-7}N$ (B) $1.7 \times 10^{-8}N$
(C) $9 \times 10^{-9}N$ (D) $5 \times 10^{-10}N$

28. What amount of force must the flagellum generate to propel the bacterium at a constant velocity of 0.003 m/s.
(A) $1.5 \times 10^{-10}N$ (B) $5 \times 10^{-10}N$
(C) $9 \times 10^{-7}N$ (D) $1.8 \times 10^{-7}N$

29. If the bacterium wished to accelerate at a rate of $0.001 m/s^2$, how much additional force would be necessary?
(A) $1 \times 10^{-18}N$ (B) $3 \times 10^{-18}N$
(C) $1 \times 10^{-15}N$ (D) $4 \times 10^{-15}N$

ASSERTION/REASON TYPE QUESTIONS

Each of the questions given below consist of Statement – I and Statement – II. Use the following Key to choose the appropriate answer.
(A) If both Statement- I and Statement- II are true, and Statement - II is the correct explanation of Statement– I.
(B) If both Statement - I and Statement - II are true but Statement - II is not the correct explanation of Statement – I.
(C) If Statement - I is true but Statement - II is false.
(D) If Statement - I is false but Statement - II is

true.

30. **Assertion:** The third law of motion concludes that the forces occur in pairs of action and reaction.
    Reason: The action force is more than the reaction force.
    (A) a   (B) b
    (C) c   (D) d

31. **Assertion:** In a free fall, weight of a body becomes effectively zero.
    Reason: Acceleration due to gravity acting on a body having free fall is zero.
    (A) a   (B) b
    (C) c   (D) d

32. **Assertion:** Newton's second law of motion is the main law of motion.
    Reason: Newton's first and third law are contained in second law.
    (A) a   (B) b
    (C) c   (D) d

49.3.1.4. LEVEL – 4 (Previous Years Questions)

SECTION - A (JEE MAIN)

1. Three point masses A, B and C are 66 gram each are connected as shown. The acceleration of system is 5 m/s². tension between B and C is approximately- **[2002]**

   (A) 0.33 Newton   (B) 4 Newton
   (C) 5 Newton   (D) 6 Newton

2. A person in an aeroplane which is coming, down at acceleration a releases a coin. After release, the acceleration of coin with respect to observer on ground and in aeroplane both will be respectively- **[2002]**
   (A) g and (g –a)   (B) (g – a), g
   (C) (g + a), g   (D) g, (g + a)

3. A light string passing over a smooth light pulley connects two blocks of masses m_1 and m_2 (vertically). If the acceleration of the system is g/8, then the ratio of the masses is – **[2002]**
   (A) 8 : 1   (B) 9 : 7
   (C) 4 : 3   (D) 5 : 3

4. One end of massless rope, which passes over a massless and frictionless pulley P is tied to a hook C while the other end is free. Maximum tension that the rope can bear is 360 N. With what value of minimum safe acceleration (in ms⁻²) can a man of 60 kg slide down the rope? **[2002]**

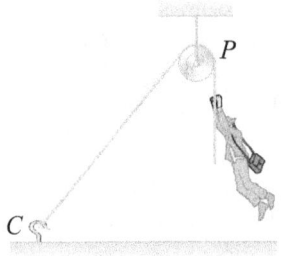

   (A) 16   (B) 6
   (C) 4   (D) 8

5. A spring balance is attached to the ceiling of a lift. A man hangs his bag on the spring and the spring reads 49 N, when the lift is stationary. If the lift moves downward with an acceleration of 5 m/s², the reading of the spring balance will be – **[2003]**
   (A) 74 N   (B) 15 N
   (C) 49 N   (D) 24 N

6. Let $\vec{F}$ be the force acting on a particle having position vector $\vec{r}$, and $\vec{\tau}$ be the torque of this force about the origin. Then – **[2003]**
   (A) $\vec{r}.\vec{\tau} \neq 0$ and $\vec{F}.\vec{\tau} = 0$
   (B) $\vec{r}.\vec{\tau} \neq 0$ and $\vec{F}.\vec{\tau} \neq 0$
   (C) $\vec{r}.\vec{\tau} = 0$ and $\vec{F}.\vec{\tau} = 0$
   (D) $\vec{r}.\vec{\tau} = 0$ and $\vec{F}.\vec{\tau} \neq 0$

7. A block of mass M is pulled along a horizontal frictionless surface by a rope of mass m. If a force P is applied at the free end of the rope, the force exerted by the rope on the block is – **[2003]**
   (A) $\frac{Pm}{M-m}$ (B) P
   (C) $\frac{PM}{M+m}$ (D) $\frac{Pm}{M+m}$

8. One end of a light spring balance hangs from the hook of the other light spring balance attached to roof and a block of mass M kg hangs from the other end. Then the true statement about the scale reading is – **[2003]**
   (A) The scale of the lower one reads M kg and of the upper one zero
   (B) The reading of the two scales can be anything but the sum of the reading will be M kg
   (C) Both the scales read M/2 kg each
   (D) Both the scales read M kg each

9. Three forces start acting simultaneously on a particle moving with velocity $\vec{v}$. These forces are represented in magnitude and direction by the three sides of a triangle ABC (as shown). The particle will now move with velocity – **[2003]**

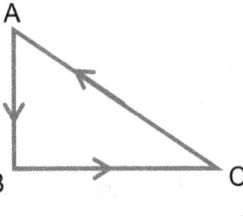

   (A) Greater than $\vec{v}$
   (B) $|\vec{v}|$ in the direction of the largest force BC
   (C) $\vec{v}$, remaining unchanged
   (D) Less than $\vec{v}$

10. A machine gun fires a bullet of mass 40 g with a velocity 1200 ms⁻¹. The man holding it can exert a maximum force of 144 N on the gun. How many bullets can he fire per second at the most? **[2004]**
(A) One (B) Four
(C) Two (D) Three

11. Two masses $m_1 = 5$ kg and $m_2 = 4.8$ kg tied to a string are hanging over a light frictionless pulley. What is the acceleration of the masses when left free to move? ($g = 9.8$ m/s²) **[2004]**
(A) 0.2 m/s²
(B) 9.8 m/s²
(C) 5 m/s²
(D) 4.8 m/s²

12. A parachutist after bailing out falls 50 m without friction. When parachute opens, it decelerates at 2 m/s². He reaches the ground with a speed of 3 m/s. At what height, did he bail out? **[2005]**
(A) 91 m (B) 182 m
(C) 293 m (D) 111 m

13. A block is kept on a frictionless inclined surface with angle of inclination 'α'. The incline is given an acceleration 'a' to keep the block stationary. Then 'a' is equal to **[2005]**

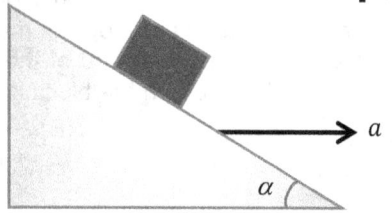

(A) $g / \tan \alpha$ (B) $g \csc \alpha$
(C) g (D) $g \tan \alpha$

14. A particle of mass 0.3 kg is subjected to a force $F = -kx$ with $k = 15$ N/m. What will be its initial acceleration if it is released from a point 20 cm away from the origin? **[2005]**
(A) 3 m/s² (B) 15 m/s²
(C) 5 m/s² (D) 10 m/s²

15. A player caught a cricket ball of mass 150 g moving at a rate of 20 m/s. If the catching process is completed in 0.1 s, the force of the blow exerted by the ball on the hand of the player is equal to – **[2006]**
(A) 30 N (B) 300 N
(C) 150 N (D) 3 N

16. A block of mass m is connected to another block of mass M by a spring (massless) of spring constant k. The blocks are kept on a smooth horizontal plane. Initially the blocks are at rest and the spring is unstretched. Then a constant force F starts acting on the block of mass M to pull it. Find the force of the block of mass m. **[2007]**
(A) $\frac{MF}{(m+M)}$ (B) $\frac{mF}{M}$
(C) $\frac{(M+m)F}{m}$ (D) $\frac{mF}{(m+M)}$

17. A body of mass $m = 3.513$ kg is moving along the x-axis with a speed of 5.00 ms⁻¹. The magnitude of its momentum is recorded as **[2008]**
(A) 17.57 kg ms⁻¹ (B) 17.6 kg ms⁻¹
(C) 17.565 kg ms⁻¹ (D) 17.56 kg ms⁻¹

18. The figure shows the position - time (x-t) graph of one-dimensional motion of a body of mass 0.4 kg. The magnitude of each impulse is **[2010]**

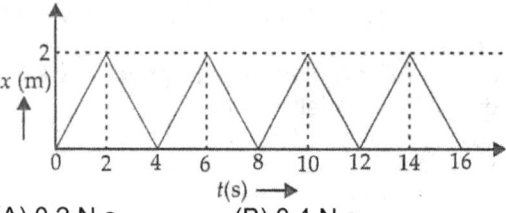

(A) 0.2 N s (B) 0.4 N s
(C) 0.8 N s (D) 1.6 N s

19. Two fixed frictionless inclined planes making an angle 30° and 60° with the vertical are shown in the figure. Two blocks A and B are placed on the two planes. What is the relative vertical acceleration of A with respect to B? **[2010]**

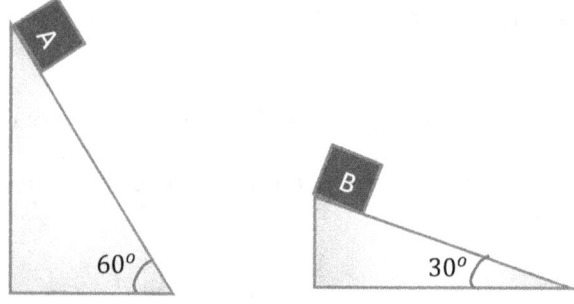

(A) 4.9 ms⁻² in vertical direction
(B) 4.9 ms⁻² in horizontal direction
(C) 9.8 ms⁻² in vertical direction
(D) zero

20. A particle of mass m is at rest at the origin at time $t = 0$. It is subjected to a force $F(t) = F_0 e^{-bt}$ in the x direction. Its speed $v(t)$ is depicted by which of the following curves? **[2012]**

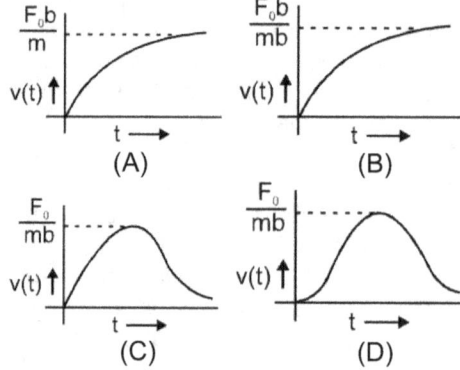

21. Two cars of masses m_x and m_2 are moving in circles of radii r_x and r_2, respectively. Their speeds are such that they make complete circles in the same time t. The ratio of their centripetal acceleration is **[2012]**
(A) $m_1 : m_2$ (B) $r_1 : r_2$

(C) 1 : 1   (D) $m_1 r_1 : m_2 r_2$

22. A block of mass m is placed on a surface with a vertical cross section given by $y = \frac{x^3}{6}$. If the coefficient of friction is 0.5, the maximum height above the ground at which the block can be placed without slipping is **[2014]**
(A) $\frac{1}{2}m$ (B) $\frac{1}{6}m$
(C) $\frac{2}{3}m$ (D) $\frac{1}{3}m$

SECTION - B (JEE ADVANCED)

1. A car is moving in a circular horizontal track of radius 10m with a constant speed of 10m/s. A plumb bob is suspended from the roof of the car by a light rigid rod of length 1.00 m. The angle made by the rod with the track is - **[1992]**
(A) zero   (B) 30°   (C) 45°   (D) 60°

2. The magnitude of the force(in newtons) acting on a body varies with time t(in microseconds) as shown in the figure. AB, BC and CD are straight line segments. The magnitude of the total impulse of the force on the body from $t = 4$ μs to $t = 16$μs is **[1994]**

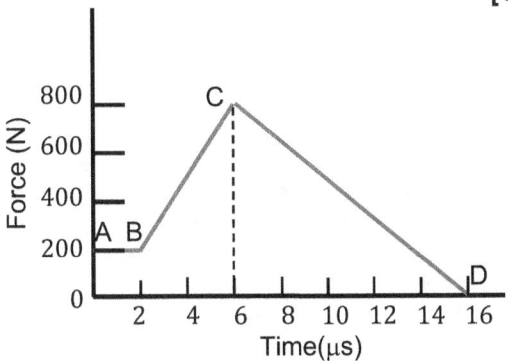

(A) .005 Ns   (B) .004 Ns
(C) .003 Ns   (D) None of these

3. A rod of weight w is supported by the parallel knife edges A and B is in equilibrium in a horizontal position. The knives are at a distance d from each other. The centre of mass of the rod is at distance x from A. The normal reaction on A is .....and on B is ... **[1997]**
(A) $(d-x)/d$, xw/d (B) $(x-d)/d$, wx/d
(C) $(d-x)$, xw (D) None of these

4. In the system shown if the inextensible string connecting 2m and m is cut, the accelerations of mass m and 2m are **[2006]**
(A) $\frac{g}{2}, \frac{g}{2}$
(B) $g, \frac{g}{2}$
(C) $\frac{g}{2}, g$
(D) g, g

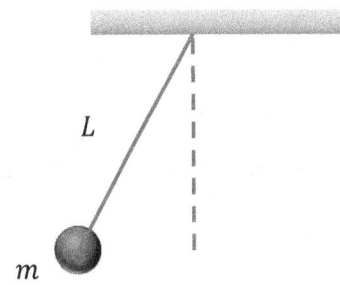

5. Two particles of mass 'm' each are tied at the ends of a light string of length 2a. The whole system is kept on a frictionless horizontal surface with the string held tight so that each mass is at a distance 'a' from the centre P (as shown in the figure). Now, the mid-point of the string is pulled vertically upwards with a small but constant force F. As a result, the particles move towards each other on the surface. The magnitude of acceleration, when the separation between them becomes $2x$, is **[2007]**

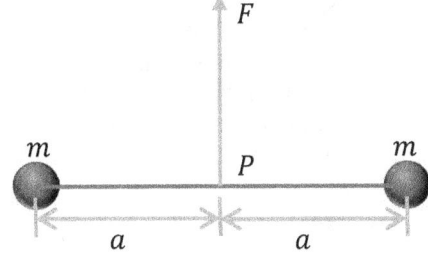

(A) $\frac{F}{2m}\frac{a}{\sqrt{a^2-x^2}}$ (B) $\frac{F}{2m}\frac{x}{\sqrt{a^2-x^2}}$
(C) $\frac{F}{2m}\frac{x}{a}$ (D) $\frac{F}{2m}\frac{\sqrt{a^2-x^2}}{x}$

6. A particle moves in the X –Y plane under the influence of a force such that its linear momentum is $\vec{p}(t) = A[\hat{i} \cos k t - \hat{j} \sin k t]$ where A and k are constants. The angle between the force and the momentum is **[2007]**
(A) 0°   (B) 30°
(C) 45°   (D) 90°

7. A piece of wire is bent in the shape of a parabola $y = kx^2$ (y-axis vertical) with a bead of mass m on it. The bead can slide on the wire without friction. It stay at the lowest point of the parabola when the wire is at rest. The wire is now accelerated parallel to the x axis with a constant acceleration a. The distance of the new equilibrium position of the bead, where the bead car stay at rest with respect to the wire, from the y-axis **[2009]**
(A) $\frac{a}{gk}$ (B) $\frac{a}{2gk}$
(C) $\frac{2a}{gk}$ (D) $\frac{a}{4gk}$

8. A ball of mass (m) 0.5 kg is attached to the end of a string having length (L) 0.5 m. The ball is rotated on a horizontal circular path about vertical axis. The maximum tension that the string can bear is 324 N. The maximum possible value of angular velocity of ball (in radian/s) is **[2011]**

(A) 9   (B) 18
(C) 27   (D) 36

9. A flat plate is moving normal to its plane through a gas under the action of a constant force F. The gas is kept at a very low pressure. The speed of the plate v is much less than the average speed u of the gas molecules. Which of the following options is/are true? **[2017]**
   (A) The resistive force experienced by the plate is proportional to v
   (B) The pressure difference between the leading and trailing faces of the plate is proportional to uv.
   (C) The plate will continue to move with constant non-zero acceleration, at all times
   (D) At a later time the external force F balances the resistive force.

49.3.2. FRICTION

49.3.2.1. LEVEL-1

CONCEPTUAL PROBLEMS

1. A block is placed on a rough floor and a horizontal force F is applied on it. The force of friction f by the floor on the block is measured for different values of F and a graph is plotted between them -
   (a) The graph is a straight line of slope 45°
   (b) The graph is straight line parallel to the F axis
   (c) The graph is a straight line of slope 45° for small F and a straight line parallel to the F-axis for large F.
   (d) There is small kink on the graph
   (A) c, d (B) a, d
   (C) a, b (D) a, c

2. Mark the correct statements about the friction between two bodies -
   (a) static friction is always greater than the kinetic friction
   (b) coefficient of static friction is always greater than the coefficient of kinetic friction
   (c) limiting friction is always greater than the kinetic friction
   (d) limiting friction is never less than static friction
   (A) b, c, d (B) a, b, c
   (C) a, c, d (D) a, b, d

3. A block A kept on an inclined surface just begins to slide if the inclination is 30°. The block is replaced by another block B and it is found that it just begins to slide if the inclination is 40°.
   (A) mass of A > mass of B
   (B) mass of A < mass of B
   (C) mass of A = mass of B
   (D) all the three are possible

4. Two cars of unequal masses use similar tyres. If they are moving at the same initial speed, the minimum stopping distance -
   (A) is smaller for the heavier car
   (B) is smaller for the lighter car
   (C) is same for both cars
   (D) depends on the volume of the car

5. Consider the situation shown in figure. The wall is smooth but the surfaces of A and B in contact are rough. The friction on B due to A in equilibrium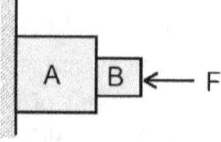
   (A) is upward
   (B) is downward
   (C) is zero
   (D) the system cannot remain in equilibrium

6. Suppose all the surfaces in the previous problem are rough. The direction of friction on B due to A -
   (A) is upward
   (B) is downward
   (C) is zero
   (D) depends on the masses of A and B

7. A body of mass M is kept on a rough horizontal surface (friction coefficient = μ). A person is trying to pull the body by applying a horizontal force but the body is not moving. The force by the surface on A is F, where -
   (A) $F = Mg$
   (B) $F = \mu Mg$
   (C) $Mg \leq F \leq Mg\sqrt{1 + \mu^2}$
   (D) $Mg \geq F \geq Mg\sqrt{1 - \mu^2}$

8. In a situation the contact force by a rough horizontal surface on a body placed on it has constant magnitude if the angle between this force and the vertical is decreased the frictional force between the surface and the body will -
   (A) increase
   (B) decrease
   (C) remain the same
   (D) may increase or decrease

9. The contact force exerted by a body A on another body B is equal to the normal force between the bodies. We conclude that –
   (a) the surfaces must be smooth
   (b) force of friction between two bodies may be equal to zero
   (c) magnitude of normal reaction is equal to that of friction
   (d) bodies may be rough
   (e) acceleration of two bodies is identical and remains constant
   (A) b, d (B) a, b (C) c, d (D) a, d

10. It is easier to pull a body than to push, because -
    (A) the coefficient of friction is more in pushing than that in pulling
    (B) the friction force is more in pushing than that in pulling
    (C) the body does not move forward when pushed
    (D) None of these

MOTION OF A BLOCK ON THE HORIZONTAL PLANE

11. A force F is required to push a crate along a rough horizontal floor at a constant speed v with friction

present. What force is needed to push this crate along the same floor at a constant speed $3v$ if friction is the same as before?
(A) A constant force $3F$ is needed.
(B) A force that gradually increases from F to $3F$ is needed.
(C) A constant force F is needed.
(D) No force is needed, since the crate has no acceleration.

12. When you're driving on the freeway it's necessary to keep your foot on the accelerator to keep the car moving at a constant speed. In this situation
(A) the net force on the car is in the forward direction.
(B) the net force on the car is toward the rear.
(C) the net force on the car is zero.
(D) the net force on the car depends on your speed

13. A worker pulls horizontally on a crate on a rough horizontal floor, causing it to move forward with constant velocity. In adjoining figure, force A is the pull of the worker and force B is the force of friction due to the floor. Which one of the following statements about these forces is correct?

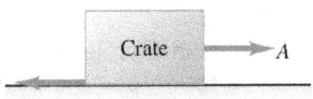

(A) $A < B$ (B) $A = B$
(C) $A > B$ (D) none of the above

14. A block of metal is lying on the floor of a bus. The maximum acceleration which can be given to the bus so that the block may remain at rest, will be -
(A) μg (B) μ/g
(C) $\mu^2 g$ (D) μg^2

15. A chain is lying on a rough table with a fraction $1/n$ of its length hanging down from the edge of the table. if it is just on the point of sliding down from the table, then the coefficient of friction between the table and the chain is -
(A) $\dfrac{1}{n}$ (B) $\dfrac{1}{(n-1)}$
(C) $\dfrac{1}{(n+1)}$ (D) $\dfrac{n-1}{(n+1)}$

16. Two masses A and B of 10 kg and 5 kg respectively are connected with a string passing over a frictionless pulley at a corner of a table as shown in the adjoining diagram. The coefficient of friction of A with the table is 0.2. The minimum mass of C that may be placed on A to prevent it from moving is equal to –

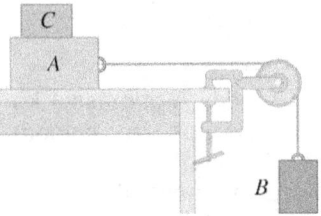

(A) 0 kg (B) 5 kg
(C) 10 kg (D) 15 kg

17. In the figure, the block A and B are of masses 3 kg and 2 kg. The coefficient of friction between the two blocks A and B is 0.3. The surface of the table is smooth. Then –

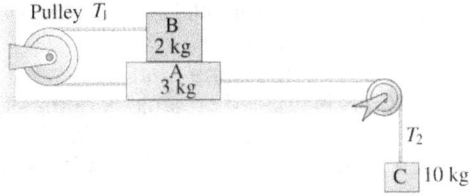

(a) The acceleration of masses is 5.75 ms^{-2}
(b) The tensions are T_1 and T_2 in the strings are 90.36 N and 17.55 N
(c) Acceleration of masses is 8.15 ms^{-2}
(d) Tension T_1 and T_2 in the strings are 17.38 N and 40.50 N
(A) a, c (B) c, d
(C) b, d (D) a, d

18. In the arrangement shown mass of A = 1 kg, mass of B = 2 kg coefficient of friction between A and B is 0.2. There is no friction between B and ground. The frictional force between A and B is: ($g = 9.8\ ms^{-2}$)

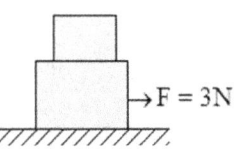

(A) 0 N (B) 2 N
(C) 1.96 N (D) 1 N

19. A block of mass M rests on a rough horizontal surface as shown. Coefficient of friction between the block and the surface is μ. A force F = Mg acting at angle θ with the vertical side of the block pulls it in which of the following cases the block can be pulled along the surface?

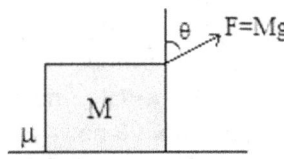

(A) $\tan\theta \geq \mu$ (B) $\tan(\theta/2) \geq \mu$
(C) $\cot\theta \geq \mu$ (D) $\cot(\theta/2) \geq \mu$

20. The coefficient of friction between car C and the inclined plane is 0.8. The maximum acceleration of the car, so that it may move up the plane inclined at an angle θ without slipping will be –

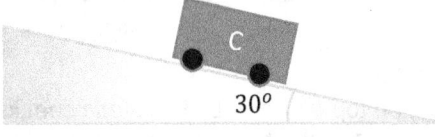

(A) 2.67 m/s^2 (B) 3.02 m/s^2
(C) 1.88 m/s^2 (D) 4.32 m/s^2

21. A block rests on a rough inclined plane as shown in figure. A horizontal force F is applied to it-
(a) Reaction on the block is $F\sin\theta + mg\cos\theta$
(b) Frictional force is zero when $F\cos\theta = mg\sin\theta$
(c) The value of limiting friction in $\mu(mg\sin\theta + F\cos\theta)$
(d) The value of limiting friction is $\mu(mg\sin\theta - F\cos\theta)$
(A) a, b (B) c, d

(C) b, d (D) b, c
22. For the arrangement shown in fig., the tension in the string to prevent it from sliding down, is -
(A) 6 N (B) 6.4 N
(C) 0.4 N (D) None of these
23. A force of 100 N is applied on a block of mass 3 kg as shown in fig. The coefficient of friction between the surface of the block is 1/4. The friction force acting on the block is
(A) 15 N downwards (B) 25 N upwards
(C) 20 N downwards (D) 20 N upwards
24. A block moves down a smooth inclined plane of inclination θ. Its velocity on reaching the bottom is v. If it slides down a rough inclined plane of some inclination, its velocity on reaching the bottom is v/n, where n is a number greater than 0. The coefficient of friction is given by -
(A) $\mu = \tan\theta\left(1 - \frac{1}{n^2}\right)$
(B) $\mu = \cot\theta\left(1 - \frac{1}{n^2}\right)$
(C) $\mu = \tan\theta\left(1 - \frac{1}{n^2}\right)^{1/2}$
(D) $\mu = \cot\theta\left(1 - \frac{1}{n^2}\right)^{1/2}$
25. A block of mass 2kg rests on a rough inclined plane making an angle of 30° with the horizontal. The coefficient of static friction between the block and the plane is 0.7. The frictional force on the block is-
(A) 0.7 × 9.8 newton
(B) 9.8 newton
(C) 0.7 × 9.8√3 newton
(D) 9.8 × √3 newton
26. A truck and a car are moving with equal velocity. On applying the brakes both will stop after certain distance, then
(A) Truck will cover less distance before rest
(B) Car will cover less distance before rest
(C) Both will cover equal distance
(D) None

49.3.2.2. LEVEL-2

1. A 40 kg slab rests on a frictionless floor. A 10 kg block rests on top of the slab. The static coefficient of friction between the block and slab is 0.60 while the kinetic coefficient is 0.40. The 10 kg block is acted upon by a horizontal force 100 N. If g = 9.8 m/s², the resulting acceleration of the slab will be –
(A) 0.98 m/s² (B) 1.47 m/s²
(C) 1.52 m/s² (D) 6.1 m/s²
2. A lift is moving downwards with an acceleration equal to the acceleration due to gravity. A body of mass M kept on the floor of the lift is pulled horizontally. If the coefficient of friction is μ, then the frictional resistance offered by the body is-
(A) Mg (B) μMg
(C) 2μMg (D) zero
3. The rear side of a truck is open and box of mass 20 kg is placed on the truck 4 meters away from rest with an acceleration of 2 m/sec² on a straight road. The truck starts from rest with an acceleration of 2 m/sec² on a straight road. The box will fall off the truck when it is at a distance from the starting point equal to (μ = 0.15) -
(A) 4 m (B) 8 m
(C) 16 m (D) 32 m
4. A uniform rope of length l lies on a table if the coefficient of friction is μ, then the maximum length of the part of this rope which can over hang from the edge of the table without sliding down is -
(A) $\frac{l}{\mu}$ (B) l
(C) $\frac{\mu l}{\mu+1}$ (D) $\frac{\mu l}{\mu-1}$
5. A particle is projected along a line of greatest slope on a rough plane inclined at an angle of 45° with the horizontal, if the coefficient of friction is 1/2, then the retardation is-
(A) $\frac{g}{\sqrt{2}}$ (B) $\frac{g}{2\sqrt{2}}$
(C) $\frac{g}{\sqrt{2}}\left(1 + \frac{1}{2}\right)$ (D) $\frac{g}{\sqrt{2}}\left(1 - \frac{1}{2}\right)$
6. An inclined plane is inclined at an angle θ with the horizontal. A body of mass m rests on it, if the coefficient of friction is μ, then the minimum force that has to be applied to the inclined plane to make the body just move up the inclined plane is -
(A) mg sin θ
(B) μ mg cos θ
(C) μmg cos θ – mg sin θ
(D) μ mg cos θ + mg sin θ
7. A cart of mass M has a block of mass m attached to it as shown in figure. The coefficient of friction between the block and the cart is μ. What is the minimum acceleration of the cart so that the block m does not fall?
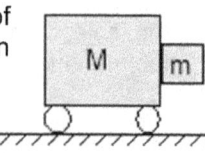
(A) mg (B) g/μ
(C) μ/g (D) $M\mu g/m$
8. A body A of mass 1 kg rests on a smooth surface. Another body B of mass 0.2 kg is placed over A as shown. The coefficient of static friction between A and B is 0.15. B will begin to slide on A, if A is pulled with a force greater than –
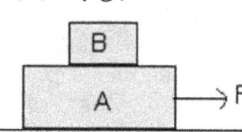
(A) 1.764 N (B) 0.1764 N
(C) 0.3 N (D) it will not slide for any F
9. A block of mass 1 kg lies on a horizontal surface in a truck. The coefficient of static friction between the block and the surface is 0.6 if the acceleration of

the truck is 5 m/s² the frictional force acting on the block is-
(A) 10 N (B) 6 N
(C) 4 N (D) 5 N

10. Starting from rest, a body slides down a 45° inclined plane in twice the time it takes to slide down the same distance in the absence of friction. The coefficient of friction between the body and the inclined plane is -
(A) 0.33 (B) 0.25
(C) 0.75 (D) 0.80

11. A heavy uniform chain lies on a horizontal table top. If the coefficient of friction between the chain and the table surface is 0.25, then the maximum fraction of the length of the chain that can hang over one edge of the table is -
(A) 20% (B) 25%
(C) 35% (D) 15%

12. A body of mass 2 kg is placed on a horizontal surface having coefficient of kinetic friction 0.4 and coefficient of static friction 0.5. If a horizontal force of 25 N is applied on the body, the frictional force acting on the body will be: ($g = 10$)
(A) 8 N (B) 10 N
(C) 20 N (D) 25 N

13. The brakes of a car moving at 20 m/s along a horizontal road are suddenly applied and it comes to rest after travelling some distance if the coefficient of friction between the tyres and the road is 0.90 and it is assumed that all four tyres behave identically, the shortest distance the car would travel before coming to a stop is -
(A) 2.27 m (B) 11.35 m
(C) 22.7 m (D) 4.54 m

14. A block rests on an inclined plane that makes an angle θ with the horizontal, if the coefficient of sliding friction is 0.50 and that of static friction is 0.75, the time required to slide the block 4 m along the inclined plane is -
(A) 25 s (B) 10 s
(C) 5 s (D) 2 s

15. Block A has a mass of 2 kg and block B has a mass of 20 kg if the coefficient of kinetic friction between block B and the horizontal surface is 0.1, and B is accelerating towards the right with a = 2 m/s², then the mass of the block C will be –

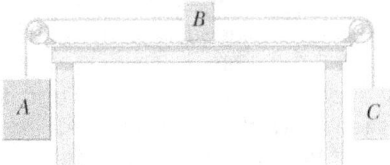

(A) 15 kg (B) 12.5 kg
(C) 5.7 kg (D) 10.5 kg

16. A force F accelerates a block of mass m on horizontal surface. The coefficient of friction between the contact surface is μ. The acceleration of m will be -

(A) $\frac{F}{m} - \mu g$
(B) zero
(C) may be (A) or (B)
(D) none of these

17. A horizontal force F is exerted on a 20 kg block to push it up an inclined plane having an inclination of 30°. The frictional force retarding the motion is 80 N. For the acceleration of the moving block to be zero, the force F must be -
(A) 206 N (B) 602 N
(C) 620 N (D) 260 N

18. With reference to the fig. shown, if the coefficient of friction at all the surfaces is 0.42, then the force required to pull out the 6.0 kg block with an acceleration of 1.50 m/s² will be –

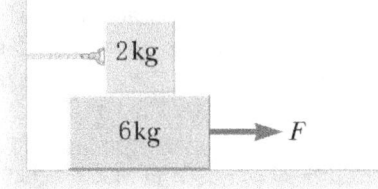

(A) 36 N (B) 24 N
(C) 84 N (D) 51 N

19. A person wants to drive on the vertical surface of a large cylindrical wooden 'well' commonly known as 'death well' in a circus. The radius of the well is R and the coefficient of friction between the tyres of the motorcycle and the wall of the well is μ_s. The minimum speed the motor cyclist must have in order to prevent slipping should be -

(A) $\sqrt{\frac{rg}{\mu}}$ (B) $tan^{-1}\mu$
(C) $\frac{\sqrt{rg}}{\mu}$ (D) $\frac{rg}{\mu}$

20. A spherical ball of mass 1/2 kg is held at the top of an inclined rough plane making angle 30° with the horizontal the coefficient of limiting friction is 0.5. If the ball just slides down the plane without rolling its acceleration down the plane is -

(A) $\left[\frac{2-\sqrt{3}}{4}\right]g$ (B) g
(C) $\left[\frac{2\sqrt{3}-1}{4}\right]g$ (D) $\left[\frac{\sqrt{3}-1}{2}\right]g$

21. An object is placed on the surface of a smooth inclined plane of inclination θ. It takes time t to reach the bottom of the inclined plane. If the same object is allowed to slide down rough inclined plane of same inclination θ, it takes time nt to reach the bottom where n is a number greater than 1. The coefficient of friction μ is given by –

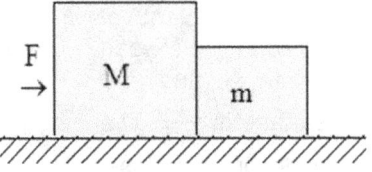

(A) $\mu = \tan\theta\left(1 - \frac{1}{n^2}\right)$

(B) $\mu = \cot\theta \left(1 - \frac{1}{n^2}\right)$

(C) $\mu = \tan\theta \left(1 - \frac{1}{n^2}\right)^{1/2}$

(D) $\mu = \cot\theta \left(1 - \frac{1}{n^2}\right)^{1/2}$

22. A given object takes n times as much time to slide down a 45° rough incline as it takes to slide down a perfectly smooth 45° incline. The coefficient of kinetic friction between the object and the incline is given by -

(A) $1 - \frac{1}{n^2}$ (B) $\frac{1}{1-n^2}$

(C) $\sqrt{1 - \frac{1}{n^2}}$ (D) $\sqrt{\frac{1}{1-n^2}}$

23. A 15 kg mass is accelerated from rest with a force of 100 N. As it moves faster, friction and air resistance create an oppositely directed retarding force given by $F_R = A + Bv$, where A = 25 N and B = 0.5 N/m/s. At what velocity does the acceleration equal to one half of the initial acceleration?

(A) 25 ms^{-1} (B) 50 m/s
(C) 75 m/s (D) 100 m/s

24. Two blocks of masses M = 3 kg and m = 2 kg, are in contact on a horizontal table. A constant horizontal force $F = 5\,N$ is applied to block M as shown. There is a constant frictional force of 2 N between the table and the block m but no frictional force between the table and the first block M, then the acceleration of the two blocks is-

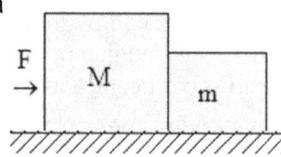

(A) 0.4 ms^{-2} (B) 0.6 ms^{-2}
(C) 0.8 ms^{-2} (D) 1 ms^{-2}

25. Starting from rest a body slides down a 45° inclined plane in twice the time it takes to slide down the same distance in the absence of friction. The coefficient of friction between the body and the inclined plane is -

(A) 0.33 (B) 0.25
(C) 0.75 (D) 0.80

26. Block A of mass M in the system shown in the figure slides down the incline at a constant speed. The coefficient of friction between block A and the surface is $\frac{1}{3\sqrt{3}}$. The mass of block B is-

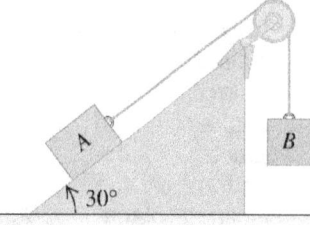

(A) M/2 (B) M/3
(C) 2M/3 (D) M/$\sqrt{3}$

27. Two blocks connected by a massless string slide down an inclined plane having angle of inclination 37°. The masses of the two blocks are $m_1 = 4$ kg and $m_2 = 2$ kg respectively and the coefficients of friction 0.75 and 0.25 respectively-

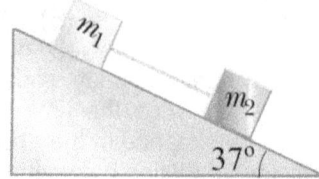

(a) The common acceleration of the two masses is 1.3 ms^{-2}
(b) The tension in the string is 14.7 N
(c) The common acceleration of the two masses is 2.94 ms^{-2}
(d) The tension in the string is 5.29 N

(A) a, d (B) c, d
(C) b, d (D) b, c

28. The coefficient of static friction between the two blocks is 0.363. What is the minimum acceleration of block 1 so that block 2 does not fall?

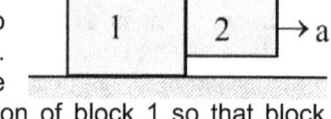

(A) 6 ms^{-2} (B) 12 ms^{-2}
(C) 18 ms^{-2} (D) 27 ms^{-2}

29. A block of mass m is placed on a rough inclined plane of inclination θ kept on the floor of the lift. The coefficient of friction between the block and the inclined plane is μ. With what acceleration will the block slide down the inclined plane when the lift falls freely?

(A) Zero
(B) $g\sin\theta - \mu g\cos\theta$
(C) $g\sin\theta + \mu g\cos\theta$
(D) None of these

30. Block A of mass 35 kg is resting on a frictionless floor. Another block B of mass 7 kg is resting on it as shown in the figure. The coefficient of friction between the blocks is 0.5 while kinetic friction is 0.4. If a force of 100 N is applied to block B, the acceleration of the block A will be (g = 10 m s^{-2})

(A) 0.8 m s^{-2} (B) 2.4 m s^{-2}
(C) 0.4 m s^{-2} (D) 4.4 m s^{-2}

31. A box of mass 8 kg is placed on a rough inclined plane of inclination θ. Its downward motion can be prevented by applying an upward pull F and it can be made to slide upwards by applying a force 2F. The coefficient of friction between the box and the inclined plane is -

(A) $\frac{1}{3}\tan\theta$ (B) $3\tan\theta$
(C) $\frac{1}{2}\tan\theta$ (D) $2\tan\theta$

32. A wooden block of mass M resting on a rough horizontal surface is pulled with a force F at an

angle φ with the horizontal. If μ is the coefficient of kinetic friction between the block and the surface, then acceleration of the block is -

(A) $\frac{F}{M}(\cos\phi + \mu\sin\phi) - \mu g$

(A) $\frac{F}{M}(\cos\phi + \mu\sin\phi) - \mu g$

(B) F sin φ/M

(C) μF cosφ

(D) μF sin φ

Each of the questions given below consist of Statement – I and Statement – II. Use the following Key to choose the appropriate answer.

(A) If both Statement- I and Statement- II are true, and Statement - II is the correct explanation of Statement– I.

(B) If both Statement - I and Statement - II are true but Statement - II is not the correct explanation of Statement – I.

(C) If Statement - I is true but Statement – II is false.

(D) If Statement - I is false but Statement - II is true.

33. **Statement I:** Pulling a lawn roller is easier than pushing it.
    Statement II: Pushing increases the apparent weight and hence force of friction.

34. **Statement I:** The force of friction in the case of a disc rolling without slipping down an inclined plane is zero.
    Statement II: When the disc rolls without slipping, friction is required because for rolling condition velocity of point of contact is zero.

35. **Statement I:** Friction is self adjusting force.
    Statement II: The magnitude of static friction is equal to the applied force and its direction is opposite to that of the applied force.

49.3.2.3. LEVEL-3

1. A block of mass 4 kg is kept over a rough horizontal surface. The coefficient of friction between the block and the surface is 0.1. At t = 0, 3 m/s ($\hat{i}$) velocity is imparted to the block and simultaneously 2N ($-\hat{i}$) force starts acting on it. Its displacement in first 5 second is (g = 10 m/s²) -
   (A) $8\hat{i}$ (B) $-8\hat{i}$
   (C) $3\hat{i}$ (D) $-3\hat{i}$

2. In the given figure the wedge is fixed, pulley is frictionless and string is light. Surface AB is frictionless whereas AC is rough. If the block of mass 3m slides down with constant velocity, then the coefficient of friction between surface AC and the block is –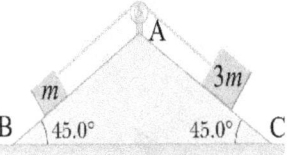

(A) $\frac{1}{3}$ (B) $\frac{2}{3}$
(C) $\frac{1}{2}$ (D) $\frac{4}{3}$

3. Mass of upper block and lower block kept over the table is 2 kg and 1 kg respectively and coefficient of friction between the blocks is 0.1. Table surface is smooth. The maximum mass m for which all the three blocks move with same acceleration is (g = 10 m/s²) –

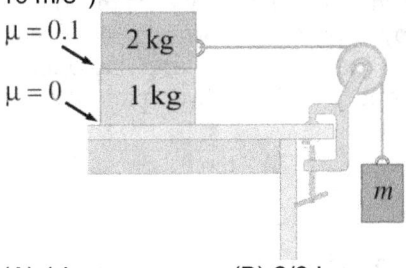

(A) 1 kg (B) 2/3 kg
(C) 1/3 kg (D) 3/4 kg

4. A block of mass m lying on a rough horizontal plane is acted upon by a horizontal force P and another force Q inclined at an angle θ to the vertical. The block will remain in equilibrium if the coefficient of friction between it and the surface is –

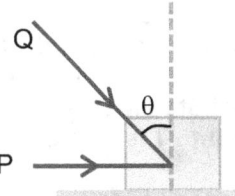

(A) $\frac{P+Q\sin\theta}{mg+Q\cos\theta}$ (B) $\frac{P\cos\theta+Q}{mg-Q\sin\theta}$
(C) $\frac{P+Q\cos\theta}{mg+Q\sin\theta}$ (D) $\frac{P\sin\theta-Q}{mg-Q\cos\theta}$

5. A block slides down an inclined surface of inclination 30° with the horizontal. Starting from rest it covers 8m in the first two seconds. Find the coefficient of kinetic friction between the two.
   (A) 0.11 (B) 0.5
   (C) 0.8 (D) 0.2

6. A body of mass 2 kg is lying on a rough inclined plane of inclination 30°. Find the magnitude of the force parallel to the incline needed to make the block move (a) up the incline (b) down the incline. Coefficient of static friction = 0.2
   (A) 13 N, 5 N (B) 13 N, 13 N
   (C) 13 N, 0 N (D) 5 N, 13 N

7. Figure shows two blocks in contact sliding down an inclined surface of inclination 30°. The friction coefficient between the block of mass 2.0 kg and the incline is μ_1, and that between the block of mass 4.0 kg and the incline is μ_2. Calculate the acceleration of the 2.0 kg block if $\mu_1 = 0.30$ and $\mu_2 = 0.20$, Take $g = 10$ m/s²

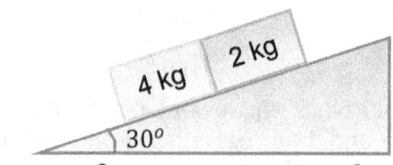

(A) 2 m/s² (B) 2.7 m/s²
(C) 4 m/s² (D) 2.4 m/s²

8. A box of mass 8 kg is placed on a rough inclined plane of inclination θ. Its downward motion can be prevented by applying an upward pull F and it can be made to slide upwards by applying a force $2F$. The coefficient of friction between the box and the inclined plane is -

(A) $\frac{1}{3}\tan θ$ (B) $3 \tan θ$
(C) $\frac{1}{2}\tan θ$ (D) $2 \tan θ$

Assertion/Reason Type Questions

Each of the questions given below consist of Statement – I and Statement – II. Use the following Key to choose the appropriate answer.
(A) If both Statement- I and Statement- II are true, and Statement - II is the correct explanation of Statement– I.
(B) If both Statement - I and Statement - II are true but Statement - II is not the correct explanation of Statement – I.
(C) If Statement - I is true but Statement - II is false.
(D) If Statement - I is false but Statement - II is true.

9. **Assertion:** A coin is placed on phonogram turn table. The motor is started, coin moves along the moving table.
Reason: Rotating table is providing necessary centripetal force to the coin.
(A) a (B) b
(C) c (D) d

10. **Assertion:** By pressing a block against a rough wall, one can balance it.
Reason: Smooth walls can not hold the block by pressing the block against the wall, however high the force is exerted.
(A) a (B) b
(C) c (D) d

11. **Assertion:** The value of dynamic friction is less than the limiting friction.
Reason: Once the motion has started, the inertia of rest has been overcome.
(A) a (B) b
(C) c (D) d

12. **Assertion:** The acceleration of a body down a rough inclined plane is greater than the acceleration due to gravity.
Reason: The body is able to slide on a inclined plane only when its acceleration is greater than acceleration due to gravity
(A) a (B) b

(C) c (D) d

Column matching Type Questions

13. A man is standing on a ladder as shown in figure. N_1 and N_2 are the normal reactions and f the force of friction. Then match the following table:

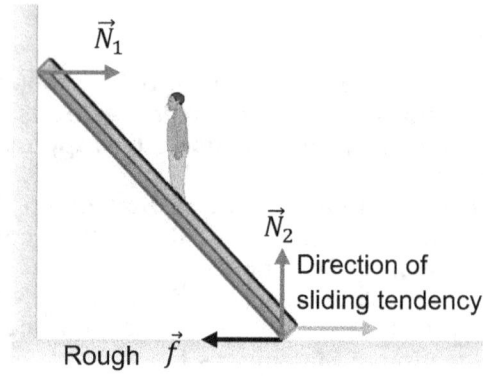

Table -1	Table -2
(A) As the man moves up the ladder	(P) N_1 will Increase
(B) If weight of ladder is increased	(Q) N_1 will decrease
	(R) N_2 will increase
	(S) N_2 will decrease
	(T) f will increase

Passage Type Questions

Two blocks of mass $m_1 = 10$ kg and $m_2 = 20$ kg are placed on a fixed inclined surface making an angle $θ = 37°$ with horizontal. One end of a light spring of spring constant $k = 100$ N/m is free and other end is connected to a support S rigidly fixed to inclined surface. The coefficient of friction between block of mass m_1 and inclined plane is $μ_1 = 0.5$ and that between block of mass m_2 and inclined plane is $μ_2 = 1$. At time $t = 0$ both blocks are released at rest from shown position. (Take $g = 10$ m/s²)

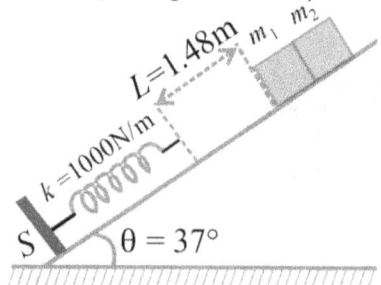

14. At the instant the speed of block of mass m₁ is maximum, the compression in the spring is
(A) 5 cm (B) 10 cm

(C) 20 cm (D) 60 cm
15. At the instant speed of block of mass m_1 is maximum, the speed of block of mass m_2 is
(A) zero (B) 2 m/s
(C) 2/3 m/s (D) 4/3 m/s
16. The maximum force exerted by block of mass m_1 on block of mass m_2 is
(A) zero (B) 40/3 N
(C) 80/3 N (D) 40 N

49.3.2.4. LEVEL-4 (Previous Years Questions)

SECTION - A (JEE MAIN)

1. A horizontal force of 10 Newton is necessary to just hold a block stationary against a wall. The coefficient of friction between the block and the wall is 0.2. The weight of block is – **[2003]**

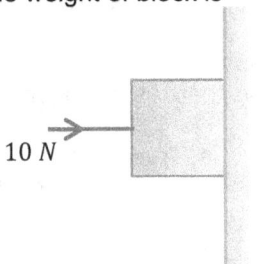

(A) 50 N (B) 100 N
(C) 2 N (D) 20 N

2. A marble block of mass 2 kg lying on ice when given a velocity of 6 m/s is stopped by friction is 10 s. Then the coefficient of friction is – **[2003]**
(A) 0.03 (B) 0.04
(C) 0.06 (D) 0.02

3. A block rests on a rough inclined plane making an angle of 30° with the horizontal. The coefficient of static friction between the block and the plane is 0.8. If the frictional force on the block is 10 N, the mass of block (in kg) is (Take g = 10 m/s²) **[2004]**
(A) 2.0 (B) 4.0
(C) 1.6 (D) 2.5

4. A smooth block is released at rest on a 45° incline and then slides a distance 'd'. The time taken to slide is 'n' times as much to slide on rough incline than on a smooth incline. The coefficient of friction is **[2005]**
(A) $\mu_k = 1 - \frac{1}{n^2}$
(B) $\mu_k = \sqrt{1 - \frac{1}{n^2}}$
(C) $\mu_s = 1 - \frac{1}{n^2}$
(D) $\mu_s = \sqrt{1 - \frac{1}{n^2}}$

5. The upper half of an inclined plane with inclination ϕ is perfectly smooth while the lower half is rough. A body starting from rest at the top will again come to rest at the bottom if the coefficient of friction for the lower half is given by **[2005]**

(A) 2 sin ϕ (B) 2 cos ϕ
(C) 2 tan ϕ (D) tan ϕ

6. Consider a car moving on a straight road with a speed of 100 m/s. The distance at which car can be stopped is [$\mu_k = 0.5$] **[2005]**
(A) 800 m (B) 1000 m
(C) 100 m (D) 400 m

7. Given in the figure are two blocks A and B of weight 20 N and 100 N, respectively. These are being pressed against a wall by a force F as shown. If the coefficient of friction between the blocks is 0.1 and between block B and the wall is 0.15, the frictional force applied by the wall on block B is: **[2015]**

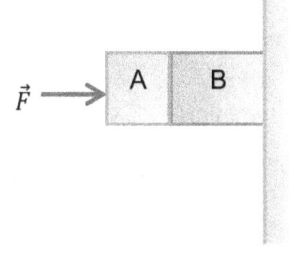

(A) 100 N (B) 80 N
(C) 120 N (D) 150 N

8. A body of mass m = 10^{-2} kg is moving in a medium and experiences a frictional force $F = -kv^2$. Its initial speed is $v_0 = 10\ ms^{-1}$. If, after 10 s, its energy is $\frac{1}{2}mv_0^2$, the value of k will be: **[2017]**
(A) $10^{-1}\ kg\ m^{-1}s^{-1}$ (B) $10^{-3}\ kg\ m^{-1}$
(C) $10^{-3}\ kg\ s^{-1}$ (D) $10^{-4}\ kg\ m^{-1}$

9. Two masses m_1 = 5 kg and m_2 = 10 kg, connected by an inextensible string over a frictionless pulley, are moving as shown in the figure. The coefficient of friction of horizontal surface is 0.15. The minimum weight m that should be put on top of m_2 to stop the motion is- **[2018]**

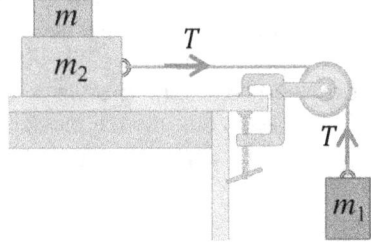

(A) 18.3 kg (B) 27.3 kg
(C) 43.3 kg (D) 10.3 kg

SECTION - B (JEE ADVANCED)

1. A block of mass 0.1 kg is held against a wall by applying a horizontal force of 5 N on the block. If the coefficient of friction between the block and the wall is 0.5, the magnitude of the frictional force acting on the block is – **[1994]**
(A) 2.5N (B) 0.98 N
(C) 4.9 N (D) 0.49 N

2. A long horizontal rod has a bead which can slide its length, and initially placed at a distance L from one end A of the rod. The rod is set in angular motion about A with constant angular acceleration a. If the coefficient of friction between the rod and the bead

is μ, and gravity is neglected, then the time after which the bead starts slipping is: **[2000]**

(A) $\sqrt{\dfrac{\mu}{\alpha}}$ (B) $\dfrac{\mu}{\sqrt{\alpha}}$

(C) $\dfrac{1}{\sqrt{\mu\alpha}}$ (D) infinitesimal

3. A block of mass $2\sqrt{3}$ kg is placed on a rough horizontal surface whose coefficient of friction is $\dfrac{1}{2}\sqrt{3}$. Then the minimum value of force F (shown in figure) for which the block starts to slide on the surface. ($g = 10$ m/s^2) **[2003]**

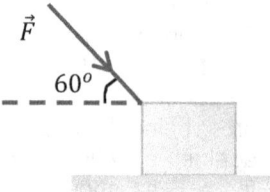

(A) 10N   (B) 5N
(C) 40N   (D)

4. Two blocks P & Q of same mass m are placed as shown in figure. Coefficient of static friction between P & Q is μ_s. There is no friction between P and horizontal surface. The block system is displaced by a distance A and released. If there is no slipping between Q & P then maximum value of friction on block P is- **[2004]**

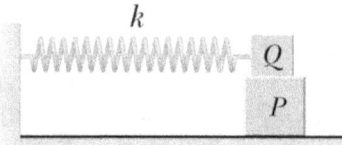

(A) 0   (B) K A
(C) KA /2   (D) μ_s mg

5. A disc having a groove as shown in kept on smooth horizontal plane. A block of mass m can slide in the groove. The disc move with acceleration $a_0 = 25\ m/s^2$ as shown in figure. The coefficient of friction between block and groove is 2/5 and $\sin\theta = 3/5$. Find the acceleration of block relative to the disc.

[2006]

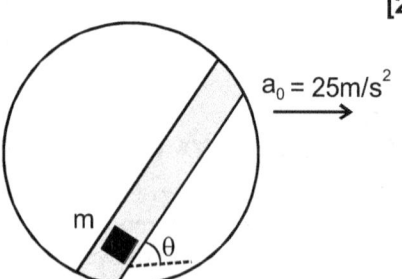

6. **STATEMENT – 1.** A block of mass m starts moving on a rough horizontal surface with a velocity v. It stops due to friction between the block and the surface after moving through a certain distance. The surface is now tilted to an angle of 30° with the horizontal and the same block is made to go up on the surface with the same initial velocity v. The decrease in the mechanical energy in the second situation is smaller than that in the first situation. Because-

STATEMENT – 2. The coefficient of friction between the block and the surface decreases with the increase in the angle of inclination.

(A) Statement-1 is True, Statement-2 is True; Statement-2 **is** a correct explanation for Statement-1
(B) Statement-1 is True, Statement-2 is True; Statement-2 **is NOT** a correct explanation for Statement-1
(C) Statement-1 is True, Statement-2 is False
(D) Statement-1 is False, Statement-2 is True.

[2007]

7. **STATEMENT –1.** A cloth covers a table. Some dishes are kept on it. The cloth can be pulled out without dislodging the dishes from the table. Because-

STATEMENT –2
For every action there is an equal and opposite reaction.

(A) Statement–1 is True, Statement–2 is True; Statement–2 **is** a correct explanation for Statement–1
(B) Statement–1 is True, Statement–2 is True; Statement–2 is NOT a correct explanation for Statement–1
(C) Statement–1 is True, Statement–2 is False
(D) Statement–1 is False, Statement–2 is True.

[2007]

8. **STATEMENT - 1**
It is easier to pull a heavy object than to push it on a level ground.
and
STATEMENT - 2
The magnitude of frictional force depends on the nature of the two surfaces in contact.

(A) STATEMENT-1 is True, STATEMENT-2 is True; STATEMENT-2 **is** a correct explanation for STATEMENT-1
(B) STATEMENT-1 is True, STATEMENT-2 is True; STATEMENT-2 is not a correct explanation for STATEMENT-1
(C) STATEMENT-1 is True, STATEMENT-2 is False
(D) STATEMENT-1 is False, STATEMENT-2 is True

[2008]

9. A block of base 10 cm × 10 cm and height 15 cm kept on an inclined plane. The coefficient of friction between them is $\sqrt{3}$. The inclination θ of this inclined plane from the horizontal plane is gradually increase, from 0°. Then **[2009]**
(A) at $\theta = 30°$, the block will start sliding down the plane
(B the block will remain at rest on the plane up to certain θ and then it will topple
(C) at $\theta = 60°$, the block will start sliding down the plane and continue to do so at higher angles

(D) at $\theta = 60°$, the block will start sliding down the plane and on further increasing θ, it will topple at certain θ.

10. A block of mass m is on an inclined plane of angle θ. The coefficient of friction between the block and the plane is μ and $\tan\theta > \mu$. The block is held stationary by applying a force P parallel to the plane. The direction of force pointing up the plane is taken to be positive. As P is varied from $P_1 = mg(\sin\theta - \mu\cos\theta)$ to $P_2 = mg(\sin\theta + \mu\cos\theta)$, the frictional force/versus P graph will look like

[2010]

50. ANSWER KEYS AND SOLUTIONS

50.1. CHECKPOINT 1

1. (D) all of the above
2. Field force
3. (B) Electromagnetic. As contact forces are EM in nature.
4. The fish exerts a contact force. It pushes on the water, and the water pushes back. The fish would also exert a gravitational force on the surrounding water, but the magnitude would be much less than the contact force.
5. Air exerts a contact force on the piece of paper
6. (D) $E > G > N$. Although, in general, $N > E > G$, but nuclear force only acts between protons and neutrons within the nucleus. Nuclear force does not act between for electrons. In other words the nuclear force is almost zero between the electrons.
7. (D) *Irrespective of the signs of the charges.*
Because force is a vector quantity, so net force zero means their vector sum is zero. It depends on the orientation of the forces on the particles.

50.2. CHECKPOINT 2

1. (D)
2. This drawing is much more like the sketch you would make when identifying forces as part of solving a problem.

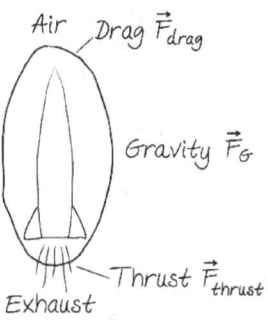

3. Identify which sources are outside the system in each case, and then decide whether these sources exert an external force on the system.
a. The spring scale, and the Earth. The elevator car and the cable are external, but are not in contact with the system, so they cannot exert a force on the system.
b. The elevator car, and the Earth The cable is not in contact with the system, so it cannot exert a force on the system.
c. The Earth and the cable are outside the system, and they both exert forces on it
4. (A), (B), and (D) Friction and the normal force are the only contact forces. Nothing is touching the rock to provide a "force of the kick."
5. (C)
6. (A) Force on the particle is parallel to its velocity, (B) Force is not parallel to velocity, (C) Force is perpendicular to the velocity.
7. (B)
8. Due to inertia. If automobile stops very slowly, then jerk can be minimized.
9. $(-2.32\hat{\imath} + 23.57\hat{\jmath})N$
10. The ball has the same mass in each instance and therefore the same inertia whether it is at rest or rolling.
11. As you have same mass in both cases, therefore the inertia will be same in both cases.
12. The heavy man is more massive and thus has more inertia

50.3. CHECKPOINT 3

1. If an object with no net force acting on it is at rest or is moving with a constant speed in a straight line (i.e., with constant velocity) relative to the reference frame, then the reference frame is an inertial reference frame. Consider sitting at rest in an accelerating train or plane. The train or plane is not an inertial reference frame even though you are at rest relative to it. In an inertial frame, a dropped ball lands at your feet. You are in a noninertial frame when the driver of the car in which you are riding steps on the gas and you are pushed back into your seat.
2. Note: An object accelerates when a net force acts on it. The fact that an object is accelerating tells us nothing about its velocity other than that it is always changing.
Yes, the object must have an acceleration relative to the inertial frame of reference. According to Newton's 1st and 2nd laws, an object must accelerate, relative to

any inertial reference frame, in the direction of the net force. If there is "only a single nonzero force," then this force is the net force.

Yes, the object's velocity may be momentarily zero. During the period in which the force is acting, the object may be momentarily at rest, but its velocity cannot remain zero because it must continue to accelerate. Thus, its velocity is always changing.

3. (C) As we saw in the solution of EXAMPLE 6, the normal force is $N = mg \cos\theta$, so unless the angle of the incline is zero, the normal force is less than mg. When the angle is zero, $N = mg$.

4. (A) An airplane cruising in a straight path at constant speed.

5. (B) Acceleration is proportional to force, so doubling the number of rubber bands doubles the acceleration of the original object from 2 m/s² to 4 m/s². But acceleration is also inversely proportional to mass. Doubling the mass cuts the acceleration in half, back to 2 m/s².

6. (a) Newton's first law states that acceleration requires at least one force; and is used to identify inertial frames. Newton's second law mathematically connects net force with inertia (mass) and acceleration.
(b) According to the second law, if the total force is zero, the object does not accelerate. This statement is consistent with the first law. It is likely that Newton stated the first law separately from the second law to address the notion of a *natural state*, which had been discussed for centuries before he published the *Principia*.

7. Attraction of the earth will produce an acceleration of g in the body. According to Newton's third law the falling body applies equal and opposite force on the earth. This will produce an acceleration of $g/2$ in the earth since mass of the earth is twice that of the body

Relative acceleration of approach = $\frac{3g}{2}$

$$\therefore \quad t = \sqrt{\frac{2H}{3g/2}} = \sqrt{\frac{4H}{3g}}$$

8. $45°$, $g/2$
9. 3.13 kg

10. (a) 6467 N, (b) 22400 N (c) 190400 N (d) T_0 and F do not change. T will increase.

50.4. CHECKPOINT 4

1. In the absence of a net force, an object moves with constant velocity, therefore, (d) is correct.
2. (C)
3. Zero
4. (a) $\frac{2F}{k}$, (b) $\frac{F}{k}$

Hint: If we replace each combination of springs by their equivalent springs, then we get infinite springs in series having spring constants, k, 2k, 4k, ... etc.
Spring constant of 2 springs in parallel = $k + k = 2k$,
Spring constant of 4 springs in parallel = $4k$,
Spring constant of 8 springs in parallel = $8k$,
and so on

(a) If $x_1, x_2, x_3, \ldots$ are the extensions on top spring, 2nd equivalent spring of spring constant 2k, third equivalent spring of spring constant 4k, and so on, then, the net displacement of top point A will be
$x_A = x_1 + x_2 + x_3 + \cdots$

$$\Rightarrow x_A = \frac{F}{k} + \frac{F}{2k} + \frac{F}{4k} + \cdots = \frac{F}{k}\left[\frac{1}{1-\frac{1}{2}}\right] = \frac{2F}{k} \quad \text{[As in series forces will be equal]}$$

(b) Displacement of B1, $x_{B1} = x_A - x_1 = \frac{2F}{k} - \frac{F}{k} = \frac{F}{k}$

5. (a) Displacement (and hence speed) of each point on the spring will be proportional to its distance from the wall.

$\therefore$ Speed of the particle. $V = 5 \times \frac{1}{3} = \frac{5}{3}$ cm/s

(b) The free end moves by 10 cm in 2.0 s

$\therefore$ Tension in the spring $= Kx = 0.6 \times 10 = 6N$

50.5. CHECKPOINT 5

1. (a) True. By definition, action-reaction force pairs cannot act on the same object. (b) False. Action equals reaction independent of any motion of the two objects.

2. The two forces exerted by the two children on the toy cannot be interaction partners (action-reaction pair) because they act on the *same* object (the toy), not on two different objects. Interaction partners act on different objects, one on each of the two objects that are interacting. The interaction partner of the force exerted by one child on the toy is the force that the toy exerts on that child.

3. $N = 12$; Tension $= \frac{F}{N} = \frac{F}{12}$

4. (a) $\frac{m_A}{m_B} = \frac{1}{3}$, (b) 0.75 m (c) $v'_A = \left(\frac{3}{2} + \frac{3}{\sqrt{2}}\right) ms^{-1}$, $v'_B = \left(\frac{3}{2} - \frac{1}{\sqrt{2}}\right) ms^{-1}$

5. No. Net external force is zero.

6. Force of the truck on the car will be equal to the force of the car on the truck? (action-reaction law)

7. If he wants to move himself to the shore, he must through away his shirt or shoe in a direction opposite to the desired direction of motion.

8. Suppose force applied by hand is F, and the force appiled by string is F_1. Since both block will move with same acceleration, therefore, we can consider both blocks in a single system of mass $(m_1 + m_2)$

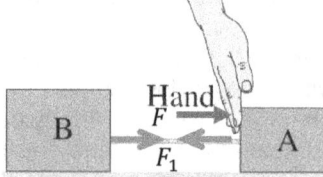

The acceleration of the system, $a = \frac{F}{m_A + m_B}$

For block B, $F_1 = m_B a = m_B \left(\frac{F}{m_A + m_B}\right)$

$\Rightarrow F_1 = \left(\frac{m_B}{m_A + m_B}F\right)$

From above it is clear that $F_1 < F$.

☞ Note that, force F is accelerating a system of mass $(m_A + m_B)$, whereas F_1 is accelerating only m_B with same acceleration a.

9. $T(x) = \frac{m}{l}(l-x)g + Mg$

10. $2\sqrt{\frac{2h}{g}}$

11. $\frac{5F}{m}$

50.6. CHECKPOINT 6

1. Determine the Concept: If there is a force on her in addition to the gravitational force, she will experience an additional acceleration relative to her space vehicle that is proportional to the net force required producing that acceleration and inversely proportional to her mass.

She could do an experiment in which she uses her legs to push off from the wall of her space vehicle and measures her acceleration and the force exerted by the wall. She could calculate her mass from the ratio of the force exerted by the wall to the acceleration it produced.

2. Determine the Concept One's apparent weight is the reading of a scale in one's reference frame.

Your apparent weight would be greater than your true weight when observed from a reference frame that is accelerating upward. That is, when the surface on which you are standing has an acceleration a such that a_y is positive: $a_y > 0$.

3. $W' = W\left(1 - \frac{a_y}{g}\right) = 50g\left(1 - \frac{9}{10}\right) = 5g = 50N$ (The weighing scale will show 5 kg)

4. (a) 80 N, (b) $\frac{640}{9}N$

Hint: (a) For B: $m_B g - T = m_B a$

$\Rightarrow (10)(10) - T = 10 \times 2 \Rightarrow T = 80N$

Thus, the tension in the string will be 80 N

Since, monkey A holds the string tightly, therefore tension will be the measure of its weight. So, A feels its weight = 80 N. Its is his real weight also.

(b) Let acceleration of string is a, in from B to A (you can also assume it from A to B). As monkey A tightly holds the string, so its acceleration will also be a in downward direction. Therefore, for monkey A, we have

$$m_A g - T = m_A a$$

or $\quad 80 - T = 8a$ ... (1)

As monkey B has downward acceleration of $4ms^{-2}$, with respect to the string, therefore its acceleration with respect to the ground will be, $(4 - a)$ (rope is slipping from his hand)

Therefore, for monkey B, we have

$$100 - T = (10)(4 - a)$$

$$60 - T = -10a \qquad ...(2)$$

Subtracting (2) in (1), we get

$$20 = 18a$$

or $\quad a = \frac{10}{9}ms^{-2}$

Calculation of T:

Using the value of a, in (1), we get

$80 - T = 8\left(\frac{10}{9}\right)$ or $T = \frac{640}{9}N$

So, in this case the apparent weight of each monkey will be $\frac{640}{9}N$.

5. (a) $\frac{8gt_0}{15}$

(b)

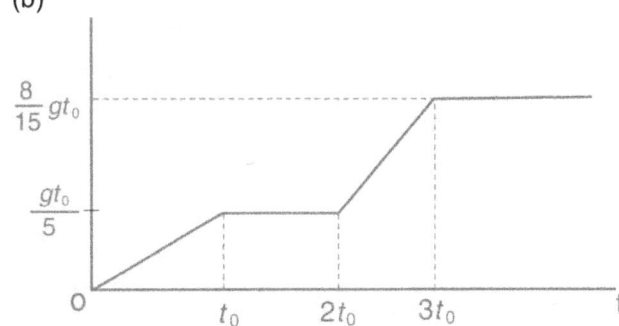

6. $v_0 = \frac{Mg}{\mu}\left[4\ln\left(\frac{4}{3}\right) - 1\right]$

7. t_2

8. $\frac{5Mg}{2k}$

Hint: Block will move up when tension in the string becomes larger than Mg.

Tension in string = Tension in cord *AB*, Tension in cord *CD* = twice the tension in string

138 CONCEPTS AND PROBLEMS IN PHYSICS

9. 2.9 sec.

10. (a) $K = 2.5$ N/cm (b) No

11. (a) The block is at height $h = 2.5$ m

(b) $v = 5\sqrt{2}$ m/s

(c) 25 m/s² (↑)

50.7. CHECKPOINT 7

1. $a_B = 0.022$ m/s² down

2. $v_A = -0.5\hat{j}$ m/s, $v_B = 3\hat{j}$ m/s

3. $v_A = 1.8$ m/s up, $a_A = 3$ m/s² down

4. $v_A = 1.333$ ft/sec up incline

5. (a) $T = N \cos \theta$, (b) $\dfrac{mg}{M \tan \theta + m \cot \theta}$

6. $t = 20$ sec

7. $2a_A + 2a_B + a_C = 0$, 2 degrees of freedom

8. $v_A = \dfrac{2\sqrt{x^2+h^2}}{x} v_B$

9. $v_A = -\dfrac{3y}{2\sqrt{y^2+b^2}} v_A$

10. $v_{B/A} = 1$ ft/sec, $a_{B/A} = 2$ ft/sec², $v_C = 4$ ft/sec (all directed up incline)

11. $4v_A + 8v_B + 4v_C + v_D = 0$, 3 degrees of freedom

12. $a_x = -\dfrac{L^2 v_A^2}{(L^2-y_2)^{3/2}}$

13. $\dfrac{44g}{205}$

14. $v_A = \dfrac{2y}{\sqrt{y^2+b^2}} v_B$

15. $v_B = -\dfrac{3x_A v_A}{2\sqrt{x_A^2+h^2}}$, $a_B = -\dfrac{3(v_A^2+x_A a_A)}{2\sqrt{x_A^2+h^2}} + \dfrac{3x_A^2 v_A^2}{2(x_A^2+h^2)^{3/2}}$ (a minus value indicates a leftward direction)

16. $h = 300$ mm,

17. $v = \dfrac{i\sqrt{4y^2+b^2}}{16y}$, $a = \dfrac{b^2 i^2}{256 y^3}$

18. $v_B = \dfrac{1}{2^n} v_A$

19. $2L/5$

20. $v_B = 62.9$ mm/s up

21. $v_B = 5.70$ m/s ↑

22. $v = 3.62$ m/s ↑

23. (a) 0.125 m/s ↑. (b) 0.5154 m/s ⦨ 14°

24. (a) 8.00 m/s ↑. (b) 4.00 m/s ↑. (c) 12.00 m/s ↑. (d) 8.00 m/s ↑

25. (a) 600 mm/s →, (b) 1200 mm/s ←,

(c) 900 mm/s ←

26. (a) $a_A = 50.8$ mm/s² →, $a_B = 25.4$ mm/s² ←.(b) $v_B = 152.2$ mm/s ←, $\Delta x_B = 458$ mm ←.

27.

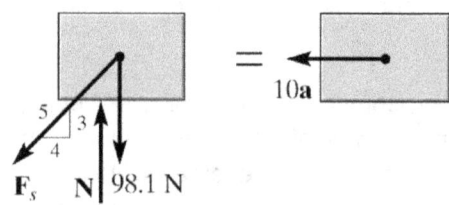

$F_s = kx = (10 \text{ N/m}) (5 \text{ m} - 1 \text{ m}) = 40$ N

$\pm F_x = ma_x; \dfrac{4}{5}(40N) = 10a$

$a = 3.2$ m/s²

28. (a) 10.00 mm/s →. (b) $a_A = 2.00$ mm/s² ↑, $a_C = 6.00$ mm/s² →.(c) 175.0 mm ↑

29. (a) $v_C = 120$ mm/s (b) $y_D - (y_D)_0 = 125.0$ mm ↑

30. (a) $v_L = 16.67$ mm/s↑ (b) $v_{B/L} = 16.67$ mm/s↑

31. $a = 4.46$ m/s² up incline.

32. (a) Force between the wall and the middle ball is maximum. It is 4 mg (b) Force between upper ball and wall is least. It is $\dfrac{4}{3}mg$.

33. (a) $g \sin \theta$, (b) $f = \dfrac{1}{2} mg \sin 2\theta$, (c) $t = \sqrt{\dfrac{2h(M+m \sin^2 \theta)}{(M+m) g \sin^2 \theta}}$

34. (a) $(M+m)g \sin \alpha$, (b) $\dfrac{(M+m)g \sin^2 \alpha}{m + M \sin^2 \alpha}$

35. $\dfrac{M}{m} = \dfrac{1}{5}$

36. $a_0 = \dfrac{48g}{199}$

37. $a = g/\sqrt{2}$, $N_{AB} = 0$

38. $\dfrac{6g}{47}$

39. (b) $T_P = 21.65$ N, (c) 3.05 kg

50.8. CHECK POINT 8

1. The observers in (a) and (b) are moving at constant velocities, so their accelerations are zero. Hence, they are both inertial frames. In (c) and (d), the frames are accelerating, so they are noninertial frames.

2. (a) 3.60 m/s² to the right (b) $T = 0$ (c) Someone in the car (noninertial observer) claims that the forces on the mass along x are T and a fictitious force $(-Ma)$. (c) Someone at rest outside the car (inertial observer) claims that T is the only force on M in the x direction.

3. $g(\cos\phi\tan\theta - \sin\phi)$

4. $v_{B/AC} = (a_0 \sin\theta)t$, $s = \frac{1}{2}(a_0 \sin\theta)t^2$

5. (a) 2 mg, (b) $t = \sqrt{\frac{3L}{\sqrt{2}g}}$

6. (a) $\vec{a}_{wedge} = \frac{g}{3}\hat{\imath}$, $a_{x\,block} = -\frac{g}{3}$, $a_{y\,block} = \frac{2g}{3}$

(b) The block hits the table normally.

(c) $\frac{3u^2}{16g}$

7. $N_{12} = \frac{Mg\cos^2\theta}{1+2\sin^2\theta}$, $a = \frac{3g\sin\theta}{1+2\sin^2\theta}$

50.9. CHECKPOINT 9

1. Statement is true. When horse pushes the ground by force F_{gh}, ground also exerts an equal reactional force on the horse. The horizontal component of this force moves the horse-cart system in forward direction (see adjoining figure).

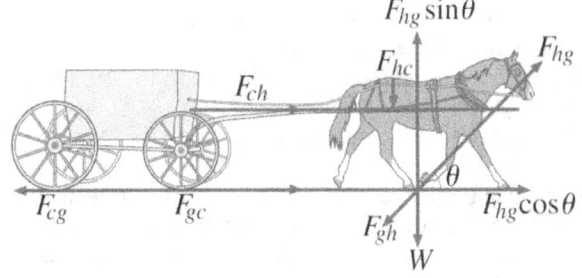

2. $v = 14.7\ m/s$

3. $T = 176\ N$

4. $s = 8.49\ m$

5. $a = 1.96\ m/s^2$

6. θ is given by $117.72 \sin\theta - 35.316 \cos\theta - 12.5 = 0$

7. 2.68 M, 1.87 s

8. 2.45 m, 1.72 s

9.   2.11 s

10. $t = 5.66$ s

11. $t = 0.519$ s

12. $x = d$ for separation.

13. 32.2 m/s

14. $P = 2mg \tan\theta$

15. $x = 51.0\ m$

16. $s = 1.710\ m$

17. (a) $8.2 \times 10^3 N$, (b) $4.8 \times 10^2 N$

18. (i) 2.5 kg, 12.5 N, (ii) $\frac{50}{3}N$, $a = \frac{5}{3}m/s^2$, (iii) $\frac{40}{3}N$, $a = \frac{5}{3}m/s^2$ (iv) $\frac{5}{6}m/s^2$

19. $\frac{\sqrt{5-2\sqrt{2}}}{\sqrt{2}-1}$

50.10. CHECKPOINT 10

1. $v = \sqrt{\frac{mg}{k}}\left[\frac{e^{2t\sqrt{mg/k}}-1}{e^{2t\sqrt{mg/k}}+1}\right]$, $v_t = \sqrt{\frac{mg}{k}}$

2. 11 0 m/s.

3. $\frac{3}{4}g$

4. $1.20\ m/s^2$, directed straight upward

50.11. PROBLEMS

1. Using the reference axes shown in Figure P1, we see that
$\Sigma F_x = T\cos 14.0° - T\cos 14.0° = 0$
and
$\Sigma F_y = -T\sin 14.0° - T\sin 14.0° = -2T\sin 14.0°$

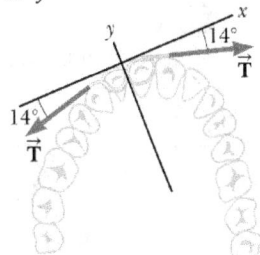

FIGURE P1

Thus, the magnitude of the resultant force exerted on the tooth by the wire brace is

$R = \sqrt{(\Sigma F_x)^2 + (\Sigma F_y)^2} = 2T\sin 14.0°$

or $R = 2(18.0\ N)\sin 14.0°$

2. Mass is defined to be $m = \dfrac{1}{\text{slope of the acceleration-versus-force graph}}$

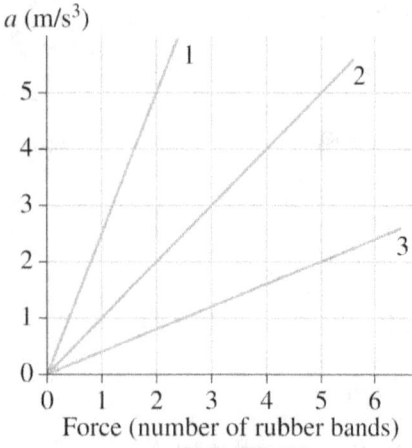

FIGURE P2

A larger slope implies a smaller mass. We know $m_2 = 0.20$ kg, and we can find the other masses relative to m_2 by comparing their slopes. Thus,

$\frac{m_1}{m_2} = \frac{1/slope\ 1}{1/slope\ 2} = \frac{slope\ 2}{slope\ 1} = \frac{1}{5/2} = \frac{2}{5} = 0.40$

$\Rightarrow m_1 = 0.40\ m_2 = 0.40 \times 0.20\ kg = 0.08\ kg$

Similarly,

$\frac{m_3}{m_2} = \frac{1/slope\ 3}{1/slope\ 2} = \frac{slope\ 2}{slope\ 3} = \frac{1}{2/5} = \frac{5}{2} = 2.50$

$\Rightarrow m_3 = 2.50\ m_2 = 2.50 \times 0.20\ kg = 0.50\ kg$

Assess: From the initial analysis of the slopes we had expected $m_3 > m_2$ and $m_1 < m_2$. This is consistent with our numerical answers.

3. Newton's second law tells us that $F = ma$ Compute F for each case:

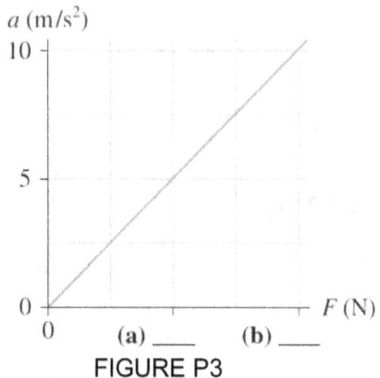

FIGURE P3

(a) $F = (0.200\ kg)(5\ m/s^2) = 1\ N$
(b) $F = (0.200\ kg)(10\ m/s^2) = 2\ N$

Assess: To double the acceleration we must double the force, as expected.

4. (a) Newton's second law is $F = ma$.
When $F = 2\ N$, we have $2\ N = (0.5\ kg)a$, hence $a = 4\ m/s^2$.

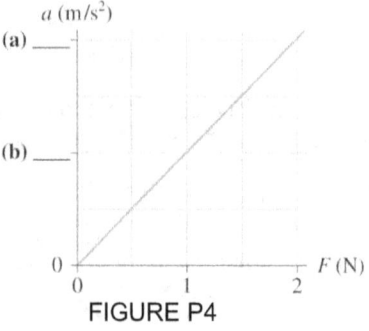

FIGURE P4

(b) When $F = 1\ N$, we have $1\ N = (0.5\ kg)a$, hence $a = 2\ m/s^2$.
After repeating this procedure at various points, the above graph is obtained

5. $F = ma = (300)(5) = (100)(15) = 1500\ N$

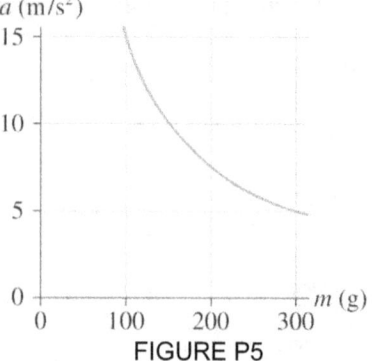

FIGURE P5

6. The force and acceleration are related by $\Sigma F_x = ma_x$, where ΣF_x is the net force. and $m = 4\ 50$ kg.

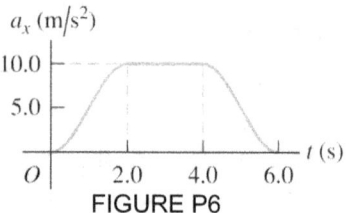

FIGURE P6

(a) The maximum net force occurs when the acceleration has its maximum value.
$\Sigma F_x = ma_x = (4.50\ kg)(10.0\ m/s^2) = 45.0\ N$
This maximum force occurs between 2.0 s and 4.0 s.
(b) The net force is constant when the acceleration is constant. This is between 2.0 s and 4.0 s.
(c) The net force is zero when the acceleration is zero. This is the case at $t = 0$ and $t = 6.0\ s$.
A graph of ΣF_x versus t would have the same shape as the graph of a_x versus t.

7. The system is accelerating so we use Newton's second law.
The acceleration of the entire system is due to the 100-N force, but the acceleration of box B is due to the force that box A exerts on it. $\Sigma F = ma$ applies to the two-box system and to each box individually.
For the two-box system:
$a_x = \frac{100\ N}{25\ kg} = 4.0\ m/s^2$

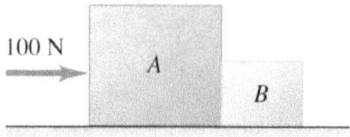

FIGURE.P7

Then for box B, where F_A is the force exerted on B by A, $F_A = m_B a = (5.0\ kg)(4.0\ m/s^2) = 20\ N$
CONCLUSION: The force on B is less than the force on A.

8. Use Newton's second law to relate the acceleration and forces for each crate.

(a) Since the crates are connected by a rope, they both have the same acceleration, 2.50 m/s²
(b) The forces on the 4.00 kg crate are shown in FIGURE. P8(a).

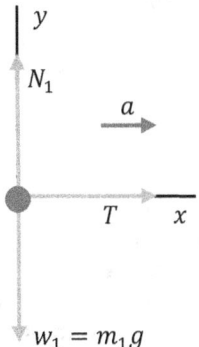

FIHURE P8 (a)

$\Sigma F_x = ma_x$
$\Rightarrow T = m_1 a = (4.00\ kg)(2.50\ m/s^2) = 10.0\ N$
(c) Forces on the 6.00 kg crate are shown in FIGURE. P8(b).

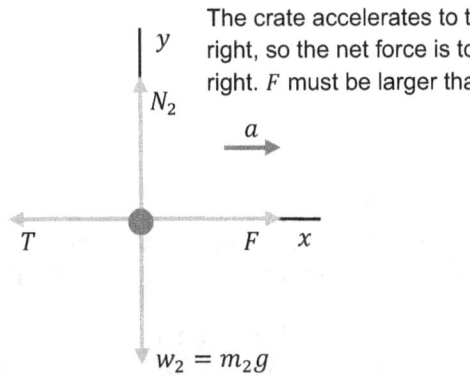

The crate accelerates to the right, so the net force is to the right. F must be larger than T

FIGURE. P8(b)

(d) $\Sigma F_x = ma_x$ gives $F - T = m_2 a$
or $F = T + m_2 a$
$= 10.0\ N + (6.00\ kg)(2.50\ m/s^2)$
$= 10.0\ N + 15.0\ N = 25.0\ N$
We can also consider the two crates and the rope connecting them as a single object of mass $m = m_1 + m_2 = 10.0\ kg$. The free-body diagram is sketched in FIGURE. P8(c).

$\Sigma F_x = ma_x$
$F = ma = (10.0\ kg)(2.50\ m/s^2)$
$= 25.0\ N$
This agrees with our answer in part (d)

FIGURE. P8(c)

9. Note that in this problem the mass of the rope is given, and that it is not negligible compared to the other masses. Apply $\Sigma \vec{F} = m\vec{a}$ to each object to relate the forces to the acceleration.
(a) The free-body diagrams for each block and for the rope are given in FIGURE P9 (a).
6.00 kg block

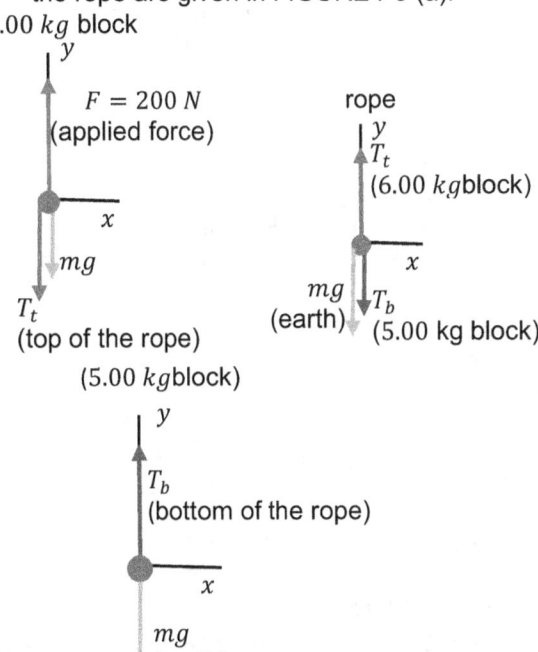

FIGURE P9 (a)
T_t is the tension at the top of the rope and T_b is the tension at the bottom of the rope.
(b) Treat the rope and the two blocks together as a single object, with mass $m = 6.00\ kg + 4.00\ kg + 5.00\ kg = 15\ kg$. Take $+y$ upward, since the acceleration is upward. The free body diagram is given in FIGURE P9 (b).

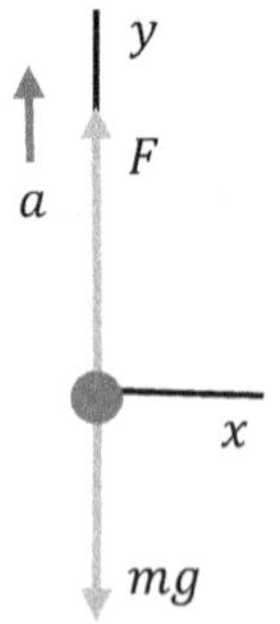

$\Sigma F_y = ma_y$
$F - mg = ma$
or $a = \frac{F-mg}{m}$

$a = \frac{200\ N - (15\ kg)(9.80\ m/s^2)}{15.0\ kg}$
$= 3.53\ m/s^2$

FIGURE P9 (b)

(c) Consider the forces on the top block ($m = 6.00$ kg), since the tension at the top of the rope T_t will be one of these forces

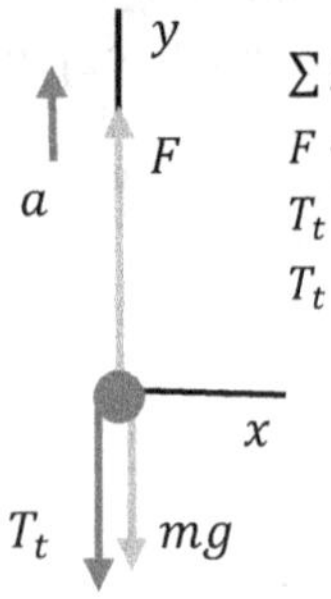

$\Sigma F_y = ma_y$
$F - mg - T_t = ma$
$T_t = F - m(g + a)$
$T_t = 200\ N - (6.00\ kg)(9.80\ m/s^2 + 3.53\ m/s^2) = 120\ N$

FIGURE P9 (c)

Alternatively, can consider the forces on the combined object rope plus bottom block ($m = 9.00$ kg):

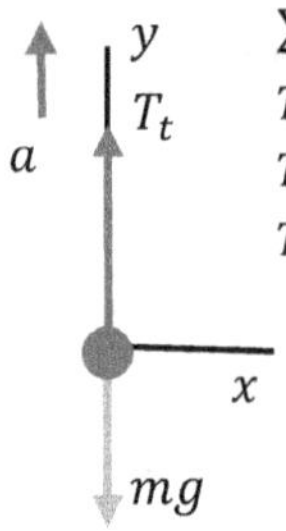

$\Sigma F_y = ma_y$
$T_t - mg = ma$
$T_t = m(g + a)$
$T_t = (9.00\ kg)(9.80\ m/s^2 + 3.53\ m/s^2) = 120\ N$

FIGURE P9 (d)

(d) One way to do this is to consider the forces on the top half of the rope ($m = 2.00$ kg). Let T_m be the tension at the midpoint of the rope.

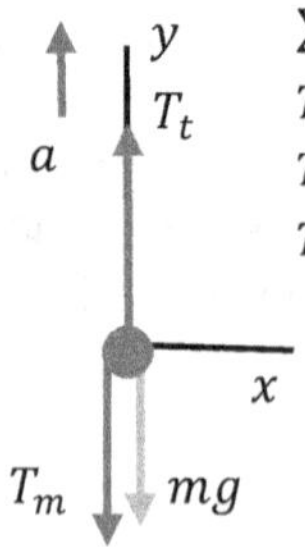

$\Sigma F_y = ma_y$
$T_t - T_m - mg = ma$
$T_m = T_t - m(g + a)$
$T_t = 200\ N - (2.00\ kg)(9.80\ m/s^2 + 3.53\ m/s^2) = 93.3\ N$

FIGURE P9 (e)

To check this answer we can alternatively consider the forces on the bottom half of the rope plus the lower block taken together as a combined object ($m = 2.00\ kg + 5.00\ kg = 7.00\ kg$)

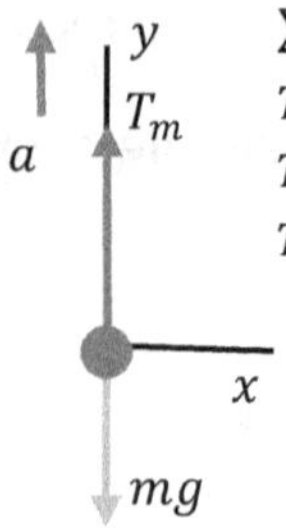

$\Sigma F_y = ma_y$
$T_m - mg = ma$
$T_m = m(g + a)$
$T_t = (7.00\ kg)(9.80\ m/s^2 + 3.53\ m/s^2) = 93.3\ N$

FIGURE P9 (f)

☞ The tension in the rope is not constant but increases from the bottom of the rope to the top. The tension at the top of the rope must accelerate the rope as well the 5.00-kg block. The tension at the top of the rope is less than F; there must be a net upward force on the 6.00-kg block

10. Apply $\Sigma \vec{F} = m\vec{a}$ to the balloon and its passengers and cargo, both before and after objects are dropped overboard.

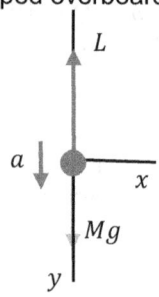

FIGURE P10 (a)

When the acceleration is downward take $+y$ to be downward and when the acceleration is upward take $+y$ to be upward.

(a) The free-body diagram for the descending balloon is given in FIGURE P10. L is the lift force.

(b) $\Sigma F_y = ma_y$ gives
$Mg - L = M(g/3)$
$\Rightarrow L = 2Mg/3$

(c) Now $+y$ is upward,
so $L - mg = m(g/2)$, where m is the mass remaining. $L = 2Mg/3$,
$\therefore 2Mg/3 - mg = m(g/2)$,
or $m = 4M/9$
$\therefore$ Mass $M - 4M/9 = 5M/9$ must be dropped overboard.

☞ In part (b) the lift force is less than the total weight and in part (c) the lift force is greater than the total weight.

11. The system is accelerating, so we apply Newton's second law to each box and can use the constant acceleration kinematics for formulas to find the acceleration.

First use the constant acceleration kinematics for formulas to find the acceleration of the system. Then apply $\Sigma F = ma$ to each box.

(a) The kinematics formula for $y(t)$ gives
$a_y = \frac{2(y-y_0)}{t^2} = \frac{2(12.0\ m)}{(4.0\ s)^2} = 1.5\ m/s^2$

For box B, $mg - T = ma$

$\Rightarrow a_y = \frac{T}{g-a} = \frac{36.0 \, N}{9.8 \, m/s^2 - 1.5 \, m/s^2} = 4.34 \, kg$

(b) For box A, $T + mg - F = ma$

$\Rightarrow m = \frac{F-T}{g-a} = \frac{80.0 \, N - 36.0 \, N}{9.8 \, m/s^2 - 1.5 \, m/s^2} = 5.30 \, kg$

☞ The boxes have the same acceleration but experience different forces because they have different masses.

12. **APPROACH:** Calculate $\vec{a}$ from $\vec{a} = \frac{d^2\vec{r}}{dt^2}$. Then $\vec{F}_{net} = m\vec{a}$, $w = mg$

    SOLUTION Differentiating twice, the acceleration of the helicopter as a function of time is
    $a = (0.120 \, m/s^3)t\hat{i} - (0.12 \, m/s^2)\hat{k}$ and at $t = 5.0s$, the acceleration is
    $a = (0.60 \, m/s^2)\hat{i} - (0.12 \, m/s^2)\hat{k}$
    The force is then
    $F = ma = \frac{w}{g}a$
    $= \frac{(2.75 \times 10^5 N)}{(9.80 \, m/s^2)}[(0.60 \, m/s^2)\hat{i} - (0.12 \, m/s^2)\hat{k}]$
    $= (1.7 \times 10^4 N)\hat{i} - (3.4 \times 10^3 N)\hat{k}$

☞ The force and acceleration are in the same direction. They are both time dependent.

13. **APPROACH:** $x = \int_0^t v_x dt$ and $v_x = \int_0^t a_x dt$ and similar equations apply to the y-component.
    In this situation, the x-component of force depends explicitly on the y-component of position. As the y-component of force is given as an explicit function of time, v_y and y can be found as functions of time and used in the expression for $a_x(t)$.

    SOLUTION $a_y = (k_3/m)t$, so $v_y = (k_3/2m)t^2$ and $y = (k_3/6m)t^3$ where the initial conditions $v_{0y} = 0$, $y_0 = 0$ have been used. Then, the expressions for a_x, v_x and x are obtained as functions of time:
    $a_x = \frac{k_1}{m} + \frac{k_2 k_3}{6m^2}t^3$, $v_x = \frac{k_1}{m}t + \frac{k_2 k_3}{24 m^2}t^4$
    and $x = \frac{k_1}{2m}t^2 + \frac{k_2 k_3}{120 \, m^2}t^5$
    In vector form,
    $\vec{r} = \left(\frac{k_1}{2m}t^2 + \frac{k_2 k_3}{120 \, m^2}t^5\right)\hat{i} + \left(\frac{k_3}{2m}t^3\right)\hat{j}$
    and $\vec{v} = \left(\frac{k_1}{m}t + \frac{k_2 k_3}{24 \, m^2}t^4\right)\hat{i} + \left(\frac{k_3}{2m}t^2\right)\hat{j}$

☞ a_x depends on time because it depends on y, and y is a function of time.

14. **APPROACH:** Apply $\Sigma F = ma$ to the object and to the knot where the cords are joined.
    Let $+y$ be upward and $+x$ be to the right.
    SOLUTION (a) $T_C = w$,

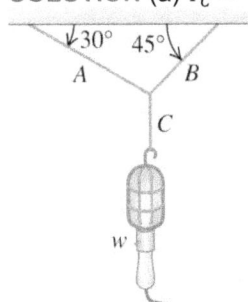
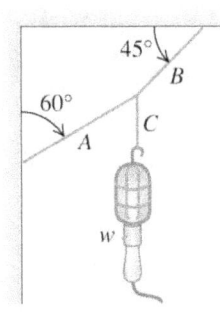

(a) (b)

FIGURE P14

$T_A \sin 30° + T_B \sin 45° = T_C = w$
and $T_A \cos 30° - T_B \cos 45° = 0$. Since $\sin 45° = \cos 45°$, adding the last two equations gives
$T_A(\cos 30° + \sin 30°) = w$, and so
$T_A = \frac{w}{1.366} = 0.732 \, w$.

and, $T_B = T_A \frac{\cos 30°}{\cos 45°} = 0.897 \, w$

(b) Similar to part (a), $T_C = w$, $-T_A \cos 60° + T_B \sin 45° = w$ and $T_A \sin 60° - T_B \cos 45° = 0$
Adding these two equations,
$T_A = \frac{w}{(\sin 60° - \cos 60°)} = 2.73w$ and $T_B = T_A \frac{\sin 60°}{\cos 45°} = 3.35w$

☞ In part (a), $T_A + T_B > w$ since only the vertical components of T_A and T_B hold the object against gravity. In part (b), since T_A has a downward component T_B is greater than w.

15. **APPROACH:** Apply Newton's first law to the hanging weight and to each knot. The tension force at each end of a string is the same.
    (a) Let the tensions in the three strings be T, T', and T'', as shown in FIGURE P15 (a).

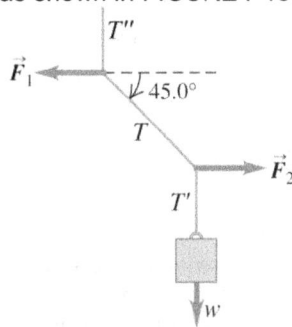

FIGURE P15 (a)

The free-body diagram for the block is given in FIGURE P15(b).

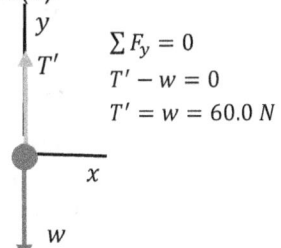

$\Sigma F_y = 0$
$T' - w = 0$
$T' = w = 60.0 \, N$

FIGURE P15 (b)

The free-body diagram for the lower knot is given in FIGURE P15 (c)

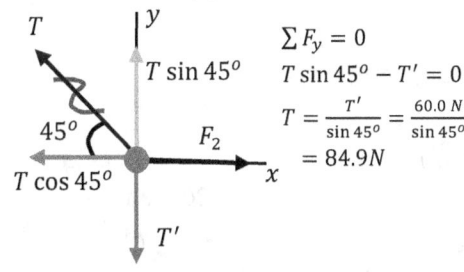

$\Sigma F_y = 0$
$T \sin 45° - T' = 0$
$T = \frac{T'}{\sin 45°} = \frac{60.0 \, N}{\sin 45°}$
$= 84.9 N$

FIGURE P15 (c)

(b) Apply $\Sigma F_x = 0$ to the force diagram for the lower knot:
$\Sigma F_x = 0$
$\Rightarrow F_2 = T \cos 45° = (84.9\ N) \cos 45° = 60.0\ N$
The free-body diagram for the upper knot is given in FIGURE P14(d).

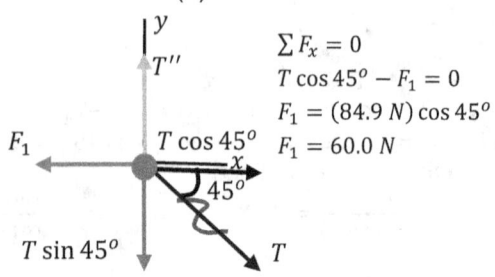

FIGURE P15(d)
Note that $F_1 = F_2$
SOLUTION Applying $\Sigma F_y = 0$ to the upper knot gives $\quad T'' = T \sin 45° = 60.0\ N = w$
If we treat the whole system as a single object, the force diagram is given in FIGURE 15(e)

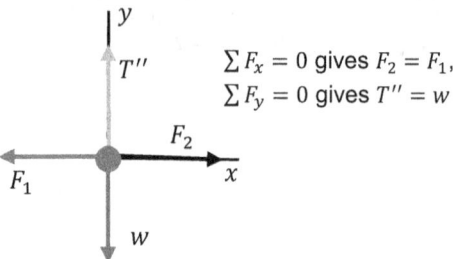

FIGURE P15(e)

16. **APPROACH:** Apply Newton's second law to the three sleds taken together as a composite object and to each individual sled. All three sleds have the same horizontal acceleration a.

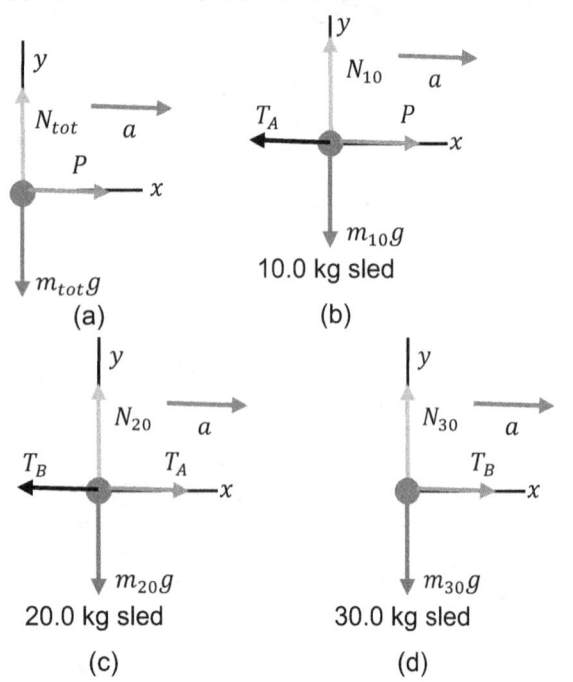

FIGURE P16

The free-body diagram for the three sleds taken as a composite object is given in Figure 5.16a and for each individual sled in Figures 5.16b–d. Let $+x$ be to the right, in the direction of the acceleration. $m_{tot} = 60.0$ kg.
SOLUTION (a) $\Sigma F_x = ma_x$ for the three sleds as a composite object gives $P = m_{tot} a$ and
$a = \dfrac{P}{m_{tot}} = \dfrac{190\ N}{60.0\ kg} = 3.17\ m/s^2$
(b) $\Sigma F_x = ma_x$ applied to the 10.0 kg sled gives $P - T_A = m_{10} a$ and
$T_A = P - m_{10} a = 190\ N - (10\ kg)(3.17\ m/s^2)$
$\qquad = 158\ N$
$\Sigma F_x = ma_x$ applied to the 30.0 kg sled gives
$T_B = m_{30} a = (30.0\ kg)(3.17\ m/s^2) = 95.1\ N$.
☞ If we apply $\Sigma F_x = ma_x$ to the 20.0 kg sled and calculate a from TA and TB found in part (b), we get
$T_A - T_B = m_{20} a$,
$\Rightarrow a = \dfrac{T_A - T_B}{m_{20}} = \dfrac{158\ N - 95.1\ N}{20.0\ kg} = 3.15\ m/s^2$, which agrees closely with the value we calculated in part (a), the difference being due to rounding.

17. (a) $\phi = \sin^{-1}\left(\dfrac{10.4}{49}\right)$ (b) 1.59 m/s
18. (a) 47.0 N, (b) 17.0 N, (c) 0
19. (a) 31N (bottom), 63 N(top); (b) 35N (bottom), 71 N(top)
20 The tension in both upper cords is 69 N. The tension in the cord between A and B is 49 N. The tension in both lower cords is 57 N. The tension in the cord attached to the mass is 98 N
21. (a) $8.76 \times 10^4\ N$, (b) $1.14 \times 10^4\ N$, (c) $1.14 \times 10^4\ N$ downward on the helicopter.
22. (a) 2.96 m/s², (b) 191 N
23. $a_1 = \dfrac{g(m_2 + m_3 - m_1)}{(m_1 + m_2 + m_3)} = 0.69\ m/s^2$ downward
24. (a) 350 N, (b) 1.5 m/s²,
25. (a) $\phi = \sin^{-1}\left(\dfrac{a}{g}\right) = \sin^{-1}\left(\dfrac{2.08}{9.8}\right)$, (b) $1.59\ m/s$
26. 2.50 m/s², (b) 1.37 kg, (c) The weight of the hanging block is $mg = (1.37\ kg)(9.80\ m/s^2) = 13.4N$. This is greater than the tension in the rope; $T = 0.75\ mg$.
27. (a) 160 m (b) 6000 N
28. (a) 0.832 m/s², (b) 17.3 s
29. (c) 3.56 N
30. (a)
Upper pulley: Lower pulley:

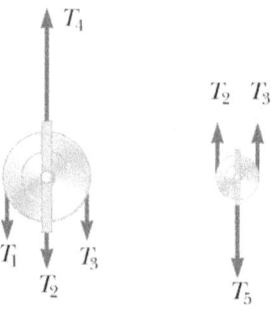

(b) $Mg/2$, $Mg/2$, $Mg/2$, $3Mg/2$, Mg (c) $Mg/2$

31. $a_2 = \frac{m_2 g}{4m_1+m_2}$ and $a_1 = 2a_2 = \frac{2m_2 g}{4m_1+m_2}$, If $m_2 = 2m_1$, then $a_2 = g/3$ and $a_1 = 2g/3$

32. (a) $M = 3m \sin \theta$, (b) $T_1 = 2mg \sin \theta$, $T_2 = 3mg \sin \theta$,
(c) $a = \frac{g \sin \theta}{1+2 \sin \theta}$;
(d) $T_1 = 4mg \sin \theta \left(\frac{1+\sin \theta}{1+2 \sin \theta}\right)$, $T_2 = 6mg \sin \theta \left(\frac{1+\sin \theta}{1+2 \sin \theta}\right)$
(e) (e) $M_{max} = 3m(\sin \theta + \mu_s \cos \theta)$; (f) $M_{min} = 3m(\sin \theta - \mu_s \cos \theta)$;
(g) $T_{2,max} - T_{2,min} = M_{max} g - M_{min} = 6\mu_s mg \cos \theta$

33. $(m_1 \cos^2 \theta + m_2) g$

34. (a)

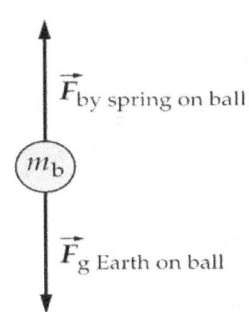

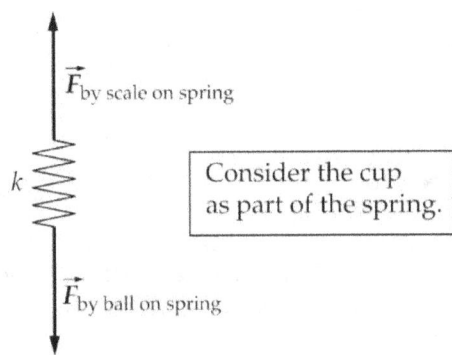

(c) $F_{by\ scale} = (m_{ball} + m_{plateform}) g$

35. (a) $A = 1.50\ m/s^2$, $B = 0.50\ m/s^3$, (b) $a_y = 5.50\ m/s^2$, (c) $T = 3.89 \times 10^4 N$, $T = 1.56w$, (d) $2.87 \times 10^4 N$

36. (a) 14.0 m, (b) 18.0 m/s

37. $k = 2Sdg$
Hint: Mass of water added when the level of water rises by x (with respect to container) $m = 2Sxd$ The spring stretches by x so that the level of water does not change with respect to ground.
$\therefore\ kx = 2Sxdg$ or $k = 2Sdg$

38. 122N

39. $\theta = \tan^{-1} \mu \approx 50°$

40. (a) The friction is static for $P = 0$ to $P = 75.0N$. The friction is kinetic for $P > 75.0N$.
(b) $\mu_s = 0.556$, $\mu_k = 0.370$
(c) When the block is moving the friction is kinetic and has the constant value $f_k = \mu_k N$, independent of P. This is why the graph is horizontal for $P > 75N$. When the block is at rest, $f_s = P$ since this prevents relative motion. This is why the graph for $P < 75\ N$ has slope +1.
(d) max f_s and f_k would double. The values of f on the vertical axis would double but the shape of the graph would be unchanged.

41. (a) If there is no applied force, no friction force is needed to keep the box at rest.
(b) $f_s^{max} = \mu_s N = (0.40)(40\ N) = 16\ N$. If a horizontal force of 6.0 N is applied to the box, then $f_s = 6\ N$ in the opposite direction.
(c) The monkey must apply a force equal to f_s^{max}, 16.0 N.
(d) Once the box has started moving, a force equal to $f_k = \mu_k N = 8.0\ N$ is required to keep it moving at constant velocity.
(e) $f_k = 8.0\ N$. $a = (18.0\ N - 8.0\ N)/(40.0\ N/9.80\ m/s^2) = 2.45\ m/s^2$

42. (b) If θ is greater than $\tan^{-1}\left(\frac{1}{\mu_s}\right)$, motion is impossible.

43. (a) $\mu_s = 0.710$, $\mu_k = 0.472$; (b) 258 N, (c) (i) 51.8 N (ii) 4.97 m/s².

44. (a) The person must apply a force of 57.1 N, directed up the ramp. (b) 146 N directed up the ramp

45. 4 71 s

46. (a) 52.5m, (b) 16 m/s

47. (a) $\mu_k = 0.556$, (b) $2.13\ m/s^2$. The acceleration is upward and the block B slows down.

48. (a) $F = \mu_k(m_A + m_B) g$, (b) $T = \mu_k m_A g$

49. (a) $\alpha = \tan^{-1} 0.35$, (b) $0.92\ m/s^2$, (c) 3 m/s.

50. (a) 21.8 cm/s, (b) 11.7 N

51. (a) $F = \frac{\mu_k mg}{\cos \theta - \mu_k \sin \theta}$, (b) $\mu_s = \frac{1}{\tan \theta}$

52. (a) If the ball is moving up, the frictional force is down, so the magnitude of the net force is $(5/4)w$ and the acceleration is $(5/4)g$, g down.
(b) While moving down, the frictional force is up, and the magnitude of the net force is $(3/4)w$ and the acceleration is $(3/4)g$, down.

53. (a) $2T \sin \theta = mg$ or $T = \frac{mg}{\sin \theta}$ $\Rightarrow$ $T = 882\ N$
(b) $\sin \theta = \frac{mg}{2T}$ $\Rightarrow \theta = \sin^{-1}(0.018)$

54. The tension in the lower chain balances the weight and so is equal to w.. The lower pulley must have no net force on it, so twice the tension in the rope must be equal to w and the tension in the rope, which equals F, is $w/2$. Then, the downward force on the upper pulley due to the rope is also w, and so the upper chain exerts a force w on the upper pulley, and the tension in the upper chain is also w.

55. (a) $T = 470\ N$ (b) 163 N directed to the right

56. $N = \frac{mg}{\cos \theta}$ (c) $T = N \sin \theta = \frac{mg}{\cos \theta} \times \sin \theta = mg \tan \theta$

57. 121 N

58. (a) $T_B = mg/\cos \phi$, (b) $T_A = 2T_B \cos \phi = 2mg$

59. 10.4 kg

60. Low pressure, $\mu_r = 0.0259$, high pressure $\mu_r = 0.00505$

61. $T(x) = F\left(1 - \frac{mx}{L(M+m)}\right)$

62. (a) $m_2 = m_1(\sin \alpha + \mu_k \cos \alpha)$ (b) $m_2 = m_1(\sin \alpha - \mu_k \cos \alpha)$
(c) $m_{largest} = m_1(\sin \alpha + \mu_k \cos \alpha)$, $m_{smallest} = m_1(\sin \alpha - \mu_k \cos \alpha)$,

63. (a) 12 N, (b) 15 N
64. (a) 1.80 N, (b) 2.52 N
65. (b) $F = \frac{\mu_s mg}{\cos\theta + \mu_s \sin\theta}$, for minimum force put $\frac{dF}{d\theta} = 0$, you will get the required result.
(c) once the block is moving the coefficient of friction will decrease, so the angle can be decreased.
66. $F = \frac{\mu_s mg}{\cos\theta - \mu_s \sin\theta}$, For minimum force denominator should ne maximum. It holds for $\theta = 0$. $F_{min} = \mu_s mg$.
67. (a) 80 N (b) 600 N, 680 N (c) 6.80 m/s².
68. 72.0 N
69. $4.90 \times 10^2 N$
70. (a) 0.931 m/s² (b) From a value of 0.625 m/s² for large x, the acceleration gradually increases, passes through a maximum, and then drops more rapidly, becoming negative and reaching −2.10 m/s² at $x = 0$ (c) 0.976 m/s² at $x = 25.0$ cm (d) 6.10 cm
71. (a) 2.13 s (b) 1.66 m
72. 490 N
73. (a) 16.8 s, $7.35 \times 10^5 N$, (b) 2770 m
74. 920 N
75. 5.9 m/s
76. (a) 11 5 m/s (b) 7 54 m/s.
77. The fraction that hangs over is $\frac{\mu_s}{1+\mu_s}$.
78. (a) 88.0 N northward, (b) 78 N, southward
79. 2 221 s, 5 43 m.
80. 3.0 N
81. (a) 12 9 kg (b) $T_{AB} = 47.2 N$, $T_{BC} = 100.6 N$
82. (a) The 100 kg block will slide down with acceleration $0.67g$.
(b) $a = 0.067 (9.80 \, m/s^2) = 0.658 \, m/s^2$.
(c) 424 N
83. 39 kg
84. (a) $a_1 = 5.17 \, m/s^2$, $a_2 = 3.43 \, m/s^2$ (b) $T = 35.3 N$
85. 2.60 kg
86. (a) $3.34 \, m/s^2$ and $6.57 \, m/s^2$
87. height reached will be 1.46 m above the floor, which is 0.860 m above its initial height
88. 105 N
89. $a = g/\mu_s$, An observer on the cart sees the block pinned there, with no reason for a horizontal force on it because the block is at rest relative to the cart. Therefore, such an observer concludes that normal reaction $N = 0$ and thus $f_s = 0$, and he doesn't understand what holds the block up against the downward force of gravity. The reason for this difficulty is that $\Sigma \vec{F} = m\vec{a}$ does not apply in a coordinate frame attached to the cart. This reference frame is accelerated, and hence not inertial. The smaller μ_s is, the larger a must be to keep the block pinned against the front of the cart.
90. (a) 11 kg of sand was added (b) 0.88 m/s².
91. (a) $a = 2.21 \, m/s^2$, (b) $T = 2.27 \, N$, (c) The string will be slack. The 4.00-kg block will have $a = 2.78 \, m/s^2$ and the 8.00-kg block will have $a = 1.93 \, m/s^2$, until the 4.00-kg block overtakes the 8.00-kg block and collides with it.

92. 0.452
93. (a) $\tan\theta = \frac{a}{g}$, (b) $\tan^{-1}\frac{1}{6}$, (c) As m_1 becomes much larger than m_2, $a \to g$ and $\tan\theta \to 1$, so $\theta \to 45°$.
94. (a) θ, (b) $\tan^{-1}\left(\frac{\sin\theta\cos\theta}{2-\sin^2\theta}\right)$
95. (a) 8.8 N, (b) 30.8 N, (c) 1.54 m/s²
96. (a) For the monkey to move up, $T > mg$. The bananas also move up
(b) The bananas and monkey move with the same acceleration and the distance between them remains constant.
(c) Both the monkey and bananas are in free fall. They have the same initial velocity and as they fall the distance between them doesn't change.
(d) The bananas will slow down at the same rate as the monkey. If the monkey comes to a stop, so will the bananas
97. (a) 6.00 m/s². (b) 3.80 m/s². (c) 7.36 m/s. (d) 8.18 m/s. (e) 1.38 m/s², (f) 3.14 s
98. (a) 1.84m, 0.61 s, (b) 0.66m, 0.283 s
99. (a) $A = -\frac{-gm}{(M+m)\tan\alpha + (M/\tan\alpha)}$, $a_x = \frac{gM}{(M+m)\tan\alpha + (M/\tan\alpha)}$, $a_y = \frac{-g(M+m)\tan\alpha}{(M+m)\tan\alpha + (M/\tan\alpha)}$
(b) When $M \gg m$, $A \to 0$, $a_x \to g\sin\alpha\cos\alpha$
(c) The trajectory is a straight line with slope $-\left(\frac{M+m}{M}\right)\tan\alpha$.
100. $F = (M+m)g\tan\alpha$
101. (b) Newton's second law is then $ma = mg - Dv^2$. Initially, when $v = 0$, the acceleration is g, and the speed increases. As the speed increases, the resistive force increases and hence the acceleration decreases. This continues as the speed approaches the terminal speed.
(c) $v_t = \sqrt{\frac{mg}{D}}a$
(d) $v = v_t \tanh(gt/v_t)$
102. (a) $a_3 = g\frac{-4m_1m_2 + m_2m_3 + m_1m_3}{4m_1m_2 + m_2m_3 + m_1m_3}$,
(b) The acceleration of the pulley B has the same magnitude as a_3 and is in the opposite direction.
(c) $a_1 = g\frac{4m_1m_2 - 3m_2m_3 + m_1m_3}{4m_1m_2 + m_2m_3 + m_1m_3}$,
(d) $a_2 = g\frac{4m_1m_2 + m_2m_3 - 3m_1m_3}{4m_1m_2 + m_2m_3 + m_1m_3}$,
(e),(f) Once the accelerations are known, the tensions may be found by substitution into the appropriate equation of motion, giving
$T_A = g\frac{4m_1m_2m_3}{4m_1m_2 + m_2m_3 + m_1m_3}$,
$T_C = g\frac{8m_1m_2m_3}{4m_1m_2 + m_2m_3 + m_1m_3}$
(g) If $m_1 = m_2 = m$ and $m_3 = 2m$, all of the accelerations are zero, $T_C = 2mg$ and $T_A = mg$. All masses and pulleys are in equilibrium, and the tensions are equal to the weights they support, which is what is expected.
103. (a) $F/2 = 62 N$, which is insufficient to raise either block; $a_1 = a_2 = 0$.
(b) $F/2 = 147 N$. The larger block (of weight 196 N) will not move, so $a_1 = 0$ but the smaller block, of weight 98 N, has a net upward force of 49 N applied to it, and so will accelerate upward with $a_2 = \frac{49 \, N}{10.0 \, kg} = 4.9 \, m/s^2$.

(c) $F/2 = 212\ N$, so the net upward force on block A is 16 N and that on block B is 114 N, so
$a_1 = \frac{16\ N}{20.0\ kg} = 0.8\ m/s^2$ and $a_2 = \frac{114\ N}{10.0\ kg} = 11.4\ m/s^2$.

104. Before the horizontal string is cut, the ball is in equilibrium, and the vertical component of the tension force must balance the weight, so $T_A \cos\beta = w$ or $T_A = w/\cos\beta$. At point B, the ball is not in equilibrium; its speed is instantaneously 0, so there is no radial acceleration, and the tension force must balance the radial component of the weight, so $T_B == w\cos\beta$ and the ratio $T_B/T_A = \cos^2\beta$
☞ At point B the net force on the ball is not zero; the ball has a tangential acceleration.

105. The sand will slide if the cone makes an angle greater than θ where $\mu_s = \tan\theta$. The volume of the cone is then $Ah/3 = \pi R^2 h/3 = \pi R^3 \tan\theta/3 = \pi\mu_s R^3/3$.

107. As m is also moving down along the incline with M, we can find the net acceleration of m using vector addition of two acceleration in m, shown in FIGURE. Thus, we have

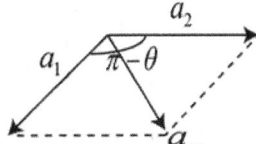

$a_m = \sqrt{a_1^2 + a_2^2 - 2a_1 a_2 \cos\theta} = 4.35 m/s^2$

108. As the weight of the man get cancelled by normal reaction, therefore the only force acting on the man is the frictional force which provides the motion along the belt. Therefore, the net force on the man is given by,
$\sum F = f = ma = 65a = 65 \times 1 = 65\ N$
Alternate method:
Pseudo force on man $F = ma = 65 \times 1 = 65N$
∴ $65N$ force is required to keep the man stationary with respect to belt
Corresponding to maximum acceleration of belt, the person will be in limiting equilibrium, therefore
$F_{max} = \mu_s N = ma_{max}$
or $\quad a_{max} = \frac{\mu_s N}{m} = \frac{\mu_s mg}{m} = \mu_s g = 0.2 \times 10 = 2ms^{-2}$

109. $\frac{[2m - \mu_2(M+m)]g}{M + m[5 + 2(\mu_1 - \mu_2)]}$

110. (A) False (normal reaction and friction due to the small block is not shown).
(B) False (acceleration will be 2a).

111. $\frac{\mu(M+m)}{1+\mu}$

112. (a) $t = 4\sqrt{\frac{l}{3g}}$ (b) $S_A = 4v\sqrt{\frac{l}{3g}} - \frac{2}{3}l$,
$S_B = 4v\sqrt{\frac{l}{3g}} - \frac{5}{3}l$

113. 0.5 sec.

114. (a) $3g, 0, 0$, (b) $0, g, g$

115. $\frac{4+\pi}{3\pi}g$

116. $\frac{mg\sin\theta}{k}$

117. (a) $a_A = \frac{3g}{11}$ (downward) $a_B = \frac{5g}{11}$ (upward) $a_C = \frac{7g}{11}$ (downward)
(b) $T_A = T_C = \frac{8mg}{11}$, $T_B = \frac{16mg}{11}$

118. $\frac{2g(2n - \sin\alpha)}{(4n+1)}$

119. $a_1 = 4\ m/s^2, a_2 = 0, a_3 = 2m/s^2$

120. $\frac{mg\sin\alpha}{M + 2m(1 - \cos\alpha)}$

121. $a = (1 - \cos\theta)\hat{\imath} + a\sin\theta\,\hat{\jmath}$

122. $\left(\frac{dx}{dt}\right) = -\frac{v}{2x}\sqrt{x^2 + b^2}$

123. g (upwards), g (downwards)

124. (i) $\frac{g}{3}\hat{\imath} - \frac{g}{3}\hat{\jmath}, \frac{g}{3}\hat{\imath}$ (ii) $\frac{2mg}{3}$ (iii) $\frac{mg}{3}$

125. $a_B = \frac{a\cos\alpha_1}{\cos\alpha_2}$

126. $-b\hat{\imath} - 4b\hat{\jmath}$

127. $a\hat{\imath} - 2(a+b)\hat{\jmath}$

128. $-4\hat{\imath} + 8\hat{\jmath}$

129. $a_B = 2\ m/s^2\ (\uparrow)$

130. (a) $a_A = g; a_B = 0; x = mg/k$; (b) $a_A = g; x = 0$

131. (a) $\left|\frac{d^2 x_2}{dt^2}\right| = \frac{3g}{2}$; (b) $\left|\frac{d^2 x_1}{dt^2}\right| = 2g$, $\left|\frac{d^2 x_2}{dt^2}\right| = \frac{3g}{2}$; (c) $\left|\frac{d^2 x_1}{dt^2}\right| = \frac{g}{2}$, $\left|\frac{d^2 x_2}{dt^2}\right| = \frac{3g}{2}$

132. $\left(\frac{r}{R}\right)_{min} = \frac{\sqrt{1+\mu^2} - \mu}{\sqrt{1+\mu^2} + \mu}$

133. 32.34 N

134. $F = \frac{Mg}{1-\mu^2}\left[\mu + \frac{\mu\cos\theta + \sin\theta}{\cos\theta - \mu\sin\theta}\right]$

135. $\mu_s = 0.362$

50.12. MULTIPLE CHOICE PROBLEMS

50.12.1. LAWS OF MOTION

50.12.1.1. LEVEL 1

Q. No.	1	2	3	4	5	6	7	8	9
Ans	A	A	A	C	B	C	B	B	C
Q. No.	10	11	12	13	14	15	16	17	18
Ans	B	C	B	D	B	D	C	D	D
Q. No.	19	20	21	22	23	24	25	26	27
Ans	C	C	A	C	C	C	D	D	B
Q. No.	28	29	30	31	32	33	34	35	36
Ans	B	D	A	B	C	A	A	A	C
Q. No.	37	38	39	40	41	42	43	44	45
Ans	A	C	D	B	B	B	B	D	D
Q. No.	46	47	48	49	50	51	52	53	54
Ans.	B	A	A	A	B	C	A	A	C
Q. No.	55	56	57	58	59	60	61	62	63
Ans.	A	C	B	C	B				

LEVEL 2

Q. No.	1	2	3	4	5	6	7	8	9

Ans	D	B	D	B	B	C	C	A	B
Q. No.	10	11	12	13	14	15	16	17	18
Ans	B	C	C	C	C	B	B	B	B
Q. No.	19	20	21	22	23	24	25	26	27
Ans	B	D	B	C	B	B	B	A	

LEVEL 3

Q. No.	1	2	3	4	5	6	7	8
Ans	C	A	A	B	A	A, C	D	D
Q. No.	9	10	11	12	13	14	15	16
Ans	B	C	B	C	A	C	A	A
Q. No.	17	18	19	20	21	22	23	24
Ans	B	A	D	C	D	A	C	C
Q. No.	25	26	27	28	29	30	31	32
Ans	C	B	D	B	A	A	C	D

LEVEL 4
SECTION A

Q. No.	1	2	3	4	5	6	7	8
Ans	A	A	B	C	D	C	C	D
Q. No.	9	10	11	12	13	14	15	16
Ans	C	D	A	C	D	D	A	A
Q. No.	17	18	19	20	21	22	23	24
Ans	C	A	B	B	B			

SECTION B

Q. No.	1	2	3	4	5	6	7	8
Ans	C	A	A	B	B	D	B	D
Q. No.	9	10	11	12	13	14	15	16
Ans	A, B, D							

50.12.2. FRICTION
50.12.2.1. LEVEL 1

Q. No.	1	2	3	4	5	6	7	8	9

Ans	A	A	D	C	D	A	C	B	A
Q. No.	10	11	12	13	14	15	16	17	18
Ans	B	C	C	B	A	B	D	D	D
Q. No.	19	20	21	22	23	24	25	26	
Ans	D	C	A	D	C	A	B	B	

50.12.2.2. LEVEL 2

Q. No.	1	2	3	4	5	6	7	8	9
Ans	A	D	C	C	C	D	B	A	D
Q. No.	10	11	12	13	14	15	16	17	18
Ans	C	A	A	C	D	D	C	A	D
Q. No.	19	20	21	22	23	24	25	26	27
Ans	A	A	A	A	C	B	C	B	A
Q. No.	26	29	30	31	32	33	34	35	36
Ans	D	A	A	A	A	A	D	D	

50.12.2.3. LEVEL 3

Q. No.	1	2	3	4	5	6	7	8
Ans	C	B	D	A	A	C	D	D
Q. No.	9	10	11	12	13			
Ans	A	B	A	D	A→(P, T) B→(P, T, R)			
Q. No.	14	15	16	17	18	19	20	21
Ans	C	A	A					

50.12.2.4. LEVEL 4
SECTION A

Q. No.	1	2	3	4	5	6	7	8	9
Ans	C	C	A	A	C	B	C	D	B

SECTION-B

Q. No.	1	2	3	4	5	6	7	8
Ans	C	A	C	C	10	B	B	B
Q. No.	9	10	11	12	13	14	15	16
Ans	B	A						

www.ingramcontent.com/pod-product-compliance
Lightning Source LLC
Chambersburg PA
CBHW080457220526
45465CB00006B/2294